Modelling and Calculation of Raw Material Industry

Modelling and Calculation of Raw Material Industry

Editors

Krzysztof Czajka
Witold Kawalec
Robert Król
Izabela Sówka

MDPI • Basel • Beijing • Wuhan • Barcelona • Belgrade • Manchester • Tokyo • Cluj • Tianjin

Editors
Krzysztof Czajka
Department of Energy
Conversion Engineering
Wroclaw University of
Science and Technology
Wroclaw
Poland

Witold Kawalec
Department of Mining
Wroclaw University of
Science and Technology
Wroclaw
Poland

Robert Król
Department of Mining
Wroclaw University of
Science and Technology
Wroclaw
Poland

Izabela Sówka
Department of Environment
Protection Engineering
Wrocław University of
Science and Technology
Wroclaw
Poland

Editorial Office
MDPI
St. Alban-Anlage 66
4052 Basel, Switzerland

This is a reprint of articles from the Special Issue published online in the open access journal *Energies* (ISSN 1996-1073) (available at: www.mdpi.com/journal/energies/special_issues/Raw_Material_Industry).

For citation purposes, cite each article independently as indicated on the article page online and as indicated below:

LastName, A.A.; LastName, B.B.; LastName, C.C. Article Title. *Journal Name* **Year**, *Volume Number*, Page Range.

ISBN 978-3-0365-5212-5 (Hbk)
ISBN 978-3-0365-5211-8 (PDF)

Contents

About the Editors

Krzysztof Czajka

Krzysztof Czajka, DSc, PhD, is a power engineer, lecturer, and researcher in the Faculty of Mechanical and Power Engineering at the Wroclaw University of Science and Technology, Poland. His research topics are related to processes of the thermochemical conversion of solid fuels (such as torrefaction, pyrolysis, gasification, and combustion), chemical kinetics and reaction mechanisms, design and construction of experimental reactors, and the development of thermal analysis techniques. He is the author and co-author of 75 scientific papers and consultancy reports.

Witold Kawalec

Witold Kawalec, Dr Eng., is a mathematician and mining engineer, lecturer, and researcher in the Faculty of Geoengineering, Mining and Geology at the Wroclaw University of Science and Technology, Poland. His main fields of interest are the mathematical modeling of systems in the mining industry with a special focus on the applications of computer methods in mining engineering (applications of integrated geological and mining packages, optimization of mine planning and design, and production scheduling), and the advanced computation of belt conveyors with the use of fuzzy numbers and object modeling approaches. He is an author and co-author of approximately 260 papers and consultancy reports.

Robert Król

Robert Król has been professionally associated with the Faculty of Geoengineering, Mining and Geology of Wrocław University of Science and Technology WUST since 1999. Currently, he is the Head of the Department of Mining and the Head of the Scientific Discipline Council of Environmental Engineering, Mining and Energy in WUST. He was appointed by the Minister of Education and Science to the expert group for auditing scientific research and evaluating the impact of scientific developments on the economy and society. His research interests include the assessment of mining facilities and systems (energy consumption, durability, and reliability), the management of exploitation of mining machines and devices, and modelling and numerical simulations in the DEM environment. His scientific activities are closely related to the socio-economic environment. He demonstrates exemplary commitment to the popularization of scientific achievements as part of the cooperation with important mining industry centers and mining and energy companies. He is the author or co-author of approximately 280 scientific papers, including 12 associated patents granted, 2 registered utility models, and over 100 documented expertise reports conducted for the Polish mining industry.

Izabela Sówka

Izabela Sówka has been active at the Wrocław University of Science and Technology, where she is currently employed as a professor in the Department of Environment Protection Engineering. She is the initiator and president of the Center for Sustainable Development and Climate Protection. She is also a member of many workgroups and committees at academic, regional, and national levels. Among her other achievements, one ought to mention her membership of the Environmental Engineering Committee of the Polish Academy of Sciences and vice-chairmanship of the State Council for Environmental Protection and Scientific Council at the Institute of Environmental Engineering of the Polish Academy of Sciences in Zabrze. Her research topics are related to identifying air pollution and its sources, air quality assessments, analysis of the health impacts of air pollution, carbon footprints, global atmospheric phenomena assessments, and RES.

energies

MDPI

Editorial

Modelling and Calculation of Raw Material Industry

Krzysztof Czajka [1,*], Witold Kawalec [2,*], Robert Król [2,*] and Izabela Sówka [3,*]

1 Department of Energy Conversion Engineering, Faculty of Mechanical and Power Engineering, Wroclaw University of Science and Technology, 27 Wybrzeże Wyspiańskiego Street, 50-370 Wroclaw, Poland
2 Department of Mining, Faculty of Geoengineering, Mining and Geology, Wroclaw University of Science and Technology, Na Grobli 15, 50-421 Wroclaw, Poland
3 Department of Environment Protection Engineering, Faculty of Environmental Engineering, Wrocław University of Science and Technology, 27 Wybrzeże Wyspiańskiego Street, 50-370 Wroclaw, Poland
* Correspondence: krzysztof.czajka@pwr.edu.pl (K.C.); witold.kawalec@pwr.edu.pl (W.K.); robert.krol@pwr.edu.pl (R.K.); izabela.sowka@pwr.edu.pl (I.S.); Tel.: +48-71-320-23-24 (K.C.); +48-71-320-68-48 (W.K.); +48-71-320-48-76 (R.K.); +48-71-320-25-60 (I.S.)

Abstract: Scientific and technical issues related to the extraction and processing of raw materials are inextricably linked with environmental concerns. The extraction, transportation and processing of raw materials and the creation of new products place a heavy burden on the environment. Therefore, the development of new technologies for the extraction and processing of raw materials which meet the demand for specific products while respecting environmental resources and saving energy can be considered one of the key challenges of modern science. The development of methods to optimize the course of certain processes related to the raw materials industry, limiting its impact on the environment, and the use of modern measurement techniques or modeling are key areas of research and development for the economy. The aim of this Special Issue was to identify certain important issues, including those related to the raw materials industry and the optimization of its processes, obtaining energy from alternative fuels and research on environmental aspects of industrial activities. The results of the research and analyses presented in the articles show that meeting the objectives in the context of sustainable raw materials industry requires: the optimization of the use of mine deposits and the recovery of materials, reductions in energy consumption, minimizations in emissions of pollutants, the perfection of quieter and safer processes and the facilitation of the recovery of materials-, water- and energy-related modern techniques and technologies.

Keywords: sustainable mining; heating and energy processes; raw material sustainable-use fossil fuels; energy conversion and storage; energy recovery; air pollution; emission reduction methods; purification and removal techniques

Citation: Czajka, K.; Kawalec, W.; Król, R.; Sówka, I. Modelling and Calculation of Raw Material Industry. *Energies* **2022**, *15*, 5035. https://doi.org/10.3390/en15145035

Received: 29 June 2022
Accepted: 8 July 2022
Published: 10 July 2022

Publisher's Note: MDPI stays neutral with regard to jurisdictional claims in published maps and institutional affiliations.

1. Introduction

The raw material mining and processing industry belongs to the most energy-exhausting and environmentally aggressive yet socially controversial branches of the economy. Therefore, its optimization should be targeted with regard to both its overall efficiency and the mitigation of its environmental impact; otherwise, the effects of mining- and fossil-fuel-based power generation would spread further than within the most developed countries. The development of modeling and calculation methods incorporating advanced cloud computing, artificial intelligence algorithms and sophisticated sensor technology allows comprehensive digital twins of objects, systems and processes to be built, representing the whole chain of the raw material mining and processing processes.

One of the most challenging industry processes in the modern economy is sustainable power generation in its transition course from being fossil-fuel-based to being based on "green" solutions to adhere to carbon and toxic emissions limits and to curb other environmental impacts such as dust, noise and vast areas of terrain degradation. Because this industry is particularly capitally and operationally costly, complex investigations that could return substantial savings measured in money, energy consumed and social benefits need to be carried out. Therefore, bearing in mind the interdisciplinary nature of the issues and the development of cooperation within the Polish scientific discipline defined in national law as environmental engineering, power engineering and mining, the Guest Editors defined the topics of this Special Issue of *Energies* as: 1. The sustainable processing of raw materials for the optimization of mining, energy production and environmental engineering. 2. The improved efficiency of raw material mining and processing processes with the use of quality control tracking from deposit modeling through the modeling of mining, transport and final processing operations with regard to environmental issues. 3. Raw material solutions for sustainable energy conversion and storage with "green power" sources. 4. Resource recycling, efficiency and secondary raw materials' management as a contribution to climate change mitigation. 5. Strategies, the identification of sources, dispersion modeling and methods of reducing and limiting air pollutant emissions to the atmosphere in the heating, mining and energy industry. This article is a synthetical summary of the articles that are part of the Special Issue, entitled "Modelling and Calculation of Raw Material Industry".

The whole set of contributions of the Special Issue can be separated into three main research topics:

(1) The generation of energy from alternative fuels [1–4];
(2) Studies on environmental aspects of activities in key sectors of the modern economy [5–9];
(3) Improvements in raw material industry processes [10–14].

The research methods applied to solve the investigated issues reflect the nature of these main topics. In the papers that deal with fuels or wastes, laboratory tests were most commonly used—either with the use of standard procedures [4], with the application of typical equipment innovatively used for specific tests [1,2,7] or with the use of dedicated testing facilities, developed or tuned for the investigations that were performed [3,5,6,8]. The analysis of the test results was confirmed using already-known theoretical equations [1,3,4], experimentally identified dependencies (linear and non-linear regression) [1,3], advanced methods of picture analysis [2] and spectral analyses [6,8]. The comparison of developed models of investigated processes was supported with statistical analyses [3,7]. The spatial distribution of measured parameters in situ was modeled with the use of the Arc-GIS Pro tools [5,6].

Papers that address the topic of improvements in raw material industry processes employed digital modeling of the technological equipment rather than laboratory tests. Depending on the given scope of the research, the "digital twin" could be either of a subassembly of a complex machinery [10], of a single object [11] or of a whole technological system [12]. These digital representations of the analyzed objects were built either on the basis of the specialized software environment implementing FEM [10], discrete simulation methods [10,12] or on the basis of the diversified family of in-house software solutions that are typical for aiding the design and analysis of less common technological objects [11,14]. The provision of better solutions for the more sustainable raw material processing does not always require novel developments. A great deal of knowledge has already been gathered by raw material industry companies. The formulation of the BAT guidelines could be successfully supported with the help of thee proper classification of objects, methods and procedures which have been truthfully verified against the versatile operational conditions of the analyzed operations [13]. Physical experiments were also used in these papers, e.g., tests were performed using a large-scale model of heavy-duty mining equipment [14].

2. Brief Characteristics of Directions and Research Areas under Study

Traditional fuels and raw materials, hard coal, lignite and biomass still account for 28% of primary energy sources [15]. The abandonment of fossil fuels and the transition to environmentally neutral energy sources will take decades; thus, it is vitally important to process and use traditional raw materials as efficiently, cleanly and safely as possible. Additionally, the retrofitting of existing solid fuels' infrastructures is crucial for successful adaptation to the transfer to clean energy systems, in which, for example, solid recovered fuels or hydrogen will play an important role. The use of solid fuels for energy and electricity production raises numerous concerns related to the environmental cost, among other concerns related to the emission of gaseous pollutants such as nitrogen oxides, NO_x, sulfur dioxide, SO_2, and carbon dioxide CO_2, which contribute to smog, acid rain, the greenhouse gas effect and respiratory illnesses.

Among the important issues related to the assessment of the impact of processes—in different areas of the economy—on the environment, there are issues related to, inter alia, the characteristics of their course and conditions, the identification of emission sources, impact assessment and technologies for limiting and reducing the pollution introduced to the environment with particular emphasis on resource savings or their reuse. Further issues are related to the modeling and calculation of the raw material industry.

Some technological processes in the raw material industry (such as sieving screens) are made with the use of vibrating machines. These machines consume relatively large amounts of energy and are prone to excessive wear. The innovative design of a vibrating subassembly built of one motor and two coaxial unbalanced masses has been analyzed, modeled and tuned to provide the realization of complex trajectories of motion that are more technologically efficient for variable parameters of the treated media and energy saving in sieving screens and other vibrating machines [10].

Continuous surface mining technology is declining in Europe due to the approaching cessation of lignite fueled power generation. However, there is still a room to improve the overall energy efficiency of mining processes. Under some technological arrangements in which the overburden is transported several dozen meters down to a spreader operating on a lower located dumping level, there it is possible to convert the potential gravitational energy of conveyed-down overburden masses into electric energy [11]. The accurate calculations of the motion resistances of the downhill belt conveyors prove that depending on the actual schedule of removing the overburden masses onto the dumpsite, the recuperative overburden conveyor working in a generator mode can recover up to 60% of the potential gravitational energy of dozens of million tons of overburden. The recuperative overburden conveyors for downhill transport align with the targets of sustainable mining, which are understood as gaining the maximum benefits from the exploited natural resources. Investments into the installation of regenerative inverters for electric power supply would be paid off within 3–4 years.

Complex mining and transportation systems can be effectively analyzed and tuned with the help of a "digital twin" built on the basis of discrete simulation methods [12].

A model based on a cyclical transportation system in a surface limestone mine, developed in the *Haulsim* software, served to evaluate alternative material transportation scenarios depending on the quality of the deposit, with regard to mineral extraction constraints. The analysis was performed for the production period of one full year, allowing for effective work and technological downtimes. The transportation model was built on the DTM of a mine, which provided a clear visualization of mining and transportation activities for the better understanding of developed solutions.

Mining aggregates and stones are still developing due to the growing demand from roadworks and railroad investments. Natural dimension stone processing generates large volumes of stone waste, causing a significant impact on the environment, as well as on the efficiency and profitability of stone processing plants. Research into the classification of the presented characteristics of dimension stone processing and the structure of the waste production processes in analyzed mines accounts for a huge portion of the state-of-the-art mining knowledge [13]. Stone waste constitutes 10–35% in relation to the quantity of the processed stone material, with the quantity of sludge being even threefold greater than the volume of solid scraps. Reducing the volume of stone waste is possible through improved planning of stone production, while at the same time, the efficiency of stone material usage can be maximized, and the most modern processing machines can be introduced.

Hard coal mines in Poland employ longwall mining technology with powered roof support. The raising depths of mined fields causes rising pressure on the roof-supporting equipment, which is addressed by its more accurate design. The hydraulic leg is the basic element that maintains the position of a powered roof support. It is located in the structure between the canopy and the floor base. Large-scale tests regarding the hydraulic leg provided valuable results that were compared with a theoretical analysis [14]. The results of the theoretical analysis showed consistency with the experimental results.

In environmental areas, it is important to use new methods in the monitoring of pollutants, which could be a complement to the traditional techniques used to measure and model air pollution [5,6] or methods that are developments or new solutions in the field of purification and removal techniques, including organic compounds and selected metals (Zn (II) and Mn (II) from solutions [7,8] and studies regarding the physicochemical properties of products obtained from used goods/products [9].

Analyzing the aspects associated with pollution reduction techniques, considerations include the various methods of NO_x reduction; the most widely recognized are so-called primary measures such as burner optimization, air staging, fuel staging, low NO_x burners and flue gas recirculation. These methods, which come under the term low-emission combustion techniques, are the most cost-effective; however, due to combustion with air deficiencies, they cause the corrosion of boilers. Thus, fast and reliable corrosion diagnostic techniques are becoming a necessity to maintain the security of the energy supply for the power grid. Hardy et al. [2] summarized the state of the art regarding non-destructive diagnostic methods for the corrosion risk assessment of industrial-scale boilers. The authors described their developed systems of online monitoring of the boundary layer gas composition which may help indicate the areas most susceptible to corrosion. Furthermore, they reviewed fast screening techniques for corrosion effects and available automatic systems with visual data processing. In conclusion, based on the collected practical experience, they recommended the electromagnetic acoustic transducer technique (*EMAT*) as a method that shows great potential and has the possibility of making quick measurements without time-consuming surface cleaning.

The corrosion of boilers is exacerbated when combusting low-quality fuels such as biomass high in potassium, *K*, phosphorus, *P*, and chlorine, *Cl*. Due to the relatively low evaporation temperature, compounds containing these elements may evaporate in the combustion chamber and condense on the superheater tubes. This phenomenon is known as fouling and weakens the alloys that tubes are made of. In their work, Król and Nowak-Woźny [4] presented two non-standard methods, i.e., a mechanical test and pressure drop test, to predict the temperature at which ash becomes viscous and starts fouling on the surfaces of heat exchangers. The comparison of the obtained results using non-standard methods with the data obtained using the standard Leitz method showed a linear relationship between them. Nevertheless, as indicated by the authors, the sintering temperatures determined using non-standard methods were lower than those obtained using the Leitz method, which seems to indicate the good potential of these tests for practical usage to prevent the weakening of heat exchanger tubes due to fouling.

Although the problem of the negative influence of SO_2 on the environment has been known since the 17th century, an optimal desulfurization method has still not been developed. The number of proposed types of flue gas desulfurization installations is constantly increasing, and one of the techniques of particular interest is the adsorption of SO_2 on carbonaceous materials. In their work, Kisiela-Czajka and Dziejarski [3] analyzed the SO_2 removal process by adsorbing it on unburned carbon obtained from fly ash and commercial activated carbons. The authors paid particular attention to the methods of determining the kinetic parameters of the processes using linear and non-linear regression analysis. They showed that regardless of the process conditions, linear regression more accurately reflected the behavior of the adsorption system. Interestingly, it was also indicated that the commonly used approach of minimizing the determination coefficient, R^2, and/or the correlation coefficient, R, does not provide the most valuable pairs of kinetic parameters and cannot be treated as evidence of the existence of a particular mechanism of adsorption reaction.

The gaseous pollutant that is causing particularly heated discussions at the moment is CO_2. Even though the impact of anthropogenic CO_2 on the natural environment is sometimes questioned, this pollution is subject to regulations, e.g., under the climate policy of the European Union. One of the promising methods of reducing CO_2 emissions while using fossil fuels is the gasification technique. Combined with carbon capture, utilization and storage, the concept of the gasification of fossil fuels in reactors in which CO_2 is the gasifying agent deserves special attention and has been intensively developed. In his work, Czajka [1] analyzed the gasification process using thermogravimetric analyses. He paid particular attention to the influence of thermal lag induced by the finite rate of heat transfer on the progress, reaction rate and kinetics of the process. Experimental studies supported by a modeling investigation indicated that the thermal lag was of importance from the perspective of the *char*-CO_2 reaction and has an impact on the kinetic parameters which is greater than 20%. The performed analysis indicated that to obtain trustworthy results, it is desirable to carry out thermogravimetric experiments with a low heating rate or to recalculate results obtained at high heating rates with a model presented in the work, which allows the determination of the true sample temperature.

3. Conclusions

The papers included in this Special Issue of *Energies*, entitled "Modelling and Calculation of Raw Material Industry", clearly prove that the investigations that addressed problems in the raw material industry have successfully adopted novel research methods and tools. As a result, there is an opportunity to improve the formerly environmentally unfriendly industry to meet the expectations of achieving the targets of sustainable operations—the decreased use of energy, quieter and safer processes, the maximization of the use of mined deposits and materials' recovery. Based on the solutions presented, greenfield investment in the raw materials industry and the use of environmentally friendly modern technological solutions can avoid most of the drawbacks and reduce the progressive degradation of the environment.

Author Contributions: Conceptualization, K.C., R.K., W.K. and I.S.; methodology, K.C., R.K., W.K. and I.S.; formal analysis, K.C., R.K., W.K. and I.S.; investigation, K.C., R.K., W.K. and I.S.; resources, K.C., R.K., W.K. and I.S.; writing—original draft preparation, K.C., R.K., W.K. and I.S.; writing—review and editing K.C., R.K., W.K. and I.S. All authors have read and agreed to the published version of the manuscript.

Funding: This research received no external funding.

Conflicts of Interest: The authors declare no conflict of interest.

References

1. Czajka, K.M. Gasification of Coal by CO_2: The Impact of the Heat Transfer Limitation on the Progress, Reaction Rate and Kinetics of the Process. *Energies* **2021**, *14*, 5569. [CrossRef]
2. Hardy, T.; Arora, A.; Pawlak-Kruczek, H.; Rafajłowicz, W.; Wietrzych, J.; Niedźwiecki, Ł.; Vishwajeet; Mościcki, K. Non-Destructive Diagnostic Methods for Fire-Side Corrosion Risk Assessment of Industrial Scale Boilers, Burning Low Quality Solid Biofuels—A Mini Review. *Energies* **2021**, *14*, 7132. [CrossRef]
3. Kisiela-Czajka, A.M.; Dziejarski, B. Linear and Non-Linear Regression Analysis for the Adsorption Kinetics of SO_2 in a Fixed Carbon Bed Reactor—A Case Study. *Energies* **2022**, *15*, 633. [CrossRef]
4. Król, K.; Nowak-Woźny, D. Application of the Mechanical and Pressure Drop Tests to Determine the Sintering Temperature of Coal and Biomass Ash. *Energies* **2021**, *14*, 1126. [CrossRef]
5. Cichowicz, R.; Dobrzański, M. 3D Spatial Analysis of Particulate Matter (PM_{10}, $PM_{2.5}$ and $PM_{1.0}$) and Gaseous Pollutants (H_2S, SO_2 and VOC) in Urban Areas Surrounding a Large Heat and Power Plant. *Energies* **2021**, *14*, 4070. [CrossRef]
6. Górka, M.; Bezyk, Y.; Sówka, I. Assessment of GHG Interactions in the Vicinity of the Municipal Waste Landfill Site—Case Study. *Energies* **2021**, *14*, 8259. [CrossRef]
7. Sobianowska-Turek, A.; Grudniewska, K.; Maciejewski, P.; Gawlik-Kobylińska, M. Removal of Zn(II) and Mn(II) by Ion Flotation from Aqueous Solutions Derived from Zn-C and Zn-Mn(II) Batteries Leaching. *Energies* **2021**, *14*, 1335. [CrossRef]

8. Urbanowska, A.; Kabsch-Korbutowicz, M.; Aragon-Briceño, C.; Wnukowski, M.; Pożarlik, A.; Niedzwiecki, Ł.; Baranowski, M.; Czerep, M.; Seruga, P.; Pawlak-Kruczek, H.; et al. Cascade Membrane System for Separation of Water and Organics from Liquid By-Products of HTC of the Agricultural Digestate—Evaluation of Performance. *Energies* **2021**, *14*, 4752. [CrossRef]
9. Urbańska, W.; Osial, M. Investigation of the Physico-Chemical Properties of the Products Obtained after Mixed Organic-Inorganic Leaching of Spent Li-Ion Batteries. *Energies* **2020**, *13*, 6732. [CrossRef]
10. Gursky, V.; Kuzio, I.; Krot, P.; Zimroz, R. Energy-Saving Inertial Drive for Dual-Frequency Excitation of Vibrating Machines. *Energies* **2021**, *14*, 71. [CrossRef]
11. Kawalec, W.; Król, R. Generating of Electric Energy by a Declined Overburden Conveyor in a Continuous Surface Mine. *Energies* **2021**, *14*, 4030. [CrossRef]
12. Krysa, Z.; Bodziony, P.; Patyk, M. Discrete Simulations in Analyzing the Effectiveness of Raw Materials Transportation during Extraction of Low-Quality Deposits. *Energies* **2021**, *14*, 5884. [CrossRef]
13. Strzałkowski, P. Characteristics of Waste Generated in Dimension Stone Processing. *Energies* **2021**, *14*, 7232. [CrossRef]
14. Szurgacz, D. Dynamic Analysis for the Hydraulic Leg Power of a Powered Roof Support. *Energies* **2021**, *14*, 5715. [CrossRef]
15. BP. Statistical Review of World Energy: 70th Edition. 2021. Available online: https://www.bp.com/en/global/corporate/energy-economics/statistical-review-of-world-energy.html (accessed on 16 April 2022).

energies

Article

Generating of Electric Energy by a Declined Overburden Conveyor in a Continuous Surface Mine

Witold Kawalec * and **Robert Król**

Faculty of Geoengineering, Mining and Geology, Department of Mining, Wroclaw University of Science and Technology, Na Grobli 15, 50-421 Wroclaw, Poland; robert.krol@pwr.edu.pl
* Correspondence: witold.kawalec@pwr.edu.pl

Abstract: Exploitation of lignite in continuous surface mines requires removing masses of overburden, which are hauled to a dumpsite. There are some technological arrangements where the overburden is transported several dozen meters down to a spreader operating on a lower located dumping level. Depending on an angle of a declined transportation route, there is a possibility to convert the potential gravitational energy of conveyed down overburden masses into electric energy. To recover the maximum percentage of stored energy, an energy-effective and fully loaded belt conveyor should work in a generator mode. Due to the implementation of such a solution, a lignite continuous surface mine, which is a great electric energy consumer, can obtain the status of an electricity prosumer and reduce its environmental impact, in particular demonstrating significant savings in primary energy consumption. Though lignite surface mining is phasing out in Europe, the recuperative, overburden conveyors for downhill transport match up the targets of sustainable mining, understood as getting the maximum benefits from the exploited natural resources. According to the analyzed case study, an investment into the installation of regenerative inverters for the electric power supply of the declined overburden conveyor would pay off within 3–4 years.

Keywords: belt conveyor; prosumer; downhill transport of overburden; specific energy consumption; recuperation; energy recovery rate

Citation: Kawalec, W.; Król, R. Generating of Electric Energy by a Declined Overburden Conveyor in a Continuous Surface Mine. *Energies* **2021**, *14*, 4030. https://doi.org/10.3390/en14134030

Academic Editor: Rouhi Farajzadeh

Received: 26 May 2021
Accepted: 1 July 2021
Published: 4 July 2021

1. Introduction

Belt conveyors are extensively used in industrial bulk material transportation worldwide. There are estimations that more than 2.5 million conveyors are in use [1,2]. Surface lignite mines are arguably the users of the largest and the most complex belt conveyor systems. The Energy Efficiency Act issued in 2016 [3], in line with the relevant directives of the European Parliament [4], imposes on such large enterprises the obligation to perform periodic energy audits, which should document the improvements introduced to reduce energy consumption, decreasing the "carbon footprint". In surface mines, one of the directions of this activity is recovering the potential gravitational energy in each case of transporting overburden downhill. Belt conveyors, driven by electric motors, can be easily converted into electricity-generating machines [5]. A flagship example of an energy recovery belt transportation is a system of innovative belt conveyors in the Los Pelambres copper ore mine in Chile. It generates up to 25 MW power on a 13 km route with a decline of 1310 m and supplies this electric energy to the entire mine [6,7].

Articles [8,9] have introduced the concept of a potential gravitational energy deposit, which exists as an accompanying deposit in the case of a mineral location in a mountain area, from which the extracted material is transported down to a storage yard by a public road or railway siding. For an aggregate deposit with an operational resource of 5 million tons, from which the output is transported to a depot lying 200 m below the level of the excavation, the potential gravitational energy resource is substantial 2.7 GWh (230 toe-equivalent oil, in which fossil fuel savings are counted [3]). The percentage of yield of this resource (a potential gravitational energy recovery) depends on the angle of declination of a conveyor route, its energy efficiency [8], and the implementation of the potential energy recovery-oriented haulage technology, consisting of avoiding its idle or low-efficiency working mode [8]. In the aggregate mines, the movement of heavy mining vehicles (exhaust gases, vibrations, destruction of roads) has a negative impact on the environment and may even prevent the exploitation of the mineral. It has been identified as a reason for the "annihilation of resources" even in barely populated areas in Norway [10]. The replacement of heavy trucks with belt conveyors allows eliminating these problems.

Unlike rock mining, belt conveyor transportation systems have always been commonly used in continuous surface lignite mines. Though these mines are mostly located on a flat area, due to their depth and specific technological arrangements, there are various possibilities to convey large amounts of overburden downhill by declined conveyors.

Continuous lignite surface mining is phasing out in Europe due to the EU climate action and the European Green Deal [11]. However, some 380 million tons of lignite is still mined annually in open cast operations [12]. Approximately 1 billion cubic meters (cbm) of overburden has to be removed annually to allow these lignite operations. Some of the overburden volumes (and masses) are conveyed downhill, which will be even more extensive during the oncoming closure works. Due to the implementation of such a solution, a lignite continuous surface mine, which is a great electric energy consumer, can obtain the status of an electricity prosumer and reduce its environmental impact (certified by so-called "white certificates" of saved energy [3]).

The efficiency of the conveyor operation in energy recuperation mode depends on the slope angle of the descending conveyor route profile and the actual capacity of the conveyor. The maximum use of the gravity of the transported material is favored by achieving low motion resistance of the main belt conveyors equipped with energy-efficient conveyor belts, rigidly mounted, energy-saving roller sets, and proper belt guidance [1]. For the conveyor transporting downwards, the balance of the component gravity force of the excavated material and aggregated resistance to motion is crucial [13]. The goal of this paper is to reveal the opportunity to mine the potential gravitational energy reserves in the lignite surface mines during their exploitation and post-closure phases.

2. Materials and Methods

The basis of the study originated from the complexity and variability of overburden removal conveyor systems in lignite continuous surface mines, which have to be rearranged several times over the life of a mine to keep transportation costs low [14]. In Figure 1, the schematic sections of a continuous surface mine show the three typical technological layouts of removing volumes of overburden downhill to the dumping levels.

The overburden removed from the upper overburden levels is often placed down the transportation ramp to the dumping levels on the lower elevation (Figure 1a). This configuration of the haulage route "creates" a potential gravitational energy deposit, and a declined conveyor installed on the transportation ramp can work in a regenerative mode. The differences in elevations are moderate, assuming conservatively not more than the height of 3–4 mining levels, i.e., 60–80 m. However, the continuous surface mines are mostly large, and from each overburden level, at least a few million cbm of overburden is removed annually (with an average density of about 1700 kg/cbm). Therefore, this transportation arrangement "creates" a potential gravitational energy resource of 1 GWh per year on each mining level. Although the amount of energy recovered is small in comparison to that consumed by the lignite coal mine transport system, this recovery allows obtaining the status of an energy prosumer, valuable not only due to money savings but also due to the growing importance of a corporate social responsibility [15].

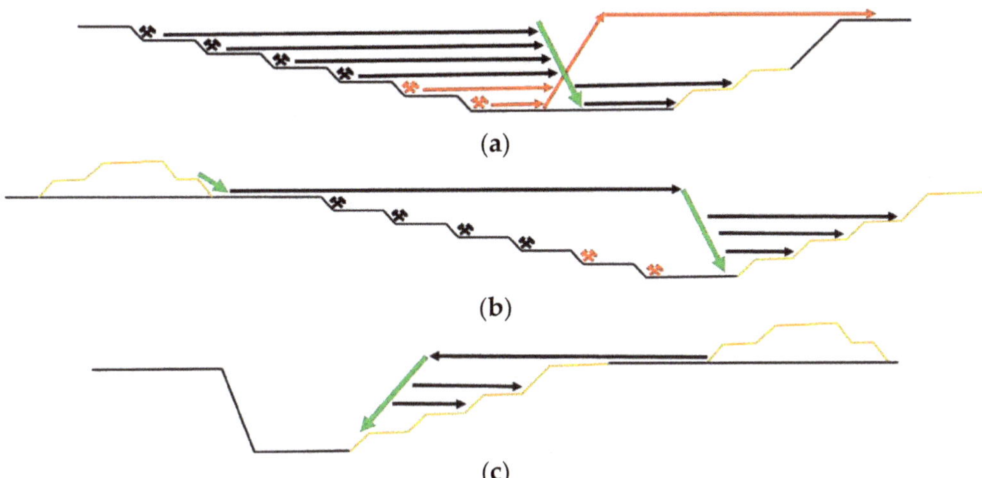

Figure 1. A schematic section through a continuous surface mine with various arrangements of overburden removal enabling the recovery of potential gravitational energy (red, black, green arrows–coal, overburden horizontal, overburden declined conveyors, respectively; yellow profile lines–dump sites, crossed hammers–exploitation levels); (**a**) Removing overburden volumes from mining levels to an in-pit dumping levels, (**b**) Translocation of overburden volumes previously temporarily dumped on a foreground to the final in-pit dumping levels, (**c**) partially liquidation of an external dump for the needs of reclamation earthworks of a post-mining void (shallowing and slopes forming)

Another technological arrangement suitable for electric energy recovery is the relocation of the overburden from a temporary dumpsite located on a foreground of the pit advance (Figure 1b) or from an external dump after the mine closure (Figure 1c). In both cases, there are anthropogenic potential gravitational energy deposits, which were previously built against a large amount of consumed energy. Both cases will become more feasible soon when, due to the European Green Deal policy, lignite mining will be put into the closure stage. Reclamation of the closed mining voids will require massive earthworks [16]. Then there is a possibility of a partial recovery of electricity previously used during the temporary dumping.

Therefore, after identifying the possibility for recovery of potential gravitational energy of the overburden in accordance with the excavation development plan (or relocation of the dumping ground), the conveyors can be properly configured on the distribution point (on the transportation ramp) with the installation of automatic drive control for maximum energy recuperation. Depending on the actual load, a declined conveyor works either in the mode of driving or braking the belt (generator operation). Though it brings increased design and operational difficulties [17], the development of automation, multi-channel monitoring of the conveyor's operation, signaling of emerging and expected threats [18], and improved calculation methods come with the help [17,19]. Regenerative drives equipped with frequency converters for energy recovery are available to control the operation of variable load drive. During braking, energy is returned to the mains instead of being converted into heat (wasted) using braking resistors (as it is still common nowadays). The recovered energy can be returned to the network in prosumer mode or used locally, e.g., to supply electricity to other nearby installed devices.

In Polish surface lignite mines, haulage of overburden is carried out mainly with high-performance B1800 or B2250 (belt width in millimeters) conveyors with steel-cord belts running with a speed of 5.25–5.9 m/s [20]. Overburden is excavated by bucket-wheel excavators with long periods of work with stable performance (Figure 2), in accordance with the current overburden removal schedule [20,21].

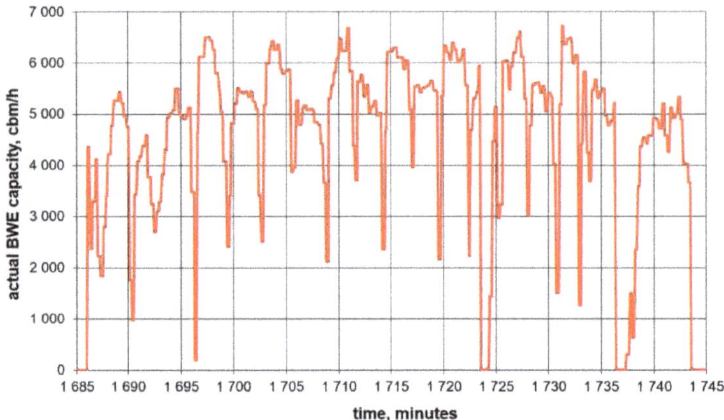

Figure 2. Actual volume capacity for the SRs 2000 bucket-wheel excavator recorded in the Belchatow surface lignite mine during the excavation of one terrace [20].

The widely adopted measure of a belt conveyor efficiency is its specific energy consumption (SEC)—an amount of energy needed to move 1 kg of conveyed material for a distance of 1 m. Having obtained measurements of actual power and capacity of a belt conveyor, it can be calculated from the equation [13]:

$$SEC = \frac{N}{\dot{M}\cdot L} \left[\frac{W\cdot s}{kg\cdot m} = \frac{N}{kg} \right] \tag{1}$$

where: N–drive power, W; $\dot{M}$–actual capacity, kg/s; L–conveyor length, m.

The drive power reflects the actual belt conveyor resistances to motion. These are divided into primary (or main) resistances that occur along the whole conveyor route, secondary (exerted at the head and tail stations–on pulleys, loading points, cleaning devices) resistance, and lift resistance. After converting (1) into the equation with resistances to the motion, we get [13]:

$$SEC = \frac{W_{main}}{M\cdot\eta} + \frac{W_{secondary}}{M\cdot\eta} + \frac{g}{\eta}\cdot sin\delta \tag{2}$$

where: W–resistance to motion, W; M–total mass of the transported material on a conveyor, kg; η–drive efficiency; g = 9.80665 m/s^2; δ–angle of inclination of a belt conveyor route

The total mass of the transported material on the conveyor equals:

$$M = \frac{L\cdot\dot{M}}{v} \tag{3}$$

where: v–conveyor belt speed, m/s

After applying (3) to (2), we get the equation of the components of specific energy consumption:

$$SEC = \frac{v}{L\cdot\dot{M}\cdot\eta}\left(W_{main} + W_{secondary} \right) + \frac{g}{\eta}\cdot sin\delta \tag{4}$$

$$or: \; SEC = SEC_{main} + SEC_{secondary} + \frac{g}{\eta}\cdot sin\delta \tag{5}$$

The SEC_{main} and $SECsecondary$ decrease with the rising actual capacity with regard to the actual parameters of a belt conveyor [9]. The $SEC_{secondary}$ decreases with the increasing conveyor length. The positive or negative value of the SEC depends on the actual balance of these components. The negative values represent the regenerative mode of operation.

In order to determine the area of possible energy recuperation with the overburden conveyor, simulation calculations of motion resistances with the necessary drive power for route variants after a fall with a moderate slope of 2 to 7 degrees were performed. A route length of about 600 m was adopted, suitable for a conveyor working on a transportation ramp (see Figure 1). In the calculations, the specialized QNK-TT software for calculating belt conveyors' motion resistances and drive power demand was used. Unlike the simplified standard algorithms, the QNK-TT software uses the more accurate method of calculating individual components of the main resistances to motion [17,19]. This is done on the basis of the theoretical analysis of the energy dissipation processes in a conveyor belt and the material load stream, and the analysis of the interaction between the belt and idlers. The investigations were validated against the numerous laboratory and industrial tests [22,23] and eventually resulted in verified algorithms for calculating motion resistances. In the QNK-TT program, a large set of data of a belt, transported bulk material, and the design characteristics of a conveyor, as well as its specific operating conditions, are used, and calculation results allow analyzing the impact of selected parameters on the belt conveyor drive power requirement.

The following conveyor parameters were adopted:

- Steel-cord belt B1800 ST 3150 (12 + 7), 79.4 kg/m, belt speed: 5.9 m/s
- Idlers: upper (carry): rigidly mounted, with a troughing angle of 45° and a spacing of 1.25 m, bottom: V-type with a spacing of 3 m; standard rotating resistance (appr. 4.5 N for each roller)
- Power units (4 × 630 or 2 × 800 kW) at the tail head station, controlled by an inverter,
- Belt take-up at the head station,
- Reference ambient temperature of 11 °C, low belt mistracking, good operating conditions
- Transported overburden: density 1700 kg/cbm; maximum capacity: 8800 cbm/h (13,500 t/h).

The lack of belt mistracking and good operating conditions are typical for main conveyors working on transportation ramps in large continuous lignite surface mines that are maintained by a highly professional staff. Motion resistance and drive power (positive or negative) were calculated for the adopted variants of the route inclination downward, assuming the belt tension force was fixed for the maximum load of transported material. The drive power map presents the results of the balance of power calculated at the drive pulley shaft (Figure 3).

The area of positive power (conveyor is driven) shown in Figure 3 above the level of 0 MW indicated that the conveyor with an angle of inclination downward up to −3° is not suitable for energy recovery. Larger angles of the route inclination downward (from −4° to −7°) allow energy recuperation in a sufficiently wide performance range, assuming that the actual conveyor capacity exceeds 10% of the maximum one. The graph above shows the isolines of calculated power values for the stabilized transported overburden stream, which is typical for overburden removal (see Figure 2). However, for uneven streams, the resistance to motion increased (up to 10% [20]), which decreased the amount of recuperated energy. In addition, any maintenance failures (e.g., worn idlers, belt mistracking) or low ambient temperature cause an increase in the respective components of motion resistances. These factors will also reduce the capability of energy recovery.

The values of power ranges of the main drive are illustrative for the technical staff. For the needs of an assessment of the recovery rate of the potential gravitational energy of conveyed down overburden masses, the map of energy recovery is more useful (Figure 4).

In Figure 4, only the area of the energy recovery rate exceeding 10% is mapped. Such a map can be used for scheduling the overburden removal in order to keep the actual conveyor capacity within the area of a high energy recovery rate.

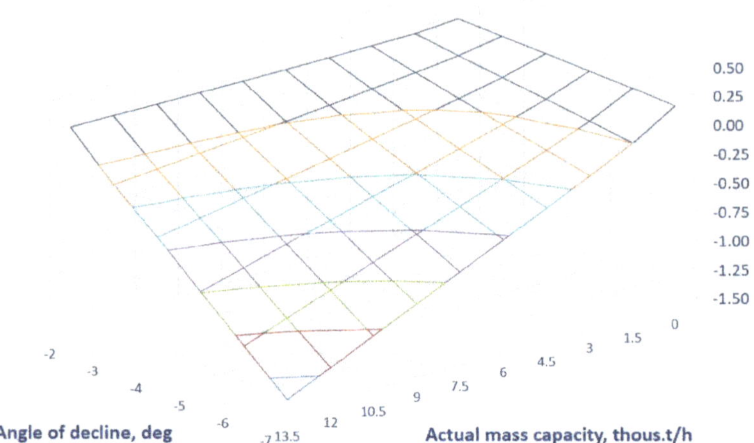

Figure 3. 3-D map of the power ranges of the main drive of the B1800 descending conveyor depending on the mass capacity and the angle of inclination downward (decline) of the conveyor route (calculations: QNK-TT).

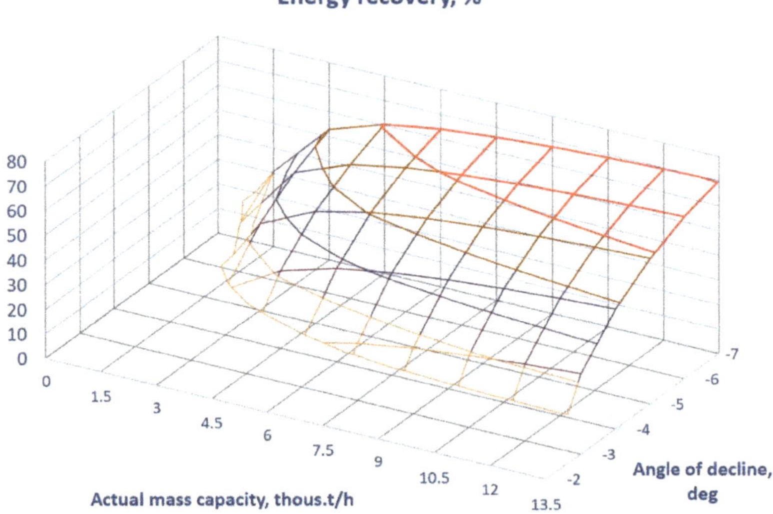

Figure 4. 3-D map of the recovery rate of the potential gravitational energy of mined overburden with the use of analyzed the B1800 descending conveyor depending on the mass capacity and the angle of inclination downward of the conveyor route (calculations: QNK-TT).

3. Results

The graph of positive or negative drive power of the overburden conveyor working on the transportation ramp presented in the previous point presents the general possibilities of electricity recuperation from overburden haulage. In each case, they should be estimated on the basis of specific plans for the advance of mining and dumping fields, containing the expected overburden volumes, which would be directed from the given overburden mining levels to the dumper operating on the lower elevation. It has to be pointed out that the overburden dumping plans are developed according to the geotechnical requirements, and the material from a given mining level can be sent to various dumping levels in order to secure the stability of the dumping slope [18]. In a large surface mine consisting of several mining and dumping levels, various configurations of dumping the overburden are used.

Assume that the collective overburden conveyor (length 565 m, average downward slope angle −7.2°) is fed on three successive overburden levels (marked A, B, C from the top; each mining level is 20 m high, the head station is positioned 70 m down from the top A level) by horizontal conveyors loaded with overburden directly from bucket-wheel excavator class KWK 1500 with a nominal capacity of 4200 cbm/h (see Figure 5). A, B, C excavators work for long periods with stable performance, according to the current disposition. The conveyor works 2000 h a year. Each excavator removes annually some 4 million cubic meters of overburden, which is transported to this conveyor. Excavators can work in different systems: all (but with limited capacity), two of three (with different levels of performance), or one of three. The actual working mode of the excavators and the actual capacity are not random as they are controlled from the mine control room by a shift dispatcher. Unlike the quality parameters driven by control of selective lignite mining in the case of mining the overburden benches, the dispatcher is obliged to follow the daily overburden removal schedule and maintain the overall energy efficiency of machinery and transportation equipment.

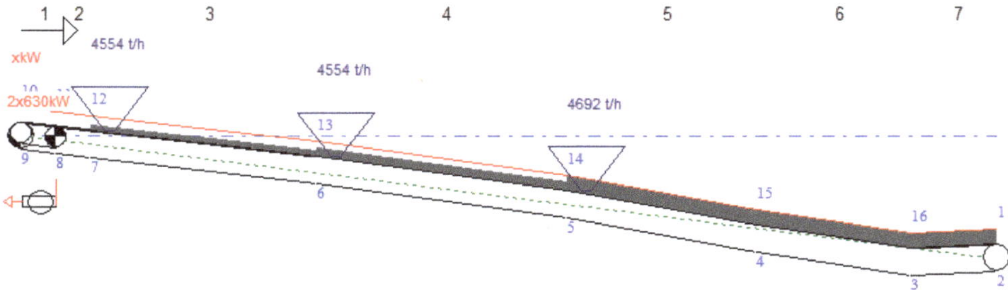

Figure 5. The profile of the study route of the overburden conveyor on the ramp, receiving overburden at three loading points on three subsequent mining levels, which is discharged at its head (source: QNK-TT).

For these assumptions, 10 modes of excavating overburden by excavators A, B, C were adopted, supplying the declined conveyor with a stabilized bulk stream controlled by the dispatcher. Modes 1–3 represent the loading by a single excavator digging at its top output, modes 4–6 and 8–10-loading by two excavators at the limited or partly limited output, while mode 7-equal loading by all three excavators (however with the limited output each). Idle or low duty work of the main conveyor is not provided. If such periods of work took place, then (according to the diagram in Figure 2) the conveyor would be driven, but then transport tasks would practically not be carried out at all. It can therefore be assumed that these would be additional hours of idle operation of the conveyor.

A hypothetical share of the working time of each of the accepted modes of work was set so as to implement the overburden removal plan. Table 1 summarizes the data on the output stream on the main, declined conveyor on the transportation ramp, and the calculated resultant drive power for each mode of operation, as well as the total annual values—the sum of the volume and mass of the overburden and the average drive power.

The declined conveyor on the transportation ramp transporting overburden from three upper mining levels to the most bottom dumping level generates approximately 790 kW (weighted average value against the presumed distribution of the supplied volumes). This means that it could produce nearly 1.6 GWh of "green" electricity per year (equivalent to some 140 toe) from the total 2.8 GWh of the combined potential gravitational energy accumulated on the overburden mining levels A, B, and C (calculated with regard to the elevation of the conveyor discharge point at its head pulley). The average energy recovery rate reaches some 54%. A similar amount of energy can be obtained from a 1 MW windmill, whose construction cost, however, is about EUR 1 million.

Table 1. Variants of loading the analyzed conveyor on a ramp with overburden stream from three excavators A, B, C at subsequent mining levels (the conveyor is discharged at its head) and the resulting drive power.

Mining Levels Supply (Volumes, Tonnages)		Mode of Loading Overburden on Levels A, B, C and Its Presumed Share of the Operation Time									
		1	2	3	4	5	6	7	8	9	10
		0.075	0.075	0.075	0.1	0.1	0.1	0.25	0.075	0.075	0.075
A	cbm/h	4200	0	0	3000	4200	0	2500	2500	0	2500
	tonnes/h	7140	0	0	5100	7140	0	4250	4250	0	4250
B	cbm/h	0	4200	0	0	3000	4200	2500	0	2500	2500
	tonnes/h	0	7140	0	0	5100	7140	4250	0	4250	4250
C	cbm/h	0	0	4200	4200	0	3000	2500	2500	2500	0
	tonnes/h	0	0	7140	7140	0	5100	4250	4250	4250	0
A + B + C	cbm/h	4200	4200	4200	7200	7200	7200	7500	5000	5000	5000
	tonnes/h	7140	7140	7140	12,240	12,240	12,240	12,750	8500	8500	8500
Calculated drive power, kW		−702	−486	−170	−847	−1229	−806	−1048	−586	−485	−788
potential energy recovery, %		52	50	29	54	60	58	60	51	52	57

For the listed above working modes of excavators on mining levels, very different levels of energy recovery were obtained. It should be assumed that in the "control room" of a given mine, an appropriate panel would appear with parameters of energy recovery by the conveyor (s) in recuperation mode so that the dispatcher could properly manage the work of removing the overburden.

The declined conveyor on the transportation ramp is often discharged on its route—not at its head pulley (see Figure 1) in order to supply the overburden to the selected dumping level. Can it still generate energy, then? The following table presents the hypothetical results of drive power demand for two variants of unloading the downhill conveyor: 1 or 2 mining levels below the second exploitation level (see Figure 6).

When a conveyor is discharged on its route, the moving discharge station is installed on the construction. The belt is lifted there from the top idlers and wrapped on two pulleys–the first is the discharging one, and the second puts the emptied belt on the top idlers again to let it go to the head pulley. The station creates additional resistances to motion that have to be taken into account for calculations of the drive power. Again, the hypothetical share of the working time of each of the accepted modes of work is implemented according to the possible overburden removal plan. Table 2 summarizes the results of simulating the transportation work of the conveyor, which supplies alternatively two dumping levels.

Under such arrangements, the declined conveyor generates less energy—approximately 530 kW (weighted average value). It equals 1.06 GWh of "green" electricity per year out of the total 2.3 GWh of the combined potential gravitational energy accumulated on the overburden mining levels A and B (calculated with regard to the actual elevation of the conveyor discharge point as showed in Figure 5). In this case, the average energy recovery rate equals 45%. Each alternative arrangement of the actual loading and discharging of the downhill conveyor can be easily checked against the available energy recovery rate. These calculations should be taken into consideration for the energy recuperating oriented planning of dumping operations. Overburden is often scattered by a spreader positioned on the upper dumping level (higher than necessary), and the dumped material falls from a big height onto the ground. It causes both the loss of the potential gravitational energy and also excessive dust contamination of the dumped material.

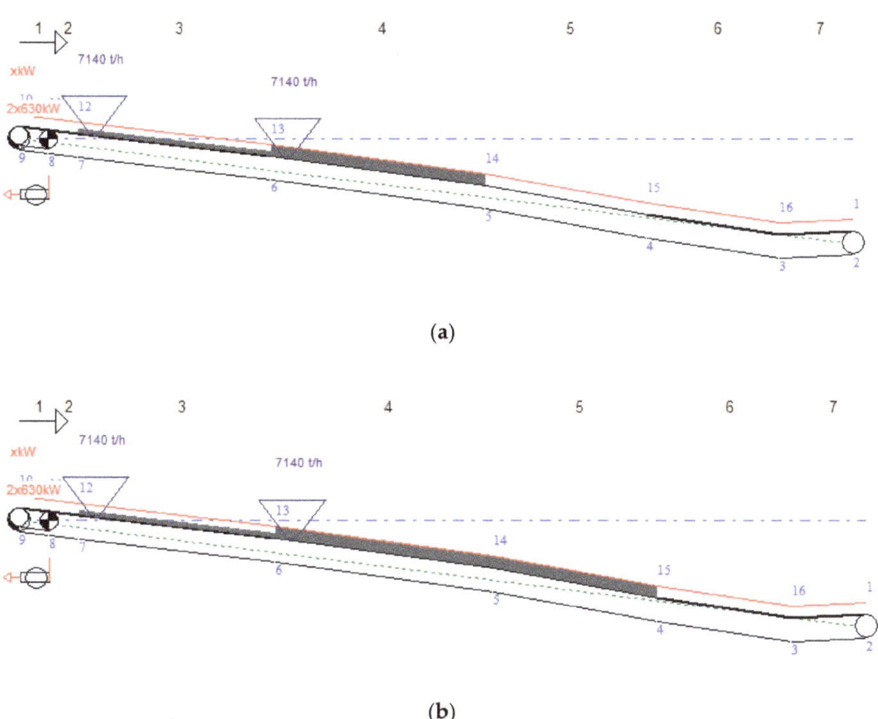

(a)

(b)

Figure 6. The profile of the study route of the overburden conveyor, loaded on two subsequent mining levels and being alternatively discharged on its route: (**a**) 1 mining level below the second loading point, (**b**) 2 mining levels below the second loading point (source: QNK-TT).

Table 2. Variants of loading the analyzed conveyor on a ramp with an overburden stream from two excavators, A and B, at subsequent mining levels (the conveyor is discharged on its route) and the resulting drive power.

Mining Levels Supply (Volumes, Tonnages)		Mode of Loading Overburden on Levels A and B and Its Presumed Share of the Operation Time									
		1	2	3	4	5	6	7	8	9	10
		0.075	0.1	0.15	0.1	0.075	0.075	0.1	0.15	0.1	0.075
A	cbm/h	4200	4200	2500	3750	0	4200	4200	2500	3750	0
	tonnes/h	7140	7140	4250	6375	0	7140	7140	4250	6375	0
B	cbm/h	0	4200	2500	3750	4200	0	4200	2500	3750	4200
	tonnes/h	0	7140	4250	6375	7140	0	7140	4250	6375	7140
A + B	cbm/h	4200	8400	5000	7500	4200	4200	8400	5000	7500	4200
	tonnes/h	7140	14,280	8500	12,750	7140	7140	14,280	8500	12,750	7140
Calculated drive power, kW		−256	−547	−251	−469	−41	−569	−1169	−623	−1024	−354
potential energy recovery, %		33	47	36	45	11	49	60	54	59	46
Position of the discharge station		1 level below the B mining level (−40 m from the level A–see Figure 5a)					2 levels below the B mining level (−60 m from level A–see Figure 5b)				

The actual removal plan can be easily applied to provide more accurate estimations of electric energy generated on the transportation ramp. If there is a decision to relocate volumes of overburden from the temporary dumpsite to the bottom of the final pit (Figure 1c), the mine can recover some electric energy after the closure of exploitation.

The cost of implementation of power supply with a regenerative inverter can be estimated to EURO 100 thousand for each drive unit. The large conveyors are usually driven by 2 or 3 drive units that give EUR 200–300 thousand of the investment costs. The rising wholesale energy price in Poland is reaching some EURO 80 per MWh, which means that the investment would pay off in 3–4 years, not to mention the value of obtained "white certificates" of saved energy [3]. After finishing the opportunity of transporting masses of overburden downhill, the regenerative inverter equipment can be easily dismantled and installed on another conveyor because large lignite belt conveyor systems are usually constructed from unified equipment, including their drive units.

Producing "green" energy is also synonymous with reducing energy production from fossil fuels, which, according to KOBIZE [24], is convertible into reducing CO_2 emissions and pollutants [8,24]. The obtained indicators provide valuable input for energy audits and reports on the reduction in environmental impact by the mine.

4. Conclusions

A declined belt conveyor is a mining machine that exploits potential gravitational energy. In a case of a continuous surface lignite mine, depending on the given technology of removing, dumping, and relocating volumes of overburden, it may be possible to recuperate the energy of the overburden transported from the upper mining levels (or from a temporary dumpsite) to the located lower spreader on the dumping level. Depending on the actual mine layout and advance, significant amounts of recovered energy can be obtained from the transported overburden massive masses for lignite mines.

Equipping the declined conveyor (operating on a transportation ramp at a distribution point) with the possibility of generator operation with appropriate automation of electricity recovery along with dedicated dispatching management procedures on given overburden levels to keep the energy recovery rate high will allow for the production of "green" electricity. The lignite mine will then become a prosumer of electricity, which has both practical (revenue from sales of energy) and public relations image significance.

According to the analyzed case study, an investment into the installation of regenerative inverters for the electric power supply of the declined overburden conveyor would pay off within 3–4 years.

The conveyor recuperating energy cannot waste its drive power on unnecessary motion resistances caused by worn elements or poor operational conditions. The calculations made in this paper were done for typical belt conveyor layout and equipment but with no allowance for additional wastes of energy. The use of an accurate method of calculating belt conveyor resistance to motion is vital to give the proper assessment of possible energy recovery. It must be noted that the efficiency of conversion of the recovered mechanical energy into the electric one has not been analyzed as it depends on the applied solution of conveyor drive units control equipment.

Lignite surface mining operations are extensive electric energy consumers. Sustainable mining operations require not only to cut all unnecessary energy consumption but also—if possible—to recover the potential gravitational energy from exploited masses of overburden. Even in the phasing out of the lignite mining in Europe, there is still an opportunity to recover energy during the reclamation works of post-mining voids.

Author Contributions: Conceptualization, W.K. and R.K.; Funding acquisition, R.K.; Investigation, W.K.; Methodology, W.K.; Project administration, R.K.; Supervision, R.K.; Validation, R.K.; Writing—original draft, W.K.; Writing—review & editing, W.K. All authors have read and agreed to the published version of the manuscript.

Funding: The research work was cofunded with the research subsidy from the Polish Ministry of Science and Higher Education granted for 2021.

Data Availability Statement: The details of calculations in the Excel spreadsheet, containing the results from the specialised QNK-TT software for calculations belt conveyors hat are presented in this study are available on request from the corresponding author

Conflicts of Interest: The authors declare no conflict of interest.

References

1. Clénet, D. *Optimising Energy Efficiency of Conveyors*; White paper WP20100601EN; Schneider Electric SA: Rue il-Malmaison, France, 2010.
2. He, D.; Pang, Y.; Lodewijks, G. Green operations of belt conveyors by means of speed control. *Appl. Energy* **2017**, *188*, 330–341. [CrossRef]
3. Available online: https://www.bialecertyfikaty.com.pl/media/Ustawa_o_efektywnosci_energetycznej.pdf (accessed on 2 July 2021).
4. Available online: https://eur-lex.europa.eu/legal-content/PL/TXT/PDF/?uri=CELEX:32018L2002&from=EN (accessed on 2 July 2021).
5. Rodríguez, J.; Pontt, J.; Becker, N.; Weinstein, A. Regenerative Drives in the Megawatt Range for High-Performance Downhill Belt Conveyors. *IEEE Trans. Ind. Appl.* **2002**, *38*, 203–210. [CrossRef]
6. Available online: http://conveyor-dynamics.com/index.php/project/los-pelambres/ (accessed on 2 July 2021).
7. Available online: https://iran.phoenix-conveyorbelts.com/pages/extreme-conveyor-belt/strength/strength_en.html (accessed on 2 July 2021).
8. Gladysiewicz, L.; Kawalec, W.; Król, R. Przenośnik taśmowy jako maszyna górnicza urabiająca energię. (Belt conveyor as a mining machine exploiting energy). *Transp. Przemysłowy Masz. Robocze.* **2019**, *4*, 5–11. (In Polish)
9. Kawalec, W.; Król, R.; Suchorab, N. Regenerative Belt Conveyor versus Haul Truck-Based Transport: Polish Open-Pit Mines Facing Sustainable Development Challenges. *Sustainability* **2020**, *12*, 9215. [CrossRef]
10. Nielsen, K. Sterilization of mineral resources by area planning and urban sprawl. In Proceedings of the International Conference on Mine Planning and Equipment Selection, Wroclaw, Poland, 1–3 September 2004.
11. European Commission. The European Green Deal. Available online: https://ec.europa.eu/info/sites/info/files/european-green-deal-communication_en.pdf (accessed on 19 May 2021).
12. Dias, P.A.; Kanellopoulos, K.; Medarac, H.; Kapetaki, Z.; Miranda-Barbosa, E.; Shortall, R.; Czako, V.; Telsnig, T.; Vazquez-Hernandez, C.; Arántegui, R.L.; et al. *EU Coal Regions: Opportunities and Challenges Ahead, Joint Research Centre (of the European Commission's) Science and Policy Report*; Publications Office of the EU: Luxembourg, 2018. [CrossRef]
13. Kulinowski, P.; Panek, P.; Rubacha, P. Aktualne kierunki poszukiwania energooszczędnych rozwiązań w konstrukcji i eksploatacji przenośników taśmowych (Current directions of seeking the energy-saving solutions of design and use of belt conveyors). *Transp. Przemysłowy Masz. Robocze.* **2013**, *3*, 7–12. (In Polish)
14. Roumpos, C.; Partsinevelos, P.; Agioutantis, Z.; Makantasis, K.; Vlachou, A. The optimal location of the distribution point of the belt conveyor system in continuous surface mining operations. *Simul. Mod. Pract. Theory* **2014**, *47*, 19–27. [CrossRef]
15. Wozniak, J.; Pactwa, K. Responsible Mining-The Impact of the Mining Industry in Poland on the Quality of Atmospheric Air. *Sustainability* **2018**, *10*, 1184. [CrossRef]
16. Stoll, R.D.; ·Niemann-Delius, C.; Drebenstedt, C.; ·Müllensiefen, K. *Der Braunkohlentagebau Bedeutung, Planung, Betrieb, Technik, Umwelt*; Springer: Berlin/Heidelberg, Germany, 2009. [CrossRef]
17. Gładysiewicz, L. Belt conveyors. In *Theory and Calculations*; Wroclaw University of Technology Publishing House: Wrocław, Poland, 2003. (In Polish)
18. Jurdziak, L.; Błażej, R.; Bajda, M. Cyfrowa rewolucja w taśmach przenośnikowych-Taśma przenośnikowa 4.0. (Digital revolution in the field of conveyor belts-Conveyor belt 4.0). *Transp. Przemysłowy Masz. Robocze.* **2018**, *2*, 6–18. (In Polish)
19. Gladysiewicz, L.; Hardygóra, M.; Kawalec, W. Determining belt resistance. *Bulk Handl. Today* **2009**, *5*, 23–28.
20. Gladysiewicz, L.; Kawalec, W. Optimised Selection Of A Belt Conveyor Loaded By A BWE. In Proceedings of the 8th International Symposium of Continuous Surface Mining ISCSM, Aachen, Germany, 24–27 September 2006.
21. Jurdziak, L. Methodology of BWE efficiency analysis for power reduction of conveyor drives. In Proceedings of the 8th International Symposium Continuous Surface Mining, Aachen, Germany, 24–27 September 2006.
22. Król, R.; Gladysiewicz, L.; Kaszuba, D.; Kisielewski, W. New Quality Standards of Testing Idlers for Highly Effective Belt Conveyors. *IOP Conf. Ser. Earth Environ. Sci.* **2017**, *95*, 42055. [CrossRef]
23. Bajda, M.; Hardygóra, M. Laboratory tests of operational durability and energy–efficiency of conveyor belts. *IOP Conf. Ser. Earth Environ. Sci.* **2019**, *261*, 12002. [CrossRef]
24. Available online: https://www.kobize.pl/pl/file/wskazniki-emisyjnosci/id/130/wskazniki-emisyjnosci-dla-energii-elektrycznej-za-rok-2017-opublikowane-w-grudniu-2018-r (accessed on 2 July 2021).

 *energies*

Article

Discrete Simulations in Analyzing the Effectiveness of Raw Materials Transportation during Extraction of Low-Quality Deposits

Zbigniew Krysa [1,*], Przemysław Bodziony [2] and Michał Patyk [2]

1 Faculty of Geoengineering, Mining and Geology, Wroclaw University of Technology, Wybrzeże Wyspiańskiego 27, 50-370 Wrocław, Poland
2 Faculty of Civil Engineering and Resource Management, AGH University of Science and Technology, Al. Mickiewicza 30, 30-059 Cracow, Poland; przembo@agh.edu.pl (P.B.); mpatyk@agh.edu.pl (M.P.)
* Correspondence: zbigniew.krysa@pwr.edu.pl

Abstract: The article presents an analysis of the influence of selected operating environment parameters on the operation of a technological system in a mine and examines the profitability of exploiting a deposit of low quality. A model based on a cyclical transportation system in a surface limestone mine, developed in the Haulsim software, served to evaluate—from an economic perspective—several material transportation scenarios depending on the quality of the deposit. A discrete simulation of machine operation allowed a detailed analysis of the technological, operating and economic parameters for selected solutions. The results may be the basis for a decision to begin or to resign from mineral extraction. The simulation results demonstrate that maintaining the operating environment in good technical condition positively influences machine operating cycle times, the required total time of the transportation task and the operating costs. The analysis was performed for the production period of one full year, allowing for the effective work and technological downtimes. This approach allowed the usefulness of the model to be demonstrated in evaluating not only the effectiveness of individual technological procedures but also its economic aspect, related to a decision on the exploitation of "difficult" deposits.

Keywords: discrete event simulation; quarry; mine machine; cost of production

Citation: Krysa, Z.; Bodziony, P.; Patyk, M. Discrete Simulations in Analyzing the Effectiveness of Raw Materials Transportation during Extraction of Low-Quality Deposits. *Energies* **2021**, *14*, 5884. https://doi.org/10.3390/en14185884

Academic Editor: Sergey Zhironkin

Received: 5 August 2021
Accepted: 9 September 2021
Published: 17 September 2021

1. Introduction

In rock raw material mining, the extraction methods and the mining and geological conditions necessitate the use of technological (machine) systems predominantly based on road vehicle transportation with adequately selected loading machines and processing devices. A rational selection of the elements of a technological system employing technological haul trucks and loading machines should be based on a detailed analysis of technical and economic aspects, allow for the deposit conditions, and most importantly reflect the influence of the operating environment on the use of a machine system [1,2].

An analysis of the profitability of deposit extraction and the included evaluation of the effectiveness of the transport system may be viewed in relation to the planned (in the case of the designed mines) or to the incurred (in the case of the already existing mines) expenditures on machine operation (both the use and the ownership), depending on the varied conditions of machine operating environments. This issue is of particular importance in the period of a positive economic situation and constantly growing demand for raw materials in construction. An increase in the price of raw materials allows deposits previously considered low quality to be profitably mined. This issue is a classic problem of natural resource economics and has been researched also for natural resources extracted with the use of mining methods, for example for crude oil [3] and metals [4,5] with the primary aim to identify an optimal production level and extraction cost or to decide on leaving an unmined fragment of the deposit.

A decision to mine new fragments of deposits, e.g., those of low quality, requires an in-depth economic analysis, as it necessitates, prior to such an activity, a reconfiguration of transport routes and a modification of the number of machines involved in the production process, in order to allow the mined material to be transported from new mining faces. An increase in the volume of the transported materials results in lowered unit costs, and thus the extraction of raw materials previously considered unprofitable may prove economical, for example by changing the streams of the mined material within the system (e.g., the limiting of waste transported to the waste heap). This article analyzes the influence of the content of the intermediate product (having a potential for further use) on the efficiency and other operating and economic parameters of the transport system, while placing an emphasis on the correlation with the operating environment, primarily in the context of the energy consumption of a transport system.

The purpose of this research was to offer a reliable simulation model of the actual machine operating conditions and to investigate the influence of selected parameters on the efficiency and cost of transporting intermediate products. The analysis also included the aspect of the reliability of a machine system and its influence on the technological parameters of road transport. The constructed models of actual operating environment and its modifications served to compare variants of the functioning of a machine system in relationship to the operating conditions. Such simulations may inform operational decisions in existing mining plants and aid the design process in the case of their expansion.

Extraction of raw materials is related to numerous technological processes, and therefore the influence of individual factors on these processes and the search of optimal solutions have been the object of interest for a number of authors and have revolved around various aspects of mining operations. The operating effectiveness of a mine is mostly influenced by the transportation costs, which account for 50–60% of the total mining costs [6]. The most frequently analyzed problems are related to the interaction between machines as part of performing a sequence of operations in the technological cycle or of selecting an appropriate type of technical equipment. Trivedi et al. [7] discussed the problem of interaction between machines as a problem typical of operations research, by analyzing the waiting time for the loading/unloading of haul trucks. The arrival and service times for successive machines define in such case a waiting line system, which can be optimized by selecting appropriate machines in order to achieve a measurable economic effect.

Technological systems in mining industry are optimized by implementing various methods based on operations research [8]. The Dijkstra algorithm for defining transportation routes in a mine was used in [9] to demonstrate the influence of selecting roads having particular parameters, e.g., inclination, on the transport cycle times. A waiting line system consisting of machines operated in a technological system of a mine was analyzed by Krause et al. [10], who suggested the application of a model based on a waiting line analysis (Finite Source Model) in designing the size of the machinery fleet in a mine. Differences in the approach to short and long-term modeling have been explored using mixed integer linear goal programming for optimal shovel and truck application in an oil sand deposit case [11].

Transportation processes may be optimized in order to meet instantaneously defined needs of a mining plant [12] or to aid short-term planning of operations. The authors of [13] demonstrated that a combination of discrete simulations of operating mining objects and process optimization models, which allow for, e.g., the limitations due to long-term planning, enables analyses of the influence of any selected factors on, e.g., the global costs, the process efficiency or the output of a mine.

Issues related to the selection of mining machines for the mining plant analyzed in this work were already discussed in [14]. The publication presents a method of multi-criteria decision support allowing for eight criteria divided into the areas of technology, environment and economy. A mixed model proposed by [15] for optimizing machinery selection involves the economic and technological factors and demonstrates the potential for reducing costs and increasing the net present value (NPV) of the project while also reducing greenhouse gas (GHG) emissions.

In the case of mining transportation systems, the problem of energy efficiency is of key importance for evaluating the effectiveness and cost-efficiency of transport. From the perspective of a global optimization of a mining plant, it will relate to the selection of the fleet of machines and their further operation [8,16]. In the case of transport systems using belt conveyors, the issue of energy-efficiency revolves around the application of certain technical solutions, such as belt management based on diagnostic systems [17], energy-efficient idler sets [18], or energy recovery by implementing inverters on conveyors transporting material in a downward direction [19].

The selection of machinery for the transportation system may be also analyzed from the perspective of the quality of the transported material [20]. A search for optimal technological solutions may result in the replacement of machines, a change of the location in the mine or a modification of a technological system. An example of such decisions is the replacement of stationary crushers with a combination of smaller mobile crushers which can be operated more elastically [21].

The failure rate of machines in the technological systems is one of the elements influencing the correct long-term simulation of the operations of the mine analyzed in this article. For one of the copper mines [22], machine failure rate was analyzed with the use of the Monte Carlo simulation, in which failure processes were modeled in order to demonstrate the key factors influencing long-term operational availability of the machines. In turn, preventive maintenance of mining equipment was described by Angeles et al. [23]. Topal et al. drew attention to the high operating costs associated with the use of wheeled transport and pointed out that production planning is very important considering the age of the machines and their availability [24].

Transportation roads were simulated and analyzed in [25], where geotechnical tests served to provide a mine with an optimal solution which allowed for the size of the employed haul trucks.

Saderowa et al. [26] developed simulated transport models, that they compared with analytical solutions and demonstrated the potential of such an approach in analyzing classic transportation systems in an aggregate mine employing haul trucks and loaders. The analysis was performed for various transport distances and operating periods, in dedicated software. Simulation models were also applied in the calculations of railway-based material transportation, and they allowed the authors to confirm the results obtained with the help of classic mathematical calculations [27]. Straka et al. [28] performed a long-term analysis of transport for an operating period of half a year in a mine, together with a description of the principles to be followed when constructing objects and their relationships, as well as when calculating the demand for railway transport. Amin et al. [29] used the Simio software to simulate transport in a limestone mine, searching for an optimal solution to minimize the number of machines. The Monte Carlo simulation was also used in an analysis of a tar sand mine, in which the machinery fleet was modeled on the basis of both the variability of the time for performing technological procedures and the production plan and the transport road network [30].

In conclusion: research into the broadly understood operational effectiveness of mining plants has been a popular area for the implementation of process modeling due to the complex character of the technological processes, variable operating conditions and the number of factors influencing the final economic effect. The authors of this article focus on yet another issue, which involves a complex evaluation of the influence of the production environment on mining operations in a low-quality deposit. Depending on the content of the valuable component, such a deposit necessitates the transportation of different material volumes, and in addition, variable factors significantly influence the profitability of the mining operations.

2. Materials and Methods

The analysis of the mining process and of the operation of the machine system was performed for a limestone deposit owned by a surface mine located in northwest Poland. The deposit has karst interlayers, and thus an average 20% of the mined rock is classified as non-technological material (NTM), which cannot be processed and which is treated as waste in the deposit. This material still needs to be transported to the waste heap, which increases the mining costs.

Figure 1 shows the actual view of the Kujawy mining plant. The central part of the figure presents the main excavation of the mine, with the successive transport ramps leading to it. The overburden and the NTM from the production processes is stored in the "Bielawy" waste heap located in the southeast.

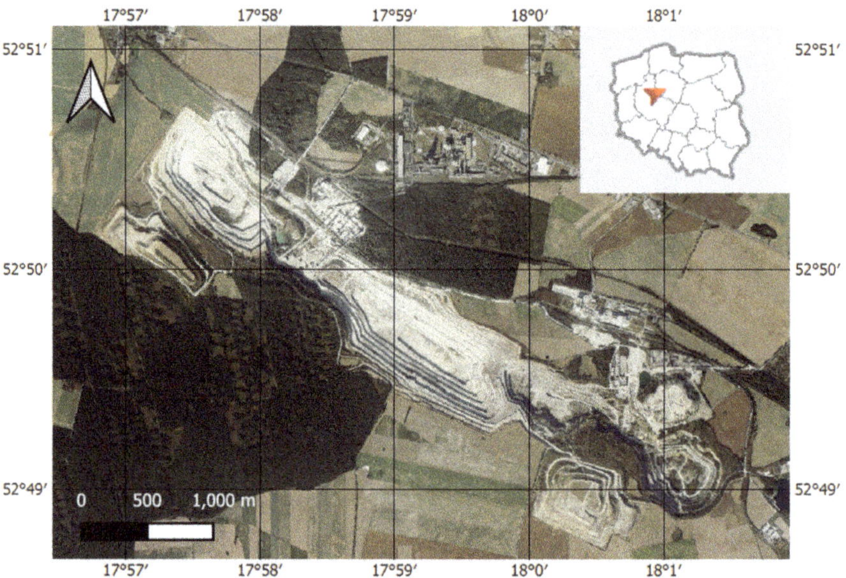

Figure 1. Limestone exploitation area and processing plant in the Kujawy mine and its location in Poland. Available online: https://www.geoportal.gov.pl/ (accessed on 7 September 2021) source: geoportal.com.

The simulation experiment was based on an assumption that depending on the price of the final product, the extraction of the deposit may be economically viable for the content of interlayers between 10% and 40%, which corresponds to the content of intermediate product in the deposit at 60–90%. An assumption was also made that the material extracted from the face should be subjected to preliminary crushing in order to remove grains containing karst 0–40 mm in size, as the fraction suitable for further use in the technological process (the intermediate product) has a grain size greater than 40 mm.

Based on the obtained ground surface model (triangle mesh) and transport road network (additional layer), the Haulsim software was used to build a spatial model comprising separately defined sections forming a network of technological roads (Figure 2). The problem was analyzed with the use of an academic license granted by RPM Software. The modeled transport road system is a coherent entity in which each end point of a road section is simultaneously a start point for another section. The geometry of the characteristic sections is constant (a straight section, a curve, an inclined plane), and the accuracy of the model ensures an increased number of marked sections. This fact allows a precise reconstruction of the road geometry, including its actual longitudinal profile. In such an approach the actual vertical curve of the section is not significantly different than the curve defined in the model. It is thus possible to discretize the model of the extraction environment corresponding to the conditions in the actual mining plant, which is of importance for the actual modeling of the movement components of the haul trucks.

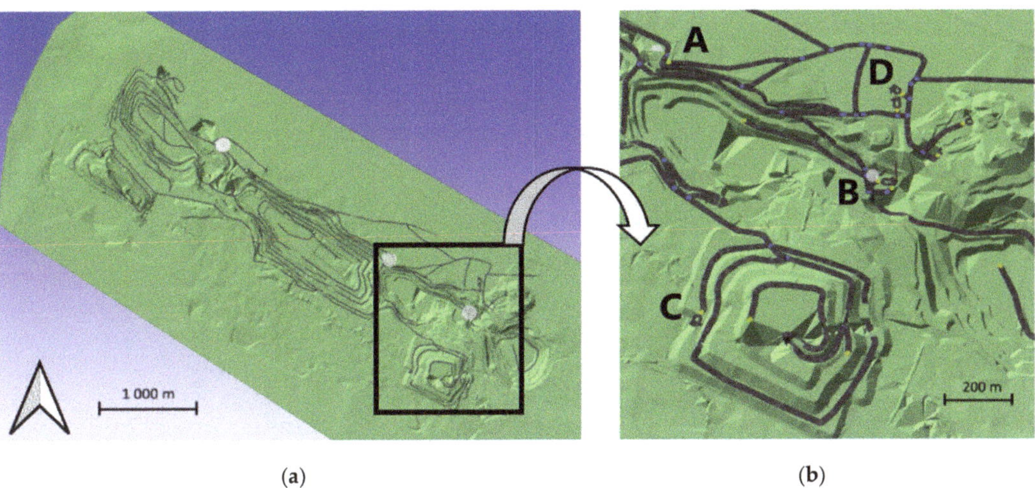

(a)   (b)

Figure 2. Spatial model of mine: top view (**a**) and simulated mine fragment with indicated transport roads and locations of basic technological procedures (**b**).

The model is additionally provided with the defined loading and unloading points, the waste heap, the crushing and storage locations of the material ready for sale, and the facilities important for the functioning of the machinery fleet, i.e., the repair station, the fuel station, etc.

The transport process was analyzed for the part of the deposit (bold selected: Figure 2a) designated with letter A (Figure 2b), in which the material is mined with the blasting technique and transported to the processing plant (B). In the processing plant, the material transported on haul trucks is fed to the Powerscreen Premiertrak 1100 × 800 crusher, where it is separated into two fractions:

- >40 mm—the product for sale—after processing it is transported via a belt conveyor to the storage site (D)
- Non-technological material (0–40 mm), which is transported on haul trucks to the external "Bielawy" waste heap (C).

The total length of the technological transport road in the analyzed part of the mine site is approximately 5000 m. All of the sections inside and outside the production excavations comprise different types of surface, its condition depending inter alia on the season of the year and atmospheric precipitation. It is maintained in good condition with the use of a subgrader machine and by filling the irregularities with the aggregate available on site.

In order to demonstrate what influence the mining environment, i.e., the type of the surface and the inclination of the road sections for the assumed deposit conditions, has on the efficiency and on the profitability of the transportation process, four simulation models were defined: a base model and two scenarios—with a road surface characterized by higher and lower rolling resistance. In the additional road network model (variant 4), the rolling resistance and inclinations of individual sections were defined as optimal for technological roads in the mine—the rolling resistance was assumed equal to that of improved roads (the rolling resistance of 2.5%), and the inclination of the transport ramps is constant along the road section. The model allows the identification of the limit for the influence of the possible changes in the road quality on the technological procedure times and as a result on the transportation costs and efficiencies.

Models of the transport processes involved two transport roads. The first haulage road extends between the face designated as Face/Hauler Loading and the mobile crusher located at a belt conveyor transporting the intermediate product (the valuable material) to the processing plant designated as ZPK Low Crusher. The road is 1089 m long and the elevation difference is 29.5 m. The second road (2542 m in length) comprises the NTM haulage cycle from the site at the mobile crusher to the "Bielawy" heap. Rolling resistances for the transport roads were identified on the basis of an in situ analysis of individual road sections. The parameters of the technological road and of the individual sections are shown in Figure 3.

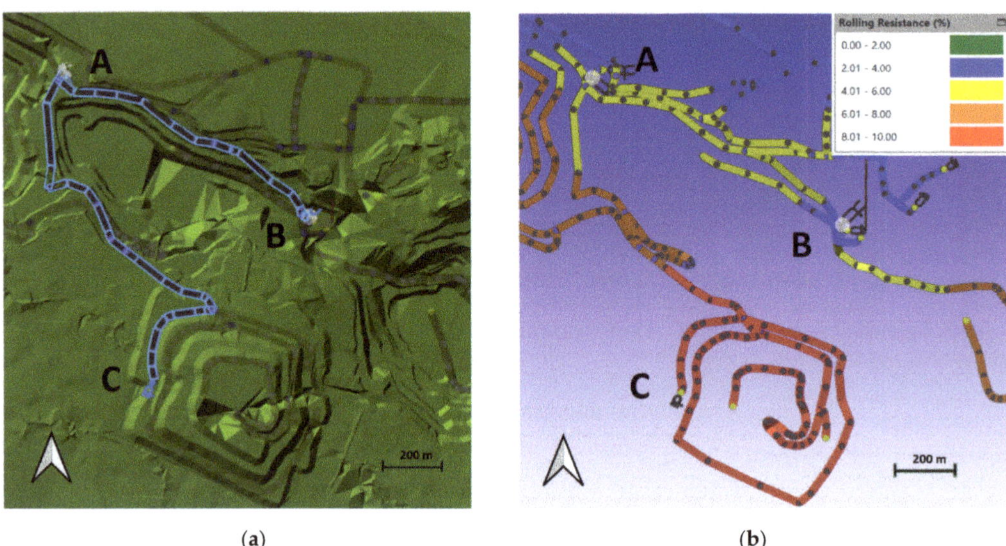

Figure 3. Identified model transport roads (**a**) and rolling resistances (**b**) for individual road sections of the mine in the base scenario.

The technological system was modeled with the actual machines used in the mine, i.e., with bucket wheel loaders Komatsu WA 600. The fleet of the machines used in transporting the run-of-mine (ROM) material to the preliminary crushing, the NTM to the waste heap and the intermediate product to the processing plant consisted of Komatsu Rigid Dump Trucks HD465-7 having a capacity of 55 Mg. The data for the machines used in the model were supplemented with the actual parameters from the operating and maintenance manuals provided by the mine.

In the Haulsim models, the analyzed movement of the technological vehicles was expressed as the energy consumption required to overcome main motion resistances of a loaded and unloaded vehicle on the road according to a certain configuration. Depending on the method for forcing the motion, the following components were identified:

- motion due to the driving force, which involves acceleration, motion at a constant speed and deceleration,
- cornering and motion on an inclined road due to the component of the gravity force.

Consideration was also paid to important aspects related to other energy-consumption components in technological road vehicle transportation, such as idling of the machine in the loading and unloading points, as well as in the waiting line. The analysis did not include the energy-consumption related to the operation of the engine needed to lift the dump body filled with ROM material during unloading, because this was considered as a constant energy output independent of the configuration of the operating environment.

In order to reconstruct real operating conditions of machines over a long period of operation, the operating times allowed for the failure process of the elements of the machine system, separately for the loaders and for the haul trucks. The reliability of the machinery fleet for the time period equal to one year (a total of 4000 operating hours was assumed) was evaluated on the basis of an analysis of the full operation period equal to 14,000 operating hours for wheel loaders (Figure 4) and haul trucks (Figure 5). The analysis included the failure intensity rate of the machinery fleet with allowance for the operation to date (6–10 thousand operating hours). The failure process of the machines in the simulation model was generated with the use of a distribution which best fitted the data from the reliability analysis, i.e., with the use of the Weibull distribution [31–34].

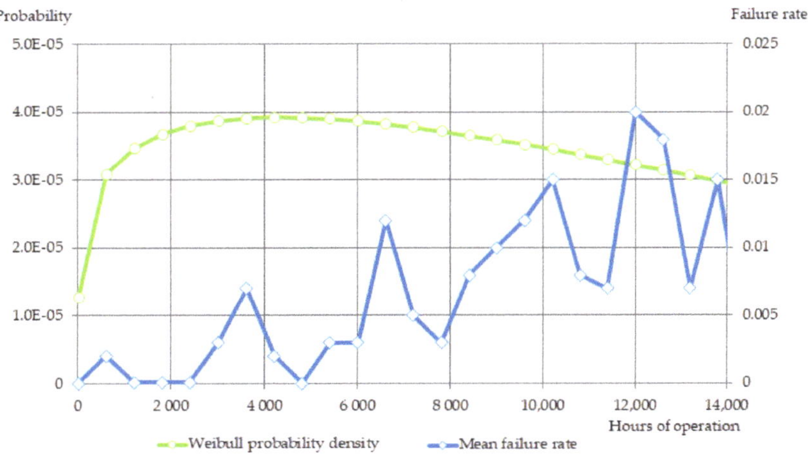

Figure 4. The failure intensity parameter for wheel loaders, for a total observation period of 14,000 h of operation [33].

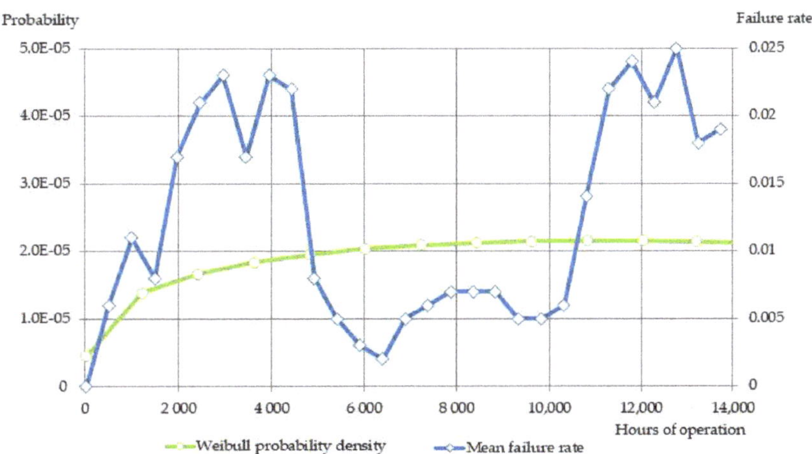

Figure 5. The failure intensity parameter for haul trucks, for the total observation period of 14,000 h of operation [33].

The distribution of failure intensity was determined for all machines based on service requests submitted to the restoration system. These requests included serious failures as well as minor defects indicated and detected by the telemetry system or the operators.

The Weibull distribution was also used to model unplanned maintenance downtime. When defining the failure intensity distribution, the mean number of events per unit of time was assumed to change with time, and in such case the number of defective objects is defined and the time to failure is random.

The study was supplemented with an additional classification and analysis of the costs representing the main components of the use and ownership of a machinery fleet, i.e.,

- costs of fuel,
- costs of lease,
- costs of the Total Maintenance and Repair (TMNR),
- costs of tires (for haul trucks),
- costs of labor—salaries.

For each configuration of the technological system, the analysis was based on an assumed extraction level of 500 Mg per hour. The machines were also assumed to be in lease and maintained by an external service partner of the manufacturer which leased the machines. The investment costs related to the machines were assigned to the category of owning costs and included as leasing costs, because the majority of machines operated in mining industry is purchased with the use of this financing method. All of the machines were operated in two working shifts, with the effective time $T_e = 7$ h. The analyzed machines were assumed to operate for 25 days per month, resulting in 350 working hours per month.

The simulations were performed in four variants of the intermediate product content, in the range 60–90% at 10% intervals. The content of the intermediate product translates directly into the volume of the transported material, as all of the NTM must be transported to the waste heap after the preliminary crushing process.

The model was validated in several stages, as the operating time of the machines required adjustment due to the duration of individual technological procedures in relation to the calendar time, which was de facto reconstructed in the simulated model. An example of visualization of simulated processes is shown in Figure 6. The validation of the model involved the following tasks:

1. Simulation of one transportation cycle, with allowance for:
   - The loading time due to the number of loading cycles,
   - The transport time involving:

     the time of maneuvering the haul truck at the loading site;
     the time of transporting the material to the unloading point;
     the time of maneuvering at the unloading point;
     the unloading time.

2. Simulation of the efficiency of the daily output, which allows for:
   - Real working hours at the mining plant;
   - Downtimes due to the cycle of production changes.

3. Simulation of the efficiency of the weekly and monthly outputs, which allows for:
   - Weekend-related downtime and average number of days in a month.
   - The need to control the process of transporting NTM from the crusher to the waste heap in order to avoid the accumulation of excess stock.

4. Simulation for the annual output allowed for:
   - Failure intensity of mining machines;
   - Necessary downtime and time out of service due to planned and unplanned maintenance.

Figure 6. The process of loading the material on the haul truck visualized against the network of mine roads (removed layer in the terrain model) in the simulated model.

The four variants of the product-NTM ratios were provided with additional three models allowing for different configurations of road qualities in the mine. In the base scenario, the roads were assumed to be identical to the direct in situ observation in the mining plant. In the modeled variants, the analysis of the operational environment took into account variation in the rolling resistance due to the effect of precipitation, as well as seasonal variation. The rolling resistance values for the individual road sections were given as average results from in situ mine road surveys. The quality of the road was not analyzed in terms of the type of material used for its construction because the input parameter in the model was the rolling resistance (which is directly influenced by the road construction material).

In a scenario referred to as weak, all of the roads were assumed to have a quality worse than in the base scenario, showing increased rolling resistance due to, e.g., transverse and longitudinal irregularities and low wheel load capacity. In the optimistic scenario, the roads were assumed to have a higher standard. In the additional scenario, the roads were assumed to have the lowest rolling resistance available for technological roads and the transport ramps—to have a constant inclination. A constantly inclined transport road minimizes the number of changes of the operating modes of the drive systems in the haul trucks, improving the transport economy. Table 1 summarizes the information on the quality of the transport roads. In the simulations, the road quality was described with the use of a basic parameter influencing transport effectiveness, i.e., with the rolling resistance of wheels on various surfaces.

Table 1. Basic parameters of the haulage roads for ROM and NTM.

Parameter	Road 1	Road 2
Length	1089 m	2542 m
Elevation difference	29.51	66.16
Average inclination	2.7%	2.6%
Average rolling resistance [1]	3.04%	8.89%

[1] Average weighted with the length of a section having a particular resistance.

For comparison, Figure 7 shows a view of the model of the transport routes divided into individual sections for the variants with low and high road surface quality. They demonstrate that the difference in the level of average rolling resistances between the "good" and the "weak" surface remains within the range of 3.0–8.6% (Table 2). The road sections having the highest surface quality in the "good" variant show a rolling resistance of 2.5%, and the sections having the lowest quality in the "weak" (degraded) variant show 9% rolling resistance for the haulage of ROM material from the face to the crusher. Mean rolling resistance along the route from the crusher to the waste heap remains within the range from 4.0% in the "good" variant to 10.5% in the "weak" variant. When considering individual sections, the differences between the rolling resistances are even greater, within a range from 2.5% to 16% (Table 3).

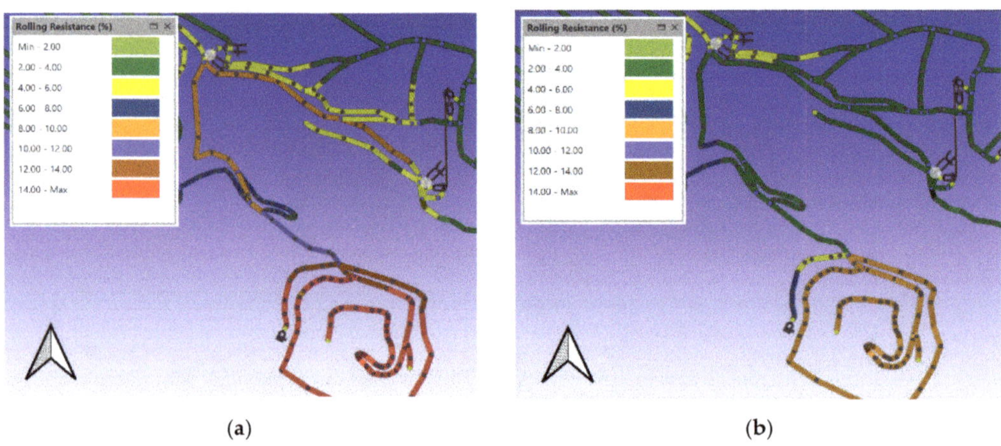

(**a**) (**b**)

Figure 7. Modeled variants of rolling resistances in the scenario for low quality roads (**a**) and good quality roads (**b**).

Table 2. Rolling resistances of road sections in different modeled transport scenarios—road 1.

Road 1					
Good		**Base**		**Weak**	
Rolling Resistance, %	**Section Lengths, m**	**Rolling Resistance, %**	**Section Lengths, m**	**Rolling Resistance, %**	**Section Lengths, m**
2.5	86.5	4	144.1	6	144.0
3	827.0	6	944.5	9	944.5
3.5	175.1				
Av. resist.	3.0%		5.7%		8.6%

Table 3. Rolling resistances of road sections in different modeled transport scenarios—road 2.

Road 2					
Good		**Base**		**Weak**	
Rolling Resistance, %	**Section Lengths, m**	**Rolling Resistance, %**	**Section Lengths, m**	**Rolling Resistance, %**	**Section Lengths, m**
2.5	51.1	4	108.7	6	108.7
3	827.0	5	1208.5	9	909.2
3.5	518.2	7	511.7	10	766.1
4	748.0	10	712.9	12	360.3
6	212.1			14	212.1
8	185.3			16	185.3
Av. resist.	4.0%		7.2%		10.5%

The estimation of the economic aspect of the modeling was possible by classifying and analyzing the costs representing the main components of the use and ownership of the machinery fleet (Table 4). The number of modeled machines was selected in accordance with the assumed parameters of the mining operations. The loading process was performed by two loaders (one at the face and one on the site at the crusher) and by two haul trucks. The costs of the loader at the crusher were assigned to the mining costs of the analyzed fragment of the deposit only to an extent to which it performed works related to the deposit (only direct loading tasks). The remaining machines performed the entire technological process, and thus 100% of their operating costs were assigned to the mining costs.

Table 4. Costs of machinery fleet (in EUR).

Cost		Loader Units WA 600	Haul Units Komatsu HD 465
Owning cost	Per hour	16.30	13.42
	Per year	142,800	117,564
Operating cost	Per hour	19.40	20.75
	Per year	169,944	181,770
Fuel	Per hour	40.00	39.40

3. Results

A total number of 16 simulation models were constructed. They allowed a multi-aspect and reliable evaluation of the product-NTM ratios, and most importantly an analysis of the influence of the mining environment on the effectiveness of the processes measured with efficiency, as well as with the total and unit cost.

The model validation process consisted of adjusting the operational availability time of the machines to the observations in the simulation, i.e., to the advancement degree of the technological process. This was performed in the dialog window of the software (see Figure 8). The base model prepared in the first stage served to construct successive variants of the models of the technological process, changing the content of the intermediate product in the mined deposit. The manipulation of the quantity of ROM material in the transported stream resulted in the change of the number of cycles along the haulage road to the waste heap and consequently in the time of the entire transport task.

As in the case of the base model, with an increased number of transport cycles to the waste heap, the successive model variants required independent (for each individual model) regulation of the ratio between the output and the NTM in order to avoid the overstock at the crusher.

The product-NTM ratio has a direct influence on the total operating costs, as it defined the volume of the processed rock mass. For example, in the case of the 90% deposit quality, 10% of NTM is transported to the waste heap, while in the first place all of the ROM material (100%) is transported from the face to the crusher by the haul trucks. Subsequently, after crushing, 90% of the material is transported to the ZPK processing plant via the belt conveyor, while the 10% of the material is transported to the waste heap with the use of wheel transportation. Thus, the total transported masses are 110%. The necessity to transport larger masses resulted in a greater number of transport cycles, which in consequence increased the duration time of both the process and its simulation. In addition, the configuration and the quality of the transport roads was demonstrated to have an influence not only on fuel consumption but also on the reduction of the cycle time and on the increase of the transport efficiency.

The total simulation time shown in Table 5 should be interpreted in relation to 10 months (300 calendar days, 250 working days) of the effective operating time of the mine per one year. This fact indicates that at 20% NTM content the target transport system (2 haul trucks + 2 loaders) is capable of performing the annual output plan only in the variant in which the transport roads are of optimal quality. Depending on the quality of the road, a greater percentage share of NTM will require a greater number of haul trucks. The simulation also demonstrated that the base (actual) road quality does not ensure the expected haulage over the planned time.

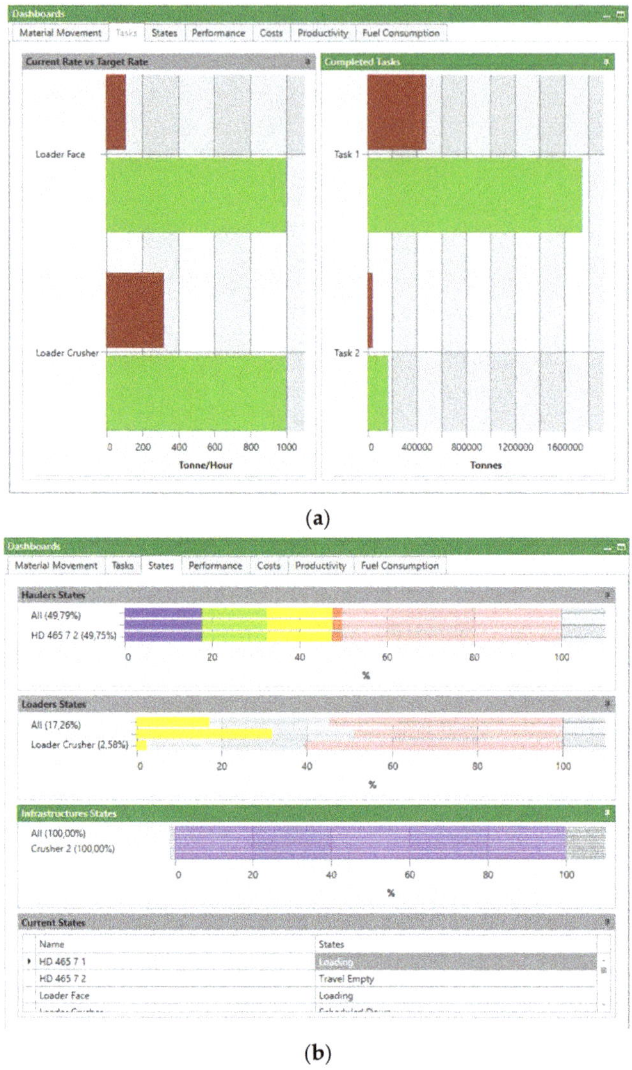

(a)

(b)

Figure 8. Source of information about the implementation of the processes in the model: (**a**) task advancement; (**b**) machine regimes.

Table 5. Operating time of the model in days (target time of 300 days).

Road Quality	Product—NTM Ratio			
	0.9	0.8	0.7	0.6
Weak	346.4	397.5	450.3	503.1
Base	317.1	362.1	403.5	450.3
Good	301.4	341.3	380.2	416.2
Optimal	269.3	303.1	333.5	367.2

The discrete simulation provided partial production costs—assigned to the unit of time and to the operated object (machine)—which were then aggregated to the total costs of the transport cycle and the time of the transportation task. Figure 9 below shows a representative model of the consumed fuel and its cost in relation to individual sections of the road in the base (actual) variant and in the optimal scenario. An analogical model was constructed for the two remaining simulation variants.

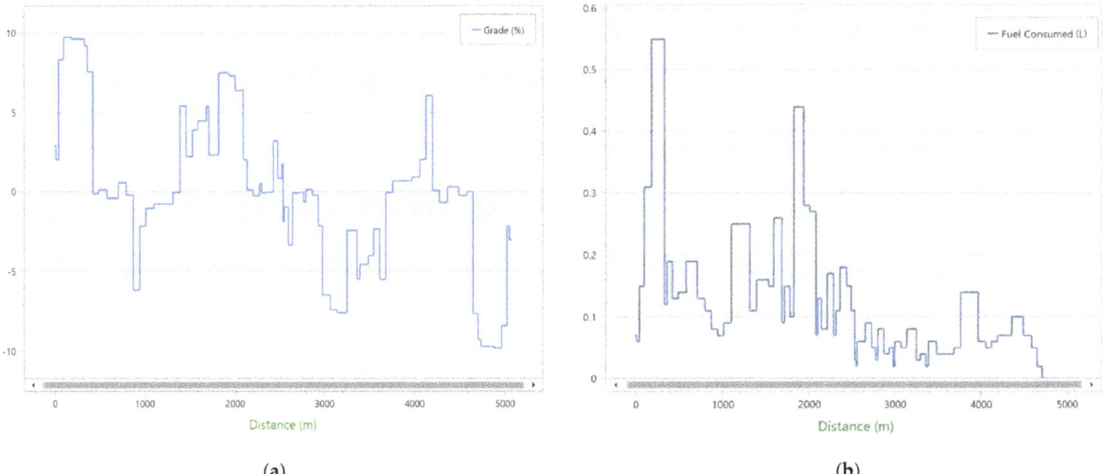

(a)   (b)

Figure 9. Longitudinal profile of the transport road (**a**) versus fuel consumption (**b**) for individual road sections modeled in the base scenario over one cycle of transporting NTM to the waste heap.

In the optimal scenario, the road inclination was defined as constant and therefore the graph in Figure 10a is a straight line. Constant road inclination is one of the factors influencing the regime of the haul trucks. The variability range of the fuel consumption is nearly two-fold lower than in the case of the base scenario (Figure 9b vs. Figure 10b), and the total fuel consumption by the haul truck per one transport cycle was 5.44 L, while in the base scenario it was 7.71 L.

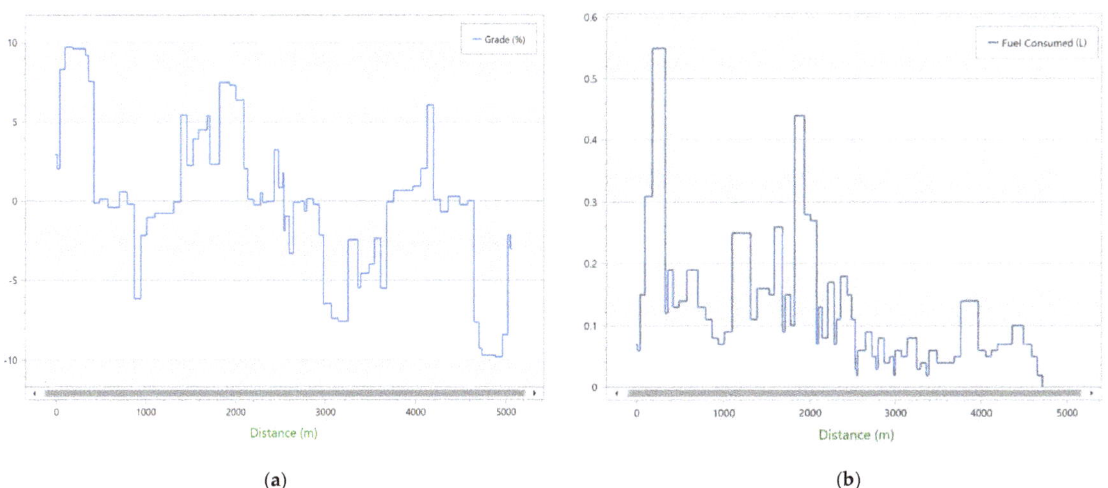

(a)   (b)

Figure 10. Longitudinal profile of the transport road (**a**) versus fuel consumption (**b**) for individual road sections modeled in the optimal scenario over one cycle of transporting NTM to the waste heap.

The costs of the transport cycle were summed for the production shift and for the total planned production volume. The results represent a complex analysis of the operational costs depending on the operating environment (see Table 6). Such calculations may become a basis for a decision to start or resign from extracting minerals from a low-quality deposit, and in combination with the sale price it may provide a reliable estimation of the future profit.

Table 6. Operational costs of transportation in the analyzed variants.

Road Quality	Product—NTM Ratio			
	0.9	0.8	0.7	0.6
Weak	325,184	389,592	446,751	514,083
Base	294,782	345,798	398,512	461,716
Good	278,393	326,877	372,317	414,012
Optimal	247,718	292,314	326,170	364,423

The unit costs summary below (Table 7) was calculated on the basis of the components shown in Table 4. In the simulation model, the production costs were calculated on the basis of the simulated values in relation to the unit of time, and subsequently aggregated to total costs. The cost structure resulting from individual models is variable and strongly correlated with the quality of the technological roads. For example, Figure 11 shows the cost structure for the variant in which NTM accounts for 10% of the ROM material. A decrease in the road quality causes an increase in the fuel consumption, as a share both in the total transportation cost and in the absolute values. The results obtained for the presented model allow an approximation that an increase in the rolling resistances by 1% results in an increase of the share of fuel costs by around 1.25%. When quantified, the fuel cost for the haul trucks changes from 56.4 thousand EUR for optimal quality roads to 139.8 thousand EUR for roads showing high rolling resistances, in the case when NTM accounted for 10% of the mined material. In the case with high NTM content (40%), the fuel cost varied in the range from 95.2 to 238.5 thousand EUR. The fluctuation range indicates the scale of the possible savings resulting from improving the road quality.

Table 7. Unit costs of transportation in the analyzed scenarios, in relation to 1 Mg of the final product.

Road Quality	Product—NTM Ratio			
	0.9	0.8	0.7	0.6
Weak	0.206	0.278	0.365	0.490
Base	0.187	0.247	0.325	0.440
Good	0.177	0.233	0.304	0.394
Optimal	0.157	0.209	0.266	0.347

In the analysis, the transportation costs remain within a range from 422 thousand EUR in the case of the variant with good quality roads to 537 thousand EUR in the case of the variant with weak quality roads. In the base variant, the sum total of the transport costs was 468 thousand EUR. The simulations did not reveal any significant changes of the cost structure in the case of changing the product-NTM ratio.

Quantitatively, the greatest cost changes were observed for the fuel used by the haul trucks, which phenomenon is related to the energy output required to overcome the motion resistances depending on the quality of the operating conditions (the transport roads and the maneuvering locations at the loading and unloading points). The range of changes in fuel costs depending on the product content and road quality is shown in Figure 12.

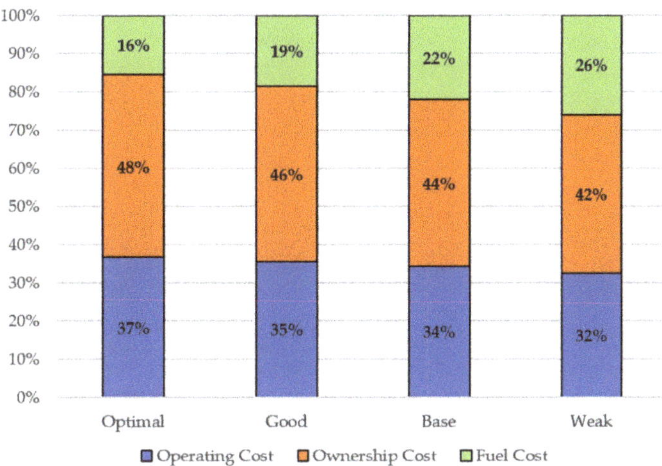

Figure 11. Cost structure for haul trucks depending on the assumed road quality scenario, for the 10% NTM content.

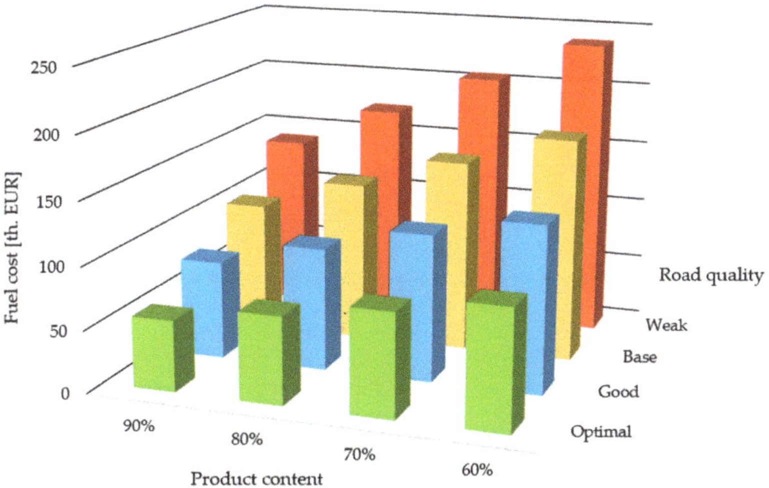

Figure 12. Fuel consumption costs for haul trucks in the 16 analyzed models of the technological system.

A decision to invest in improving the quality of the transport roads has a measurable influence on the effectiveness of the system and reduces the transport costs according to a scale presented in Table 8.

In addition to the above calculations, an observation should be made that the cost of a set of tires for the analyzed type of mining machines oscillates between 40 and 50 thousand EUR and accounts for approx. 30% of the costs associated with operating haul trucks [1]. By modifying the operating environment, it is possible not only to lower fuel consumption, but also to prolong the life of the tires, as well as of the machines. Further analyses should focus on the economic balance of investments on improving the quality of technological roads with respect to the lower costs of operating the machine system. Such an analysis may allow economically justified decisions in the perspective of the planned lifetime of both the mine and the machinery fleet. Improved road quality in a mining plant has another, unquantified effect in the form of increased work safety.

Table 8. Annual profit/loss in transport costs in the case of changing the parameters of the technological roads with respect to the base scenario.

Road Quality	Product—NTM Ratio			
	0.9	0.8	0.7	0.6
Weak	−30,402	−43,794	−48,239	−52,367
Base	0	0	0	0
Good	16,390	18,921	26,195	47,704
Optimal	47,065	53,484	72,342	97,293

4. Discussion

The results of research related to the modeling of technological processes in surface mining, as published to date and known to the authors, focus mainly on the modeling of the stream of the ROM material, on the general representation of the technological diagrams, or on short-term analyses of individual work cycles of machines.

This work offers a novel approach to the problem of long-term modeling of the operation of mining machines in a surface mine, with allowance for the aspects of energy-consumption and failure intensity of the technological system. Such a task entailed a number of difficulties related to properly defining all of the input parameters and the multi-object simulation model.

The results of the simulation experiments clearly indicate the reliability of the selected and analyzed operating and technological parameters, primarily for the estimation of the profitability of mining a low-quality deposit. It is particularly important to compare the results of the base model, which reproduces the current state, with the actual times of the production task. The difference is less than 6% (317 days in the model vs. 300 days of actual work). This fact indicates high consistency, especially if account is made of the time range of the model operation and of the projected randomness of phenomena during the operation. The running time of the model determines the costs and efficiency of the haulage process, and therefore estimating the time of individual technological operations is crucial for the credibility of the obtained results.

The discrete simulation enabled an almost costless analysis (except for the cost of work of the analyst and for the cost of the software) aiding investment decisions and allowing for a number of parameters, which are excluded in analytical models, or which are not included in short-term predictions due to the limited probability of the occurrence of particular circumstances (e.g., random events such as failures). The comparison involved total system production values obtained when considering the equipment reliability, availability and maintainability (RAM) characteristics. It is demonstrated that by considering the RAM aspects in the context of truck and wheel loaders, the total production of the system can be maximized and specific information on the production availability and productivity of its components can be obtained.

5. Conclusions

The methodology presented in this article illustrates a practical application of modeling transportation processes in surface mining, which allow haul trucks and loading machines to be evaluated in particular operating conditions of a quarry, especially in the context of low-quality deposits.

The implemented modeling allowed a systematic arrangement of alternative variants of the operating environment by both unit and total costs of machine operation, as well as by other directly influencing parameters. Specifying a universal set of reliable parameters in the modeling of the transportation and loading processes allowed a complex and long-term analysis of the functioning of a technological plant and an evaluation from the operational and economic perspective.

Such analyses enable the identification of the production potential understood as the ability of the entire system or each of its elements to perform technological processes. It is additionally possible to indicate an object characterized by the highest and the lowest failure intensity, and similar operating characteristics may serve as a basis to develop optimized operating systems, e.g., the method for performing planned maintenance and for scheduling their frequency for different time periods of the functioning of a mining plant.

As demonstrated in the analysis, in mining plants in which wheel transportation dominates, technological roads should be maintained in the best possible quality. The sole costs of fuel may increase more than two-fold in the case when the roads are not maintained properly (the decrease from the good to the weak quality).

The analyses show the structure and the influence of the technological and operating parameters, and primarily of the operating costs of a varied machinery fleet in rock raw material mining industry, relative to low-quality deposits. Further research should be performed in order to identify the profitability limit of investments on the modification of the operating environment of a machine system with respect to the potentially lower component operating costs of machines and also to their increased durability, reliability and operating life.

The proposed methodology has a universal character and therefore may be implemented in similar conditions in which vehicles and other transport machines operate subjected to a relationship of resistance resulting from the operating environment—transport roads—and overcoming these resistances in vehicle transport. The maximum service speed, the traction force profile and the rolling resistance are the unit contributions necessary to define the running frequencies in order to allow a vehicle to slot into the overall traffic pattern on a particular transport line. Collaborative analyses can be performed not only in quarrying, but in the case of any similar relationships, for example mining of metal ores and environmental engineering.

It should be noted that the analyzed model concerned several technological activities and did not simulate the operation of the entire mine. It is not yet clear at this stage whether simulations of complex systems will also provide unambiguous results, showing savings from better road quality or other investments in the mine. The reason is the difficulty in simultaneously observing the modeled simulation of discrete events produced by many objects (faces, dumps, machines) and interdependent processes (queues, blocked roads, breakdowns, etc.). In the case of a mine with several independent production processes, the analysis of complex transport systems can be divided into basic technological lines where the presented method allows the assessment of the impact of changes in the transport system/network on the haulage efficiency.

Author Contributions: Conceptualization, Z.K. and P.B.; methodology, Z.K. and P.B.; software, Z.K.; validation, M.P. and P.B.; formal analysis, M.P. and P.B.; investigation, Z.K.; resources, M.P.; data curation, M.P.; writing—original draft preparation, Z.K. and P.B.; writing—review and editing, Z.K., P.B.; visualization, Z.K.; supervision, Z.K., P.B. and M.P.; project administration, Z.K.; funding acquisition, Z.K. All authors have read and agreed to the published version of the manuscript.

Funding: This research was funded with the research subsidy from the Polish Ministry of Science and Higher Education granted for 2021.

Data Availability Statement: The detailed results of simulation tests for all presented variants are available on request from the corresponding author.

Acknowledgments: The authors thank RPM Software Pty Ltd. for granting the Haulsim license to the Wrocław University of Technology.

Conflicts of Interest: The authors declare no conflict of interest.

References

1. Czaplicki, J.M. *Shovel-Truck Systems. Modelling, Analysis and Calculation*; Taylor & Francis Group: London, UK, 2008; ISBN 978-042-920-696-2.
2. Koptev, V.; Kopteva, A.; Ivanova, T. Directions for the development of transport machines for open-pit mining. *J. Appl. Eng. Sci.* **2021**, *19*, 137–141.
3. Stiglitz, J. Monopoly and the Rate of Extraction of Exhaustible Resources. *Am. Econ. Rev.* **1976**, *66*, 655–661.
4. Chermak, J.M.; Patrick, R.H. Comparing tests of the theory of exhaustible resources. *Resour. Energy Econ.* **2002**, *24*, 301–325. [CrossRef]
5. Halvorsen, R.; Smith, T. A Test of the theory of exhaustible resources. *Q. J. Econ.* **1991**, *106*, 123–140. [CrossRef]
6. Nel, S.; Kizil, M.; Knights, P. Improving truck-shovel matching. In Proceedings of the 35th APCOM Symposium—Application of Computers and Operations Research in the Minerals Industry, Wollongong, Australia, 24–30 September 2011; pp. 381–391.
7. Trivedi, R.; Rai, P.; Nath, R. Shovel-truck optimization study in an opencast mine—A queueing approach. *Indian J. Eng. Mater. Sci.* **1999**, *6*, 153–157.
8. Shah, K.; Rehman, S. Modeling and Optimization of Truck-Shovel Allocation to Mining Faces in Cement Quarry. *J. Min. Environ.* **2020**, *11*, 21–30. [CrossRef]
9. Souza, F.; Câmara, T.; Torres, V.; Nader, B.; Galery, R. Mine fleet cost evaluation-dijkstra's optimized path. *REM-Int. Eng. J.* **2019**, *72*, 321–328. [CrossRef]
10. Krause, A.; Musingwini, C. Modelling open pit shovel-truck systems using the Machine Repair Model. *J. S. Afr. Inst. Min. Metall.* **2007**, *107*, 469–476.
11. Upadhyay, S.P.; Askari-Nasab, H.; Tabesh, M.; Badiozamani, M.M. Simulation and optimization in open pit mining. In *Application of Computers and Operations Research in the Mineral Industry, Proceedings of the 37th International Symposium, Florence, Italy, 16–24 May 2015*; Society for Mining Metallurgy & Exploration Inc. (SME): Englewood, CO, USA, 2015; pp. 532–543.
12. Sofranko, M.; Wittenberger, G.; Skvarekova, E. Optimisation of technological transport in quarries using application software. *Int. J. Min. Miner. Eng.* **2015**, *6*, 1–13. [CrossRef]
13. Upadhyay, S.P.; Askari-Nasab, H. Simulation and optimization approach for uncertainty-based short-term planning in open pit mines. *Int. J. Min. Sci. Technol.* **2018**, *28*, 153–166. [CrossRef]
14. Patyk, M.; Bodziony, P.; Krysa, Z. A Multiple Criteria Decision Making Method to Weight the Sustainability Criteria of Equipment Selection for Surface Mining. *Energies* **2021**, *14*, 3066. [CrossRef]
15. Nakousi, C.; Pascual, R.; Anani, A.; Kristjanpoller, F.; Lillo, P. An asset-management oriented methodology for mine haul-fleet usage scheduling. *Reliab. Eng. Syst. Saf.* **2018**, *180*, 336–344. [CrossRef]
16. Matsimbe, J. Optimization of Shovel-Truck Productivity in Quarries. *Int. J. Res. Advent Technol.* **2020**, *8*, 1–9.
17. Błażej, R.; Jurdziak, L.; Kawalec, W. Condition Monitoring of Conveyor Belts as a Tool for Proper Selection of Their Replacement Time. In *Advances in Condition Monitoring of Machinery in Non-Stationary Operations*; Springer: Cham, Switzerland, 2017. [CrossRef]
18. Król, R.; Kawalec, W.; Gładysiewicz, L. An Effective Belt Conveyor for Underground Ore Transportation Systems. *IOP Conf. Ser. Earth Environ. Sci.* **2017**, *95*, 042047. [CrossRef]
19. Kawalec, W.; Król, R. Generating of Electric Energy by a Declined Overburden Conveyors in a Continous Surface Mine. *Energies* **2021**, *14*, 4030. [CrossRef]
20. Šofranko, M.; Lištiaková, V.; Žilák, M. Optimizing transport in surface mines, taking into account the quality of extracted raw ore. *Acta Montan. Slovaca* **2012**, *17*, 103–110.
21. Teplická, K.; Straka, M. Sustainability of extraction of raw material by a combination of mobile and stationary mining machines and optimization of machine life cycle. *Sustainability* **2020**, *12*, 10454. [CrossRef]
22. Morad, A.M.; Pourgol-Mohammad, M.; Sattarvand, J. Application of reliability-centered maintenance for productivity improvement of open pit mining equipment: Case study of Sungun Copper Mine. *J. Cent. South Univ.* **2014**, *21*, 2372–2382. [CrossRef]
23. Angeles, E.; Kumral, M. Optimal Inspection and Preventive Maintenance Scheduling of Mining Equipment. *J. Fail. Anal. Prev.* **2020**, *20*, 1408–1416. [CrossRef]
24. Topal, E.; Ramazan, S. A new MIP model for mine equipment scheduling by minimizing maintenance cost. *Eur. J. Oper. Res.* **2010**, *207*, 1065–1071. [CrossRef]
25. Nurić, A.; Nurić, S. Numerical modeling of transport roads in open pit mines. *J. Sustain. Min.* **2019**, *18*, 25–30. [CrossRef]
26. Saderova, J.; Rosova, A.; Kacmary, P.; Sofranko, M.; Bindzar, P.; Malkus, T. Modelling as a tool for the planning of the transport system performance in the conditions of a raw material mining. *Sustainability* **2020**, *12*, 8051. [CrossRef]
27. Saderova, J.; Marasova, D.; Gallikova, J. Determining of mining transport capacity by simulation. In Proceedings of the 19th International Multidisciplinary Scientific GeoConference-SGEM, Albena, Bulgaria, 28 June–6 July 2019; Volume 19, pp. 217–224.
28. Straka, M.; Rosová, A.; Lenort, R.; Besta, P.; Šaderová, J. Principles of computer simulation design for the needs of improvement of the raw materials combined transport system. *Acta Montan. Slovaca* **2018**, *23*, 163–174.
29. Amin, I.; Adil, M.; Rehman, S.U.; Ahmad, I. Simulation of Truck-Shovel Operations Using Simio. *Int. J. Econ. Environ. Geol.* **2017**, *8*, 55–58.

30. Upadhyay, S.; Tabesh, M.; Badiozamani, M.; Askari-Nasab, H. A Simulation Model for Estimation of Mine Haulage Fleet Productivity. In *Springer Series in Geomechanics and Geoengineering, Proceedings of the 28th International Symposium on Mine Planning and Equipment Selection*; Springer: Berlin/Heidelberg, Germany, 2020; pp. 42–50.
31. Barabady, J.; Kumar, U. Reliability analysis of mining equipment: A case study of a crushing plant at Jajarm Bauxite Mine in Iran. *Reliab. Eng. Syst. Saf.* **2008**, *93*, 341–361. [CrossRef]
32. Bebbington, M.; Lai, C.D.; Zitikis, R. A flexible Weibull extension. *Reliab. Eng. Syst. Saf.* **2007**, *92*, 719–726. [CrossRef]
33. Bodziony, P.; Patyk, M. Analysis of Utilisation States and Reliability Parameters of Process-Line Configuration for Low-Quality Rock Deposits. *Adv. Sci. Technol. Res. J.* **2021**, *15*, 126–133. [CrossRef]
34. Sinha, R.S.; Mukhopadhyay, A.K. Failure rate analysis of jaw crusher using Weibull model. *Proc. Inst. Mech. Eng. Part E J. Process Mech. Eng.* **2017**, *231*, 760–772. [CrossRef]

Article

Dynamic Analysis for the Hydraulic Leg Power of a Powered Roof Support

Dawid Szurgacz [1,2]

[1] Center of Hydraulics DOH Ltd., 41-906 Bytom, Poland; dawidszurgacz@vp.pl
[2] Polska Grupa Górnicza S.A., 40-039 Katowice, Poland

Abstract: This paper presents the results of a study conducted to determine the dynamic power of a hydraulic leg. The hydraulic leg is the basic element that maintains the position of a powered roof support. It is located in the structure between the canopy and the floor base. The analysis assumes that its power must be greater than the energy of the impact of the rock mass. The energy of the rock mass is generated by tremors caused mainly by mining exploitation. The mining and geological structure of the rocks surrounding the longwall complex also have an influence on this energy generation. For this purpose, stationary tests of the powered roof support were carried out. The analysis refers to the space under the piston of the leg, which is filled with fluid at a given pressure. The bench test involved spreading the leg in the test station under a specified pressure. It was assumed that the acquisition of dynamic power would be at the point of pressure and increase in the space under the piston of the leg under forced loading. Based on the experimental studies carried out, an assessment was made with the assumptions of the methodology adopted. The results of the theoretical analysis showed consistency with the experimental results.

Keywords: power; powered roof support; hydraulic leg; bench testing; dynamic load

Citation: Szurgacz, D. Dynamic Analysis for the Hydraulic Leg Power of a Powered Roof Support. *Energies* **2021**, *14*, 5715. https://doi.org/10.3390/en14185715

Academic Editors: Robert Król, Izabela Sówka, Witold Kawalec and Krzysztof M. Czajka

Received: 9 June 2021
Accepted: 8 September 2021
Published: 10 September 2021

1. Introduction

One of the main problems associated with underground coal mining is descending depth. Therefore, the exploitation of coal deposits requires the mining industry to conduct research on technology development [1] and the monitoring of work parameters [2–6] in order to improve the safety of crews [6–9]. The development of mining determines the need for experimental [10–13] and theoretical [14–18] work, aimed at a deeper understanding and analysis of the influence of various factors and the relationships between them. Research into the dynamic loading of roof support defined future directions in [19], and it was shown, among other things, which forces are generated by the pressure on the test leg. This resulted in the theoretical identification of the parameters of susceptible shoring for gallery workings [20].

Ongoing testing of hydraulic legs under mass impact loading is intended to determine the actual resistance of the leg to dynamic loading [21–23]. The dynamic load resistance of a hydraulic leg is understood as the capacity to absorb the energy of a mass impact over a well-defined period of time without breaking or being damaged. Dynamic tests of hydraulic legs allow the assessment of their suitability for use in tremor hazard conditions [24–27]. The weight of the impactor mass used in the tests is assumed to be approximately 1/10th of the working support of the leg. Making this assumption allows for the technical implementation of the research. In mass impact studies, we relate its value to the support of a model-type leg with a constant rate of hydraulic fluid pressure build-up [28–30].

The dynamic load on a powered roof support is caused by the movement of the rock mass, causing energy to be transferred from the rock mass to the powered roof support. These loads occur during short-term tremors. The tremor-causing centres are usually thick sandstone banks located in the roof or the bottom of the mine workings [31].

Among the dynamic phenomena occurring in mines, a distinction is made between tremors, stress relief, and rockbursts. A tremor is a seismic phenomenon that manifests itself in mines by vibration of the rock mass and acoustic effects. In the case of strong tremors, the vibrations of the rock mass is felt at the surface. The relaxation of the rock mass manifests itself through its vibration, acoustic effects, and the cracking of the rock around the excavation. The phenomenon is combined with minimal slides of the support and minor damage, which does not reduce the functionality of the excavation. A rockburst causes rocks to be thrown into the excavation, vibrating the rock mass. A rockburst is usually accompanied by damage or destruction of the support and subsequent clamping and collapse of the workings [32]. Legs that are not designed to take dynamic loads are most likely to be damaged (Figure 1).

Figure 1. Area of a damaged hydraulic leg cylinder structure of a powered roof support.

The main task of a powered roof support is to maintain the roof above the workings, i.e., to provide resistance to the rock mass compressing the mine workings. Its construction has a certain load-bearing capacity, and it must be equipped with a safety valve that is placed in the construction of the hydraulic leg. The safety valve determines the behaviour of the powered roof support [33]. In a longwall system, powered roof supports play a special role, namely, the roof support set. This is because the support consists of so-called sections, which are repetitive units set along the wall with a specific pitch, most often 1.5 m. Thus, a set of supports in a wall may consist of one hundred and several dozen to two hundred sections. The support section consists of such basic elements and assemblies as the feller plate, which forms its base, the roof support floor, the hydraulic support system with hydraulic legs, and the sliding system (Figure 2).

Studies on the dynamics of hydraulic cylinders of powered roof supports can be found in the literature [34–42]. The research is based on the hydroelasticity theory taking into account the coupling between the cylinder–liquid–piston system. On the basis of this model, the problems of harmonic vibrations under force and kinematic excitation were solved. The result of this research was to determine the actual dynamics, which was proven. The study of the magnitude and nature of dynamic loads resulting from the impact of rock mass as tremors directly on powered roof supports has been the subject of numerous studies [43–48]. Many researchers from all over the world are conducting studies to identify the required dynamic resistance of the leg [49–53].

The mining exploitation in the conditions of tremor hazard sets high requirements for the producers of powered roof supports in the process of designing and testing in the test legs [54]. A powered roof support must be secured in such a way that it will not be damaged by a dynamic load. The hydraulic leg analysed in the publication had sufficient power to carry the dynamic load. The prepared leg for bench testing was used in longwalls exploited in conditions of tremor hazard (Table 1). It is on the basis of laboratory tests that it is explained how it develops its power.

The aim of this article is to propose a methodology based on bench testing to determine how much power a hydraulic leg develops as a result of dynamic loading.

(a)

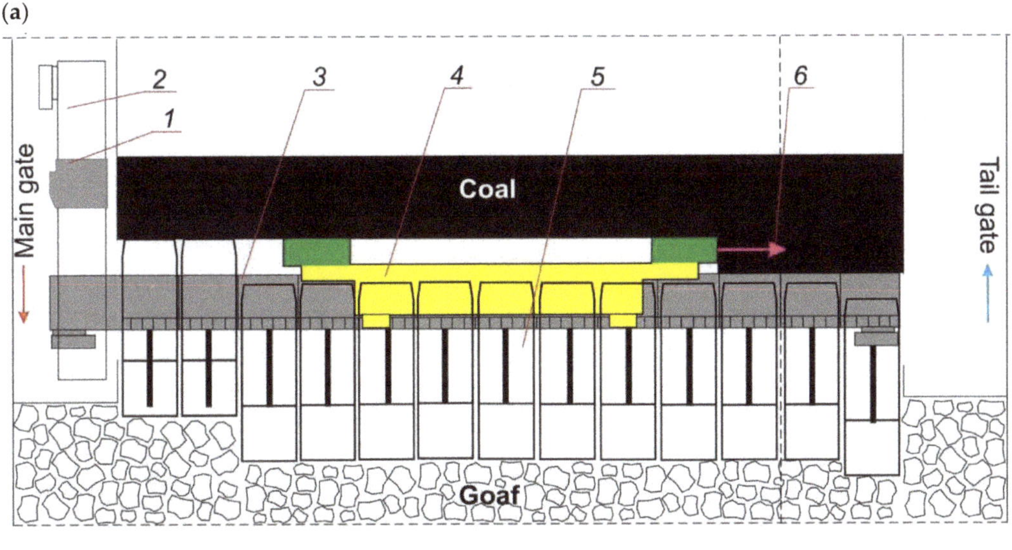

(b)

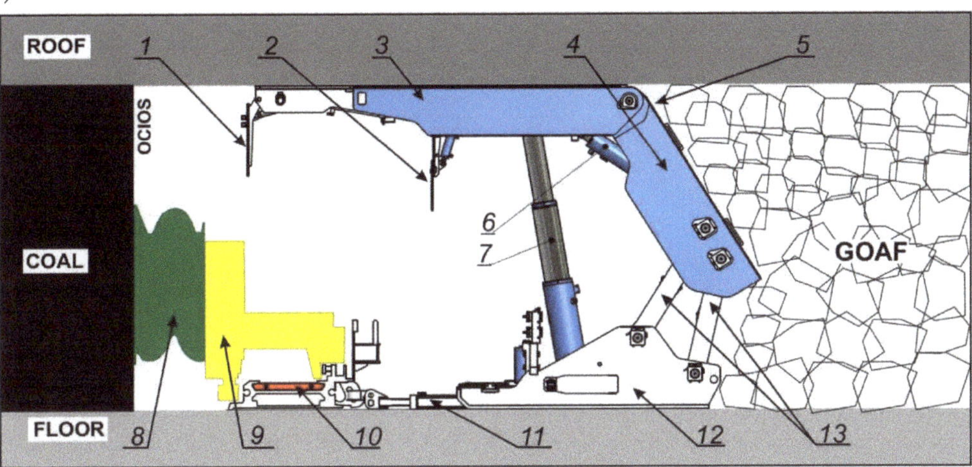

Figure 2. Example of arrangement of basic machinery of a longwall set: (**a**) longitudinal section of a longwall, where: 1—crusher, 2—main gate conveyor, 3—longwall conveyor (chock), 4—longwall shearer, 5—powered roof support, 6—mining direction; (**b**) cross-section of a longwall, where: 1—face shield, 2—transition shield, 3—canopy, 4—shield support, 5—powered roof support, 6—hydraulic support actuator, 7—hydraulic leg, 8—shearer loader, 9—shearer loader, 10—armoured face conveyor, 11—traverse system, 12—floor base, 13—linkage system.

Table 1. Calculation results for the real conditions analysed.

Longwall No.	Seam	E_p (J)	$V_{gmax(p)}$ (m/s)	Estimated Evaluation Based on the Power Method	E_p (J)	$V_{gmax(rz)}$ (m/s)	Actual Evaluation Based on the Power Method
1J	504 J	1.5×10^5	0.2	$N_{maxstoj} > D_{max\ gór}$	5×10^6	0.2	$N_{maxstoj} > D_{max\ gór}$
6	409	1.28×10^4	0.1	$N_{maxstoj} > D_{max\ gór}$	6×10^6	0.2	$N_{maxstoj} > D_{max\ gór}$
2J	504 J	9×10^7	0.2	$N_{maxstoj} > D_{max\ gór}$	5×10^6	0.2	$N_{maxstoj} > D_{max\ gór}$
3Jd	502 J	9×10^6	0.2	$N_{maxstoj} > D_{max\ gór}$	6×10^6	0.2	$N_{maxstoj} > D_{max\ gór}$
1	510 K	9×10^5	0.1	$N_{maxstoj} > D_{max\ gór}$	6×10^6	0.2	$N_{maxstoj} > D_{max\ gór}$

Note: E_p—predicted tremor energy, $V_{gmax(p)}$—predicted maximum pit clamping speed, $V_{gmax(rz)}$—actual maximum pit clamping speed, E_{rz}—actual tremor energy.

2. Materials and Methods

The basic element of a powered roof support is the hydraulic leg. It is located between the canopy and the floor base. In its sub-piston space, there is a fluid under operating pressure. When the pressure of the rock mass on the leg is constant, the working pressure is also constant. If the rock mass exerts pressure on the leg, the working pressure in its sub-piston space begins to rise until the fluid reaches the nominal pressure. The resulting pressure due to the dynamic action can lead to the destruction of the leg. For this reason, a safety valve is used for protection. Its function is to secure the under-piston space in a way that drops some of the hydraulic medium from this space to the outside. The leg is slid so that the pressure of the fluid in the sub-piston space does not rise above the nominal pressure.

2.1. Calculation Model for Actual Conditions

The power exerted by the rock mass on the leg can be calculated by knowing the speed of the force exerted on the leg and the speed of clamping the excavation using the following formula [55]:

$$N(t) = f(t)*V(t) \tag{1}$$

where:

$N(t)$—the power of the rock mass;
$f(t)$—the time course of the dynamic force acting on the support;
$V(t)$—the dynamic process of clamping the workings;
$*$—the intertwining of two time functions.

The calculation of the variation of the leg power waveform by means of Formula (1) is very difficult due to the impossibility of obtaining the corresponding variation of the leg load waveform in time as well as the pit clamping speed and the calculation of the combination of the two time functions. Based on the literature [55,56], it can be suggested that the computation of the function weave is a very complicated procedure. For the end case, i.e., the occurrence of a rockburst, the maximum power is determined by multiplying the maximum force on the leg by the maximum speed of clamping.

Ensuring safe operation of a powered roof support during strong dynamic phenomena may be achieved if the permissible load of the hydraulic leg is higher than the expected load of the leg as a result of the impact of the rock mass, which is described by the following relation:

$$N_{max\ stoj} > D_{max\ gór},\ N{\cdot}m{\cdot}s^{-1} \tag{2}$$

where:

$N_{max\ stoj}$—the maximum leg power, $N{\cdot}m{\cdot}s^{-1}$;
$D_{max\ gór}$—the maximum power that the rock mass will exert on the support, $N{\cdot}m{\cdot}s^{-1}$.

$$D_{max\ gór} = F_{gmax}{\cdot}V_{gmax},\ N{\cdot}m{\cdot}s^{-1} \tag{3}$$

where:

F_{gmax}—the maximum value of dynamic force applied to a single leg, N;
V_{gmax}—the maximum clamping speed of the excavation, $m \cdot s^{-1}$.

The maximum permissible load on the leg will be calculated for the case when the workings are clamped, and there is an outflow of fluid from the working space of the leg through the safety valve. Based on the relationships described in hydromechanics and technical parameters, the leg load can be determined using the following formula:

$$N_{max\ stoj} = Q_{zaw} \cdot k_p \cdot P_{max}, \ N \cdot m \cdot s^{-1} \tag{4}$$

where:

Q_{zaw}—the volume flow of the safety valve for a given pressure, $m^3 \cdot s^{-1}$;
P_{max}—the maximum working pressure of the leg matching the adopted allowable load-bearing capacity of the leg, N;
k_p—the leg's overload coefficient.

Introducing Equation (2) into Equations (3) and (4), the result is:

$$Q_{zaw} \cdot k_p \cdot P_{max} > F_{gmax} \cdot V_{gmax} \ N \cdot m \cdot s^{-1} \tag{5}$$

By determining the minimum value of the flows using Equation (5), it is possible to achieve a fluid discharge within the safe range for the operation of the support. This value is expressed by the following formula:

$$Q_{zaw} \geq F_{gmax} \cdot V_{gmax} / P_{max} \cdot k_p \ (m^3 \cdot s^{-1}) \tag{6}$$

2.2. Calculation Model for Test Conditions

The kinetic energy created by a moving rock mass is proportional to the mass and the square of the velocity. The velocity of the moving rock mass is very low, on the order of a few millimetres per hour, and the kinetic energy of the rock mass is also very low. This energy is transformed into the work of sliding the powered roof support, the slide being caused mainly by the opening of the safety valve built into the hydraulic leg. For bench conditions, the mass impactor was taken as the load [25]. The leg of the support slides by conducting work that can be calculated using the following formula:

$$\varphi = P \cdot \pi d^2 / 4 \cdot h_c \ (m) \tag{7}$$

where:

φ—the total work of the leg slide in the moment of dynamic loading;
P—the nominal pressure;
d—the internal diameter of the hydraulic leg cylinder;
h_c—the total leg slide in the moment of dynamic loading.

Dynamic power arises over a period of time, limited by a beginning and an end. The time in which power is created is a process of combining force and speed. Power, in this case, is defined as the energy gained over time. According to the above assumptions, this process occurring as a result of the dynamic action on the hydraulic leg will be determined by the following formula:

$$\varepsilon = E/t = \varphi/t = \delta \cdot h_c / t \ (W) \tag{8}$$

where:

ε—the dynamic power occurring in the piston sub-space of the leg;
E—the value of kinetic energy converted by the leg;
φ—the total work of the leg slide in the moment of dynamic loading;
h_c—the total leg slide in the moment of dynamic loading;
t—the time taken for the total leg slide of h_c;
δ—the hydraulic force acting on the piston of the leg at the moment of its retraction.

3. Conducted Tests

The dynamic load on the hydraulic leg of a powered roof support is characterised by high power. The energy loading the leg is transferred to the work of sliding that leg, provided that the resulting power of the leg is at least equal to the power of its load. For safe operation, the leg power should be greater than the load power. From Equations (2) and (8), we conclude that as the load period becomes shorter, the more powerful the leg must be. Consequently, a leg operating under the tremor hazard of a rock mass has to have significantly more power than the impact of the rock mass.

In bench tests, we observe the power generated as a result of dynamic loading that occurs through the work of the leg slide. In this case, we are dealing with two relations: the slide, which is the result of elastic deformation of the stator cylinder, and the slide, which is the opening of the safety valve protecting the stator sub-piston space. Very often, these two power cases occur together, and their sum is called total power. The slide that takes place during the arising power is called the total slide, and the work conducted during this slide is called total leg slide work. There may be cases during which there is a power of energy in the hydraulic leg as a result of the opening of the leg safety valve or a load as a result of elastic deformation of the leg cylinder (short-term dynamic load in which the leg safety valve fails to open) [25].

3.1. Test Bench Model

In bench tests, when a dynamic load is applied to the leg, a rapid increase in fluid pressure is observed in its sub-piston space. This may affect the enlargement of the diameter of the leg cylinder within the elastic limit, but this issue is not being tested. The load results in a leg slide. Figure 3 shows the research model adopted. In Figure 3a, the green colour indicates the fluid pressure before the load, while Figure 3b shows the leg slide and the increase in liquid pressure indicated in red.

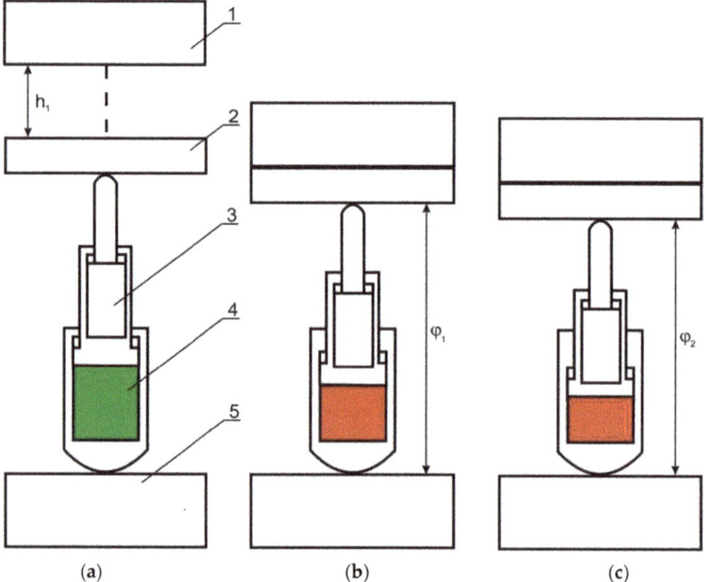

Figure 3. Schematic of the test model: (**a**) position of the leg before loading, where: 1—impact mass of the traverse m_1, 2 h_1—height for the falling mass, mass of the traverse m_2, 3—hydraulic leg, 4—working pressure, 5—foundation of the stand; (**b**) dynamic increase in fluid pressure, where φ_1—the first slide of the leg, (**c**) final increase in fluid pressure, where φ_2—the second slide of the leg.

In the process of the dynamic loading of the leg, there is a change in the fluid pressure in its sub-piston space (Figure 3). The characteristic elements of these changes are two fluid pressure increments, namely, dynamic fluid pressure increase (Figure 3b) and dynamic final fluid pressure increase (Figure 3c).

3.2. Bench Tests

According to the presented scheme (Figure 3), the test consisted of dynamically loading a hydraulic leg with an inner diameter of Ø 240 mm with a safety valve, an impact mass falling from a specified height, and a traverse of a specified mass resting on the leg. The impact weight was 20,000 kg and the traverse weight was 1800 kg. An example test for an impact mass drop height of 0.9 m is shown in Figure 4.

Figure 4. Testing of leg including safety valve (**a**) drop height h_1 = 0.9 m; (**b**) first chute φ1 under dynamic load; (**c**) second slide φ2 of the leg for final pressure build-up.

4. Results and Discussion

Based on the presented computational model for the predicted and real conditions, leg power was analysed. Based on the tremor energy, the velocity of movement of rock mass was assumed. The analysis referred to the longwalls in the Śląsk coalmine in which the powered roof supports operated. Based on the mining and geological conditions analysed, the results are presented in Table 1.

On the basis of the analysis carried out (Table 1) for five different longwalls in which the powered roof support operated, it can be concluded that the leg had an adequate reserve of operational safety.

In the bench test, taking the working pressure as the initial stage (Figures 3a and 4a), the power under consideration starts as a result of the dynamic loading (Figures 3b and 4b). In contrast, the slide under study starts at the level of the maximum pressure obtained, as the working pressure increases in the sub-piston space of the leg due to the dynamic load on the leg. Consequently, there is leg slide caused by the displacement of the impact mass. Figure 5 shows the maximum increase in working pressure loaded with an impact mass from a height of approximately 0.9 m.

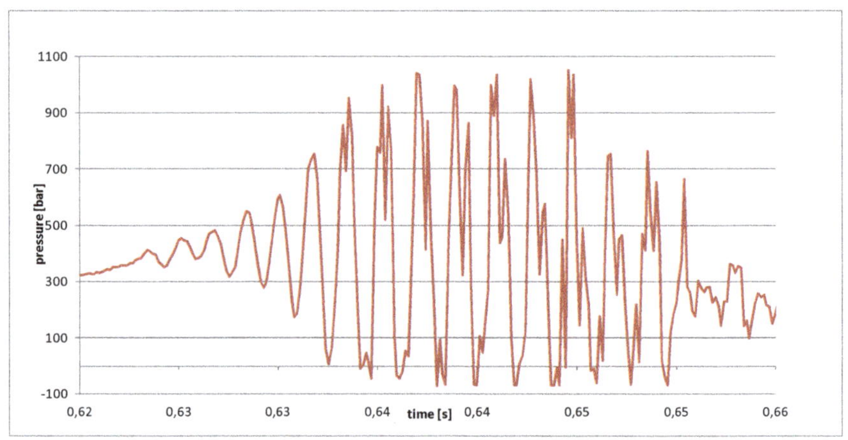

Figure 5. The resulting course of maximum fluid pressure in the sub-piston space.

The dynamic increase in pressure of the liquid in the piston cavity of the leg (Figure 5) refers to the value of the generated power as a function of the operating pressure. The dynamic increase in fluid pressure resulted in an increase in the energy loading on the leg cylinder at the time of the dynamic pulse.

In the case analysed (Figure 5), the obtained dynamic impulse lasted several tenths of a millisecond, which caused the opening of the safety valve protecting the sub-piston space. As a result, the leg slid down. The resulting dynamic pulse under bench test conditions was a high-power pulse, characterised by high kinetic energy.

During the time period indicated in the diagram (Figure 7) by the segment ($a1$), the hydraulic leg of a powered roof support is stretched in the test stand (Figures 3 and 4). It is loaded with the mass of the traverse resting on it as the fluid pressure in its sub-piston space equals the working pressure (P_{rob1}). The resulting dynamic impulse as a result of the drop in rock mass takes place at time (a_2). After time (t_2) from the occurrence of the impulse, the fluid pressure in the sub-piston space increases to the value (P_{max}), as further illustrated in Figure 8. Immediately after the activation of the dynamic impulse, the pressure of the fluid in this space decreases to the value of the working pressure (P_{rob2}) and remains at this level during the time (a_3) that the safety valve operates (Figure 6). It is only when the safety valve is opened that the pressure of the fluid in the piston cavity of the leg drops to the nominal pressure (P_{nom}). The figure also illustrates the dynamic increase in fluid pressure (a) and the dynamic increase in final fluid pressure (b). The value of the obtained power for different impact mass drop heights is shown in Table 2.

Table 2. Power occurring in the hydraulic leg during dynamic loading.

Height of Drop (m)	Energy of Impact Mass E_u (kJ)	Maximum Pressure (bar)	Leg Slide (m)	Power of the Leg (W)
0.3	58.8	550	0.03	0.5
0.4	78.4	650	0.04	0.7
0.5	98.1	730	0.05	0.9
0.6	117.7	790	0.06	1.1
0.7	137.3	830	0.08	1.3
0.8	156.9	900	0.08	1.5
0.9	176.5	990	0.09	1.7

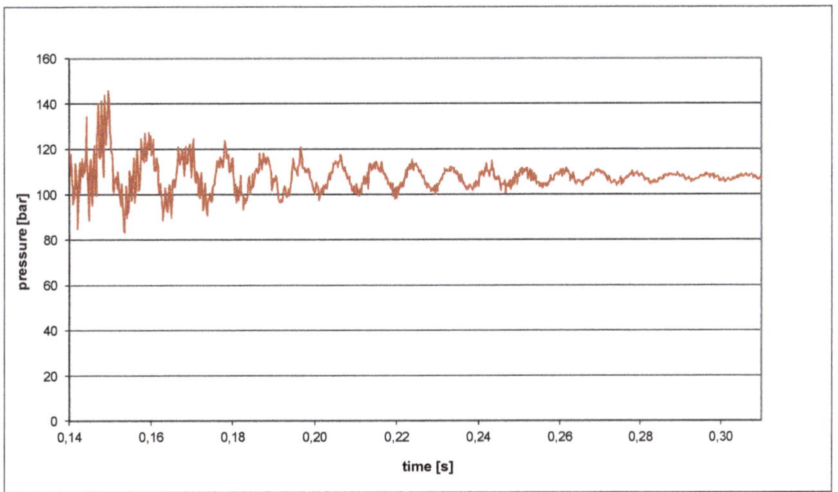

Figure 6. Opening of the safety valve.

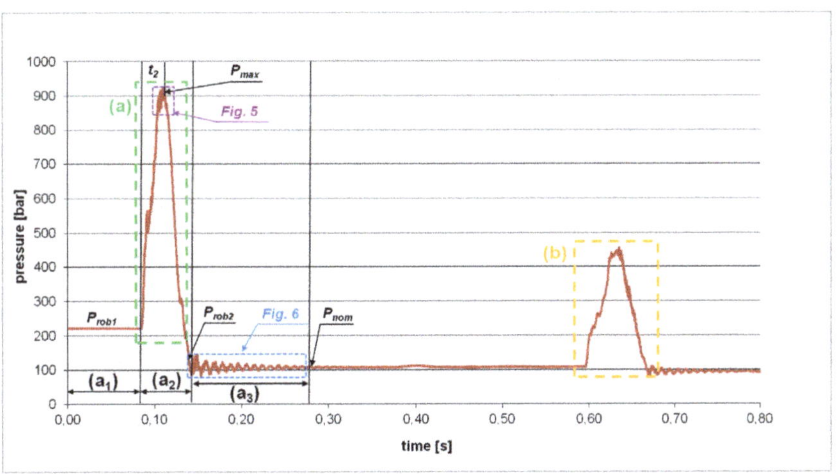

Figure 7. The course of the increase in fluid pressure for power in a specified time in the sub-piston space of a leg when it is loaded with a mass impact.

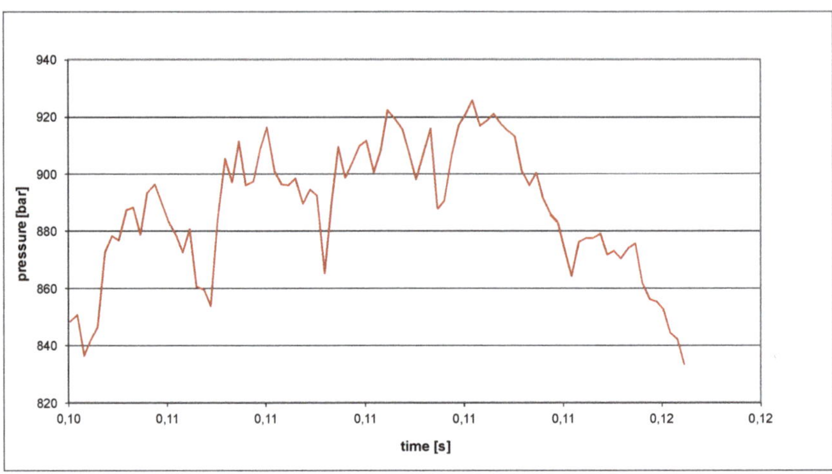

Figure 8. Increase in fluid pressure under load in progress.

5. Conclusions

The power developed by the casing must be greater than the power developed by the rock mass as a result of the tremor (Table 1). In order not to be destroyed, the powered roof support has to have greater power than the power of the rock mass. As a result of the transfer of energy from the rock mass to the powered roof support, the hydraulic legs of the support perform a slide action during the specified time in which the load energy is transferred to the powered roof support. If the powered roof support develops less power than the power developed by the rock mass, it will be damaged or destroyed. On the basis of the data (Table 1), the power of the hydraulic leg of a powered roof support operating under tremor hazard conditions was determined as well as the power that the rock mass has at the moment of the tremor.

Thirty bench tests were performed, and the tests consisted of loading the leg with an impact mass falling from a specified height (Figure 4). Each test was performed three times for a specified height of drop of the impact mass. The height of the fall in the first test was assumed to be 10 cm. For each height, three trials were carried out. The height of mass drop after three trials was increased by 10 cm. The test stand and the technical capabilities of the hydraulic leg allowed us to obtain the maximum height of mass drop to 100 cm. The paper discusses the result of the obtained test for the impact mass drop (Figure 5) from the height of 0.9 m.

When the mass impact loads the leg, it performs the work of sliding (φ), which is given by Equation (7). In the diagram (Figure 5), we have marked (a) the section in which the work of sliding is produced (φ). The resulting value of the chute work (φ) is equal to the value of the kinetic energy imparted to the leg at the time of impact. In tests, the leg was not damaged or destroyed. In contrast, if the work of the slide is less than the energy of the load, the leg could be damaged or destroyed. That is, one can conclude that if the energy created was not converted into work of the chute (φ), we would not have obtained the required leg power. The greater the sliding work (φ) a hydraulic leg can carry out, the greater the power it can achieve and the greater its resistance to dynamic loads.

The methodology presented assumes dynamic impact loading of the mass to obtain the required hydraulic leg power. This assumption makes it possible to qualify the suitability of the structure for the occurring cases of dynamic loading of the roof support's leg as a result of a rockburst. However, the research analysis presented here does not exhaust the issue of dynamic loads and only reflects the current state of knowledge and research capabilities. It is advisable to carry out further research under real and bench conditions.

Funding: This research received no external funding.

Institutional Review Board Statement: Not appliciable.

Informed Consent Statement: Not appliciable.

Data Availability Statement: Not appliciable.

Conflicts of Interest: The authors declare no conflict of interest.

References

1. Szurgacz, D.; Zhironkin, S.; Cehlár, M.; Vöth, S.; Spearing, S.; Liqiang, M. A Step-by-Step Procedure for Tests and Assessment of the Automatic Operation of a Powered Roof Support. *Energies* **2021**, *14*, 697. [CrossRef]
2. Bortnowski, P.; Gładysiewicz, L.; Król, R.; Ozdoba, M. Energy Efficiency Analysis of Copper Ore Ball Mill Drive Systems. *Energies* **2021**, *14*, 1786. [CrossRef]
3. Bajda, M.; Hardygóra, M. Analysis of Reasons for Reduced Strength of Multiply Conveyor Belt Splices. *Energies* **2021**, *14*, 1512. [CrossRef]
4. Wodecki, J.; Góralczyk, M.; Krot, P.; Ziętek, B.; Szrek, J.; Worsa-Kozak, M.; Zimroz, R.; Śliwiński, P.; Czajkowski, A. Process Monitoring in Heavy Duty Drilling Rigs—Data Acquisition System and Cycle Identification Algorithms. *Energies* **2020**, *13*, 6748. [CrossRef]
5. Ziętek, B.; Banasiewicz, A.; Zimroz, R.; Szrek, J.; Gola, S. A Portable Environmental Data-Monitoring System for Air Hazard Evaluation in Deep Underground Mines. *Energies* **2020**, *13*, 6331. [CrossRef]
6. Borkowski, P.J. Comminution of Copper Ores with the Use of a High-Pressure Water Jet. *Energies* **2020**, *13*, 6274. [CrossRef]
7. Tutak, M.; Brodny, J.; Szurgacz, D.; Sobik, L.; Zhironkin, S. The Impact of the Ventilation System on the Methane Release Hazard and Spontaneous Combustion of Coal in the Area of Exploitation—A Case Study. *Energies* **2020**, *13*, 4891. [CrossRef]
8. Góralczyk, M.; Krot, P.; Zimroz, R.; Ogonowski, S. Increasing Energy Efficiency and Productivity of the Comminution Process in Tumbling Mills by Indirect Measurements of Internal Dynamics—An Overview. *Energies* **2020**, *13*, 6735. [CrossRef]
9. Sofranko, M.; Khouri, S.; Vegsoova, O.; Kacmary, P.; Mudarri, T.; Koncek, M.; Tyulenev, M.; Simkova, Z. Possibilities of Uranium Deposit Kuriskova Mining and Its Influence on the Energy Potential of Slovakia from Own Resources. *Energies* **2020**, *13*, 4209. [CrossRef]
10. Sivák, P.; Tauš, P.; Rybár, R.; Beer, M.; Šimková, Z.; Baník, F.; Zhironkin, S.; Čitbajová, J. Analysis of the Combined Ice Storage (PCM) Heating System Installation with Special Kind of Solar Absorber in an Older House. *Energies* **2020**, *13*, 3878. [CrossRef]
11. Zhironkin, S.; Selyukov, A.; Gasanov, M. Parameters of Transition from Deepening Longitudinal to Continuous Lateral Surface Mining Methods to Decrease Environmental Damage in Coal Clusters. *Energies* **2020**, *13*, 3305. [CrossRef]
12. Beer, M.; Rybár, R.; Cehlár, M.; Zhironkin, S.; Sivák, P. Design and Numerical Study of the Novel Manifold Header for the Evacuated Tube Solar Collector. *Energies* **2020**, *13*, 2450. [CrossRef]
13. Kawalec, W.; Suchorab, N.; Konieczna-Fuławka, M.; Król, R. Specific Energy Consumption of a Belt Conveyor System in a Continuous Surface Mine. *Energies* **2020**, *13*, 5214. [CrossRef]
14. Szrek, J.; Wodecki, J.; Błażej, R.; Zimroz, R. An Inspection Robot for Belt Conveyor Maintenance in Underground Mine—Infrared Thermography for Overheated Idlers Detection. *Appl. Sci.* **2020**, *10*, 4984. [CrossRef]
15. Kozłowski, T.; Wodecki, J.; Zimroz, R.; Błażej, R.; Hardygóra, M. A Diagnostics of Conveyor Belt Splices. *Appl. Sci.* **2020**, *10*, 6259. [CrossRef]
16. Schmidt, S.; Zimroz, R.; Chaari, F.; Heyns, P.S.; Haddar, M. A Simple Condition Monitoring Method for Gearboxes Operating in Impulsive Environments. *Sensors* **2020**, *20*, 2115. [CrossRef]
17. Hebda-Sobkowicz, J.; Zimroz, R.; Wyłomańska, A. Selection of the Informative Frequency Band in a Bearing Fault Diagnosis in the Presence of Non-Gaussian Noise—Comparison of Recently Developed Methods. *Appl. Sci.* **2020**, *10*, 2657. [CrossRef]
18. Sobik, L.; Brodny, J.; Buyalich, G.; Strelnikov, P. Analysis of methane hazard in longwall working equipped with a powered longwall complex. *E3S Web Conf.* **2020**, *174*, 1011. [CrossRef]
19. Brodny, J. Determining the working characteristic of a friction joint in a yielding support. *Arch. Min. Sci.* **2010**, *55*, 733–746.
20. Brodny, J. Tests of Friction Joints in Mining Yielding Supports Under Dynamic Load. *Arch. Min. Sci.* **2011**, *56*, 303–318.
21. Budirský, S. Interaction of powered supports and the strata in coal seams under a heavy roof. *Int. J. Min. Eng.* **1985**, *3*, 113–138. [CrossRef]
22. Szurgacz, D.; Brodny, J. Analysis of the Influence of Dynamic Load on the Work Parameters of a Powered Roof Support's Hydraulic Leg. *Sustainability* **2019**, *11*, 2570. [CrossRef]
23. Stoiński, K.; Mika, M. Dynamics of Hydraulic Leg of Powered Longwall Support. *J. Min. Sci.* **2003**, *39*, 72–77. [CrossRef]
24. Szweda, S. Dynamic action of rock mass on the powered support legs. *J. Min. Sci.* **2003**, *38*, 154–161. [CrossRef]
25. Gwiazda, J.B. *Górnicza Obudowa Hydrauliczna Odporna na Tąpania*; Wydawnictwo Śląsk: Katowice, Poland, 1997.
26. Budirský, S.; Martinec, P. Influence of support resistance on roof control in single-pass thick-seam mining (4.5 m): A case history. *Min. Sci. Technol.* **1986**, *4*, 59–67. [CrossRef]
27. Klishin, V.I.; Klishin, S.V. Coal Extraction from Thick Flat and Steep Beds. *J. Min. Sci.* **2010**, *46*, 149–159. [CrossRef]

28. Buyalich, G.; Buyalich, K.; Byakov, M. Factors Determining the Size of Sealing Clearance in Hydraulic Legs of Powered Supports. *E3S Web Conf.* **2017**, *21*, 3018. [CrossRef]
29. Buyalich, G.; Byakov, M.; Buyalich, K. Factors Determining Operation of Lip Seal in the Sealed Gap of the Hydraulic Props of Powered Supports. *E3S Web Conf.* **2017**, *41*, 1045. [CrossRef]
30. Buyalich, G.; Byakov, M.; Buyalich, K.; Shtenin, E. Development of Powered Support Hydraulic Legs with Improved Performance. *E3S Web Conf.* **2019**, *105*, 3025. [CrossRef]
31. Kabiesz, J.; Konopko, W.; Patyńska, R. *Annual Report on the State of Basic Natural and Technical Hazards in Hard Coal Mining*; GIG: Katowice, Poland, 2020; pp. 97–101.
32. Stoiński, K. *Mining Roof Support in Hazardous Conditions of Mining Tremors*; The Central Mining Institute: Katowice, Poland, 2000.
33. Świątek, J.; Janoszek, T.; Cichy, T.; Stoiński, K. Computational Fluid Dynamics Simulations for Investigation of the Damage Causes in Safety Elements of Powered Roof Supports—A Case Study. *Energies* **2021**, *14*, 1027. [CrossRef]
34. Tomski, L.; Kukla, S. Dynamical response of bar-fluid-shell system simulating hydraulic cylinder subjected to arbitrary axial excitation. *J. Sound Vib.* **1984**, *92*, 273–284. [CrossRef]
35. Tomski, L.; Uzny, S. A hydraulic cylinder subjected to Euler's load in aspect of the stability and free vibrations taking into account discrete elastic elements. *Arch. Civ. Mech. Eng.* **2011**, *11*, 769–785. [CrossRef]
36. Sochacki, W.; Tomski, L. Free Vibration and Dynamic Stability of a Hydraulic Cylinder Set. *Mach. Dyn. Probl.* **1999**, *23*, 91–104.
37. Tomski, L. Forced Vibrations of Hydraulic Props of Longwall Supports. *Arch. Min. Sci.* **1979**, *24*, 19–33.
38. Tomski, L. *Dynamika Stojaków Hydraulicznych Obudów Górniczych*; Zeszyty Naukowe Politechniki Częstochowskiej, Nr 17; Politechnika Częstochowska: Czestochowa, Poland, 1979.
39. Tomski, L. Elastic Carrying Capacity of a Hydraulic Prop. *Eng. Trans.* **1977**, *25*, 247–263.
40. Tomski, L. Elastic Stability of Hydraulic Props of Longwall Supports. *Arch. Min. Sci.* **1978**, *23*, 217–231.
41. Tomski, L. Longitudinal Mass Impact of Hydraulic Servo. *Z. Angew. Math. Mech.* **1981**, *61*, 191–193.
42. Tomski, L.; Chwalba, W.; Kukla, S.; Mrowiec, A.; Posiadała, B. Dynamical Response of Hydraulic Cylinder Connected with External Shock Absorber Subjected to Axial Mass Impact. *Arch. Min. Sci.* **1987**, *32*, 439–454.
43. Yang, Z.K.; Sun, Z.Y.; Jiang, S.B.; Mao, Q.H.; Liu, P.; Xu, C.Z. Structural Analysis on impact-mechanical properties of ultra-high hydraulic support. *Int. J. Simul. Model* **2020**, *1*, 17–28. [CrossRef]
44. Uzny, S. Free Vibrations of an Elastically Supported Geometrically Nonlinear Column Subjected to a Generalized Load with a Force Directed toward the Positive Pole. *J. Eng. Mech.* **2011**, *137*, 740–748. [CrossRef]
45. Uzny, S.; Osadnik, M. Influence of longitudinal elastic support on stability of a partially tensioned column. In Proceedings of the 23rd International Conference in Engineering Mechanics, Svratka, Czech Republic, 15–18 May 2017; pp. 1006–1009.
46. Uzny, S.; Kutrowski, Ł. Strength analysis of a telescopic hydraulic cylinder elastically mounted on both ends. *J. App. Math. Comput. Mech.* **2019**, *18*, 89–96. [CrossRef]
47. Uzny, S.; Kutrowski, Ł. The Effect of the Type of Mounting on Stability of a Hydraulic Telescopic Cylinder. *Mach. Dyn. Res.* **2018**, *42*, 53–60.
48. Uzny, S. Free vibrations and stability of hydraulic cylinder fixed elastically on both ends. *Proc. Appl. Math. Mech.* **2009**, *9*, 303–304. [CrossRef]
49. Kong, D.Z.; Jiang, W.; Chen, Y. Study of roof stability of the end of working face in upward longwall top coal. *Arab. J. Geosci.* **2017**, *10*, 185. [CrossRef]
50. Michalak, M. Gradual-Randomized model of powered roof supports working cycle. *Comput. Sci. Inform. Technol.* **2015**, 15–24. [CrossRef]
51. Axin, M. Mobile Working Hydraulic System Dynamics. Ph.D. Thesis, Linköping Studies in Science and Technology, Linköping University, Linköping, Sweden, 2015.
52. Brady, B.H.G.; Brown, E.T. *Rock Mechanics for Underground Mining*; Kluwer Academic Publishers: New York, NY, USA, 2005.
53. Singh, R.; Singh, T.N. Investigation into the behavior of a support system and roof strata during sublevel caving of a thick coal seam. *Geotech. Geol. Eng.* **1999**, *17*, 21–35. [CrossRef]
54. Gil, J.; Kołodziej, M.; Szurgacz, D.; Stoiński, K. Introduction of standardization of powered roof supports to increase production efficiency of Polska Grupa Górnicza, S.A. *Min. Inform. Autom. Electr. Eng.* **2019**, *56*, 33–38. [CrossRef]
55. Szurgacz, K. An attempt to determine the dynamic powered of mechanized longwall housing leg deisgned to work in hazardous conditions of rock mass tremors—Discussion article. *Res. Rep. Min. Environ.* **2011**, *1*, 79–88.
56. Stoiński, K. *Powered Roof Support for Rock Tremors Conditions*; The Central Mining Institute: Katowice, Poland, 2018.

 *energies*

Article

Energy-Saving Inertial Drive for Dual-Frequency Excitation of Vibrating Machines

Volodymyr Gursky [1], **Igor Kuzio** [1], **Pavlo Krot** [2,*] **and Radoslaw Zimroz** [2]

[1] Institute of Engineering Mechanics and Transport, Lviv Polytechnic National University, 79013 Lviv, Ukraine;
  vol.gursky@gmail.com (V.G.); kuzo@polynet.lviv.ua (I.K.)
[2] Faculty of Geoengineering, Mining and Geology, Wroclaw University of Science and Technology,
  50-370 Wroclaw, Poland; radoslaw.zimroz@pwr.edu.pl
* Correspondence: pavlo.krot@pwr.edu.pl

Abstract: The low energy efficiency and excessive power of electric motors of large-scale vibrating machines for processing bulk materials motivated a new design of the inertial drive. This drive consists of one motor and two coaxial unbalanced masses, whose rotational frequencies are related in the ratio 2:1. This approach allows for a generation of the excitation force with variable amplitude and frequency, which changes depending on the inertial characteristics and shaft rotation frequency and does not relate to the phase difference of the unbalanced masses. Because of this, the symmetry axis of the resulting vector hodograph can be changed. The spectral composition of the exciting force up to 200 Hz contains higher harmonics, the energy share of which is 25.4% from the 2nd harmonic and 14.1% from the 3rd and higher harmonics that correspondingly improves bulk material treatment in comparison to single-frequency vibrators. The finite element model is used for checking the strength capacity of the most loaded units of a dual-frequency drive. Its use allows the realization of complex trajectories of motion that are more technologically efficient for variable parameters of the treated media and energy saving in sieving screens and other vibrating machines.

Keywords: sieving screen; inertial vibrator; dual-frequency; spectrum; FEM simulation

Citation: Gursky, V.; Kuzio, I.; Krot, P.; Zimroz, R. Energy-Saving Inertial Drive for Dual-Frequency Excitation of Vibrating Machines. *Energies* **2021**, *14*, 71. https://dx.doi.org/10.3390/en14010071

Received: 12 November 2020
Accepted: 23 December 2020
Published: 25 December 2020

1. Introduction

Vibrating screens, conveyors, rammers, and other various bulk materials processing machines are widely used in mining, metallurgical, construction, and other industries. Unfortunately, until now the energy efficiency of existing vibrating machines is low enough and the energy consumed by electric motors is spent on heating bearings and suspension units of inertial vibrators, and only small part actually results in useful work. This is because media consisting of separate particles can dynamically change their spatial distribution structure, physical and mechanical characteristics during screening, transportation or compaction.

Because of the abovementioned factors, the efficiency of vibrating machines with fixed oscillation parameters decreases. The required additional process control and changes in the technology and parameters of vibrating machines usually leads to practical difficulties, as it is necessary to ensure the uncertain values of the relevant parameters (amplitude and frequency of oscillations) within the appropriate limits.

Besides, the power of the electric motors of the most frequently used above-resonance vibrating machines needs to increase by 30–50% (depending on machine design) to pass the main resonance of the machine structure and to prevent Sommerfeld effect manifested in the form of the non-ideal (limited power) drive restraining around the critical range of rotation speed [1]. Upon passing resonance range of speed, electric drives spend excessive energy for excitation of vibrating machines, e.g., sieving screens. Additional power of the electric motors of inertial vibrators causes the installation of more expensive bearings and other elements of structure with a greater carrying capacity that increases the overall cost of machines.

Following the theory and practice of bulk media separation and transportation, it is recommended to implement higher values of accelerations and displacements. However, according to the practical observations given in a review [2], doubling of vibrator motor speed quadruples stress on elements, doubling of stroke length doubles the stress, and even a 10% increase in vibrator rotor speed halves the shaft bearings life. Hence, the trivial approaches to inertial vibrator design by increasing power and rotation velocity are impractical and lead to higher energy consumption and cause more frequent failures due to cyclic fatigue, while diagnostics is difficult due to action of high amplitude periodical excitation [3,4].

Like the solution for a problem of the increasing technological and energetic efficiency of vibrating machines, the task involves developing inertial drives with a simple design but with regulated operating modes and the possibility to provide complicated trajectories of vibrating surface motion. Those drives should provide a wider spectrum of vibrations excited without additional motors and energy consumption.

2. State of the Art in Design of Inertial Drives

The analysis of recent studies indicated the main directions in the development of new types of inertial drives for vibrating machines working in various technological chains. This primarily applies to more complex trajectories and poly-frequency oscillations of the moving bodies.

The electromagnetic or hydraulic drives are more easily controlled and preferable for use in some types of resonant vibrating machines, which, even when excited by unbalanced drives have much less energy consumption [5]. In [6,7], the asymmetric elastic piecewise-linear characteristics have been synthesized for the realization of the two-frequency resonant operation modes of the two-mass vibratory machines with pulse electromagnetic disturbance. Due to this, the corresponding operation modes with vibration impacts having a wide spectrum were obtained. However, electromagnetic drives are not a common case for a majority of industrial plants. Their implementation is rather assumed for manual vibrating tools of small power. In contrast, inertial drives are widely used in the majority of industrial vibration machines.

Depending on the technological process and design of the vibration machine, several separated vibrators may be installed [8]. There are obligatory requirements to ensure their stable synchronous rotations, which can be provided by kinematic links via gears or dynamically. In some types of machines, the in-phase rotation is required at multiple frequencies [9,10], as well as implementing the mentioned elliptical trajectories under conditions of gradual wear and cyclic fatigue of supporting springs [11].

An important point in the studies is the analysis of dynamical processes taking into account the electromechanical characteristics of the drive [12,13] for its sufficiency and avoidance of negative effects during the motor start-up and passing resonances of the machine. It should be noted that there are sufficient conditions under which stable synchronization takes place [14]. In this case, in contrast to a single-frequency system, it is more effective to provide dual-frequency oscillations of technological machines [15,16]. For this purpose, resonant machines with multiple frequencies and with clearly established values of multiple harmonics of the oscillations at the corresponding frequencies are proposed [17]. To control these machines, it is ultimately necessary to implement adaptive and synchronized drives.

Following the task of implementing complex oscillation trajectories, some studies take into account the influence of changes in frequencies and initial phases of rotation of individual motors on the trajectory of vibrating surfaces. Provision of elliptical and circular trajectories has led to the use of three [18] and even four [19] electric motors in systems. This significantly complicates the design and process of equipment management, and increases its cost. In contrast, dual-frequency systems with a single motor are more attractive.

For example, such solutions are based on the combined use of automatic balancing and synchronization [20,21]. Therefore, the use of one electric motor is a priority in any case.

The different features of double or more rotors for designing dual-frequency inertial drives are considered in [22–25]. Dynamic characteristics analysis of a single motor and coaxial dual-rotor systems with inter-shaft bearings are represented in [26,27] and issues related to friction and shaft imbalance are investigated. Excitation of poly-harmonic vibrations in single-body vibration machine with inertia drive due to non-linearity in the elastic clutch is investigated in [28]. Design of nonlinear anti-resonance vibrating screen is described in [29]. A changeable amplitude-frequency mode is implemented in [30] for the vertical concrete compactor using interfered fields of vibration waves. An energy-saving vibration unit with a poly-phase spectrum of vibrations is proposed in [31].

Searching the patents within the class B06B1 categories: 162 (making use of masses with adjustable amount of eccentricity), 161 (adjustable systems, i.e., where amplitude or direction of the frequency of vibration can be varied) and 166 (systems, where the phase-angle of masses mounted on counter-rotating shafts, can be varied) have discovered new types of vibrators and methods of bulk material fractions separation based on impulsive force generation, e.g., KROOSH screening technology [32]. Their vibrating separator [33] contains a source of single-frequency vibration excitation but special adapters generate mechanical impulses with a wide-range random spectrum. This provides continuous self-cleaning of the vibrating surfaces and intensive disaggregation of the sieved material.

In any type of rotating vibration machine, to provide reliable operation, it is necessary to analyze the most loaded elements for strength capacity and durability under operating conditions using an FEM-based CAE-calculation systems [34].

3. Methodology of Design

In this section, the new design is represented of dual-frequency unbalanced inertial drive, its kinematic and force generation functions as well as spectral components of the resulting force vector.

3.1. Structural Scheme and Design of the Drive

After considering several constructive schemes of dual-frequency inertial vibrating drives, the final design is selected and shown in Figure 1. The drive contains two independently and coaxially mounted unbalanced masses 1 and 2, which are placed in a single housing 3 through the bearing supports 4. The outer housing 3 has mounting locations for mounting on the vibrating machine. The two gears 5 and 6 rotate unbalanced masses 1 and 2 from one electric motor 7. Masses 1 and 2 can be independently installed with the corresponding phase shift $\varphi = \varphi_1 - \varphi_2$, which determines the behavior of the vibration system.

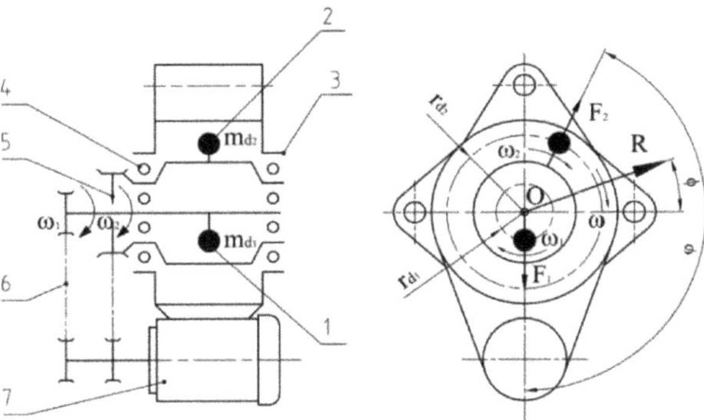

Figure 1. The design concept of the inertial dual-frequency drive.

Due to the rotation of individual unbalanced masses, the resulting vector $\vec{R}$ is created and applied to the common axis of rotation with the corresponding phase angle ϕ.

3.2. Determination of Force and Kinematic Characteristics

The resulting vector of the excitation force $\vec{R}$ is defined as the vector sum of the two inertial forces components:

$$\vec{R} = \vec{F}_{m1} + \vec{F}_{m2} \tag{1}$$

Components of inertial forces of individual unbalanced masses

$$\left|\vec{F}_{m1}\right| = F_1 \sin(\omega_1 t + \varphi_1) \tag{2}$$

$$\left|\vec{F}_{m2}\right| = F_2 \sin(\omega_2 t + \varphi_2) \tag{3}$$

where φ_1 and φ_2—the corresponding initial phase angles.

Amplitude values of inertial forces due to the rotation of unbalanced masses:

$$F_1 = m_{d1}\omega_1^2 r_{d1}; \tag{4}$$

$$F_2 = m_{d2}\omega_2^2 r_{d2}. \tag{5}$$

The vector sum (1) can be represented as the following dependence taking into account the periodic functions of inertial components:

$$R(t) = \sqrt{F_1^2 + F_2^2 + 2F_1 F_2 \cos(\omega_2 t - \omega_1 t + \varphi_2 - \varphi_1)}. \tag{6}$$

The resulting instantaneous phase $\phi(t)$ is the angle of the resulting force vector $R(t)$ rotation around the axis:

$$F_x(t) = F_1 \cos(\omega_1 t + \varphi_1) + F_2 \cos(\omega_2 t + \varphi_2); \tag{7}$$

$$F_y(t) = F_1 \sin(\omega_1 t + \varphi_1) + F_2 \sin(\omega_2 t + \varphi_2); \tag{8}$$

$$\phi(t) = angle[F_x(t), F_y(t)]. \tag{9}$$

A time-varying angular velocity $\omega(t)$ of the radius vector is defined as the derivative of the angle of rotation:

$$\omega(t) = \frac{d}{dt}\phi(t) = \frac{F_1^2\omega_1 + F_2^2\omega_2 + F_1 F_2 \cos(\omega_1 t - \omega_2 t + \varphi_1 - \varphi_2)(\omega_1 + \omega_2)}{F_1^2 + F_2^2 + 2F_1 F_2 \cos(\omega_1 t - \omega_2 t + \varphi_1 - \varphi_2)} \tag{10}$$

The obtained analytical dependences are time-varying force and kinematical parameters functions of the inertial vibrator. They functionally take into account all the design characteristics of unbalanced masses, their initial position, and velocity of shaft rotation. The technical parameters of the considered design of vibrator and corresponding amplitudes of generated inertial forces are given in Table 1. The 3D model of dual-frequency vibrator assembled with a motor is shown in Figure 2.

Table 1. Characteristics of imbalances and amplitudes of generated inertial forces.

Mass	ω, [rad/s]	m_d, [kg]	$r_d \times 10^{-3}$, [m]	F, [kN]
1	314	3.9	6.2	2.38
2	157	11.0	21.9	5.94

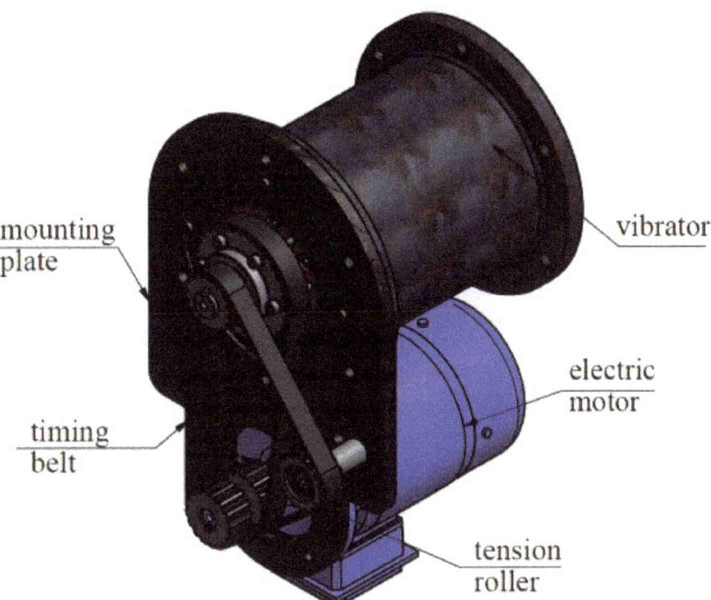

Figure 2. General view of dual-frequency vibrator assembled with a motor.

4. Results of Inertial Drive Simulation

The use of the obtained calculation Formulas (1)–(10) allows us to investigate the advantages and particular features of the proposed design solution. Taking into account the set values of unbalanced masses (see Table 1), their rotation speeds, and initial phases, it is possible to synthesize specific kinematic characteristics and trajectories that will determine the functionality of the proposed design solution.

4.1. Analysis of Force and Phase Relations

The derived functional dependences (6) and (9) can be combined into the form of a polar graph $R(\varphi)$, which allows to estimate the changes in the amplitude of the resulting force and to construct a required trajectory of oscillations about the common axis of unbalanced masses rotation. It turns out that the phase shift angle allows us to change the location of the symmetry plane of the hodograph of the resulting force vector.

Consider the influence of the phase shift angle between unbalanced masses in Figure 3. For the first case of phase difference ($\varphi_1 = 0°$ and $\varphi_2 = 90°$), perturbation force is symmetrical about the horizontal direction and for the second case ($\varphi_1 = 90°$ and $\varphi_2 = 0°$), the drive produces symmetric oscillations in the projection on the vertical axis.

Depending on the purpose of the vibrating installation, the required values of the initial phases can be selected. The asymmetric law of oscillation in the projection of the resulting vector on the corresponding axis will cause the effect of transportation along this axis. Therefore, the use of the phase shift angle like in Figure 3a will lead to the appearance of transporting the bulk media along the axis, which is required for the vibrating screens. For stationary vibrating tables, a phase shift should be used, like in Figure 3b, as there should be no asymmetric oscillations in the axial direction to avoid the movement of the media in the filled compaction form.

Regardless of the set initial value of the phase shift angle φ, developed drive produces the periodic resulting force, which varies within a range $R(t) = 3.55–8.32$ kN (see Figure 4a) and instant frequency range $\omega(t) = 51.7–202.0$ rad/s (see Figure 4b) with median value about 179.2 rad/s.

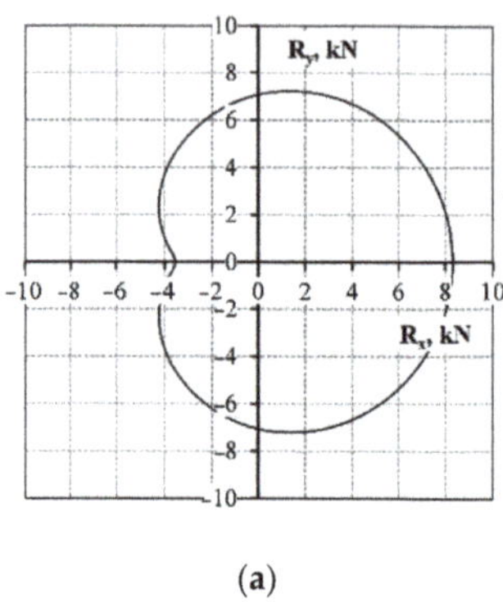

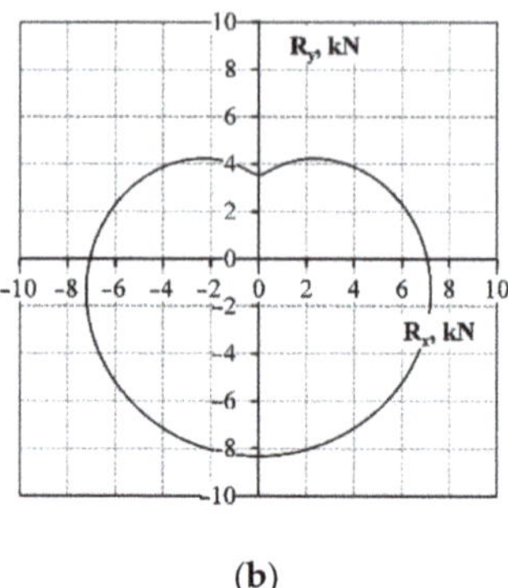

(**a**)

(**b**)

Figure 3. Projections of the resulting force vector for different phase shift angles: (**a**) $\varphi_1 = 0°$ and $\varphi_2 = 90°$; (**b**) $\varphi_1 = 90°$ and $\varphi_2 = 0°$.

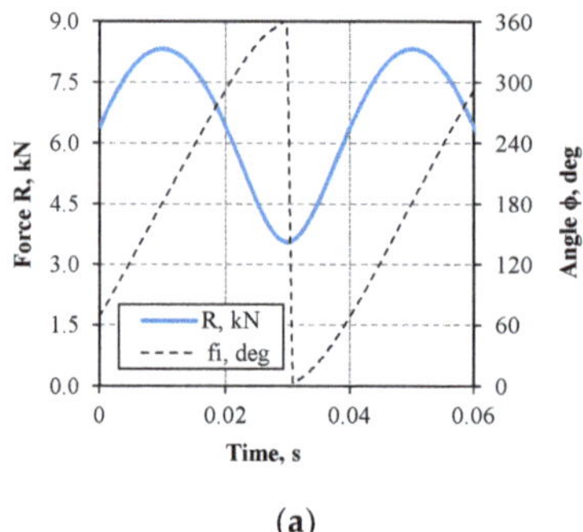

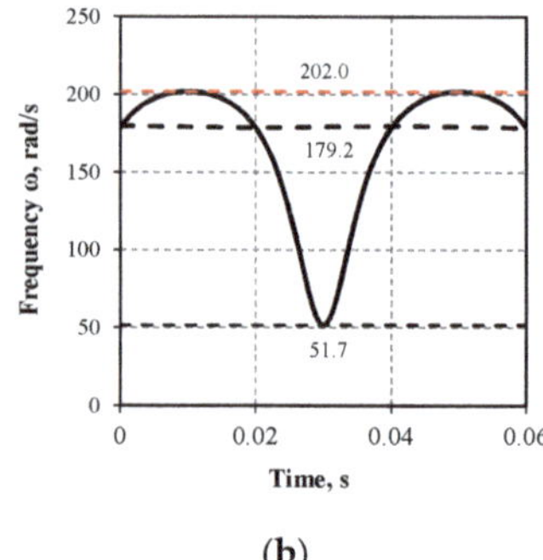

(**a**)

(**b**)

Figure 4. Time-series: (**a**) resulting force and angle of its rotation; (**b**) angular velocity.

The resulting perturbation according to Equation (2) and Figure 4a is asymmetric and periodic. However, as it turned out, its maximum value $R_{max} = 8.32$ kN does not depend on the difference in the angle of phase shift but depends on the inertial characteristics of unbalanced masses and their rotational frequencies.

The spectral composition of the resulting force is shown in Figure 5. The effective range is limited to 200 Hz for the assumed frequencies of shafts rotation. In addition to the two frequencies of excitation with the main peaks about 26 Hz and 51 Hz, the higher harmonics are also excited due to non-symmetrical waveform of resulting force.

The amplitudes and corresponding energy share of these harmonics (from 3rd to 7th) in the resulting force is about 14.1%. If to compare with a single-frequency vibrator, the contribution of the 2nd–7th harmonics increases to 25.4%. This means that for lower supplied power of the electrical motor in a vibration machine, it can demonstrate the same efficiency of media treatment. This is because of the fine fractions of bulk materials prone to aggregation require a higher frequency of, e.g., sieving screen, vibration or more amplitude of working surface displacement that supposes more energy consumption of the electrical motor. Therefore, the efficiency of the technological process can be enhanced by the proposed drive, due to a wider poly-frequency range of treated media excitation always having variable physical properties (size of particles, hardness, and water content).

The non-linear relation of resulting inertial force $R(t)$ and instant frequency $w(t)$ is shown in Figure 6. In contrast, the power characteristic of industrial single-frequency vibrators has the form of a point at this diagram. Using such graphs, designers can synthesize working characteristics of dual-frequency vibrators depending on specific requirements of technological processes.

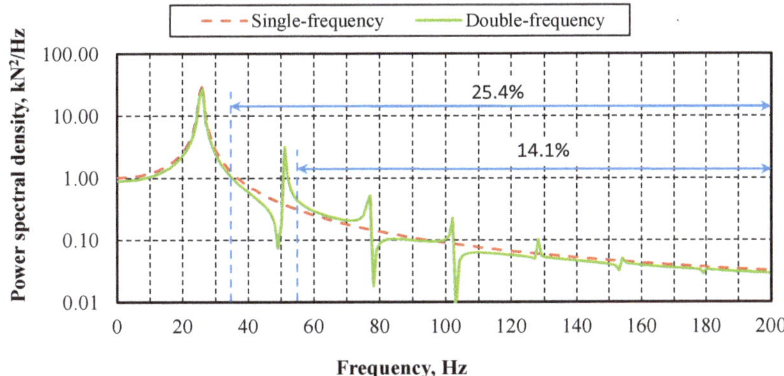

Figure 5. Power spectral density of a resulting force generated by dual-frequency vibrator and energy shares of higher harmonics.

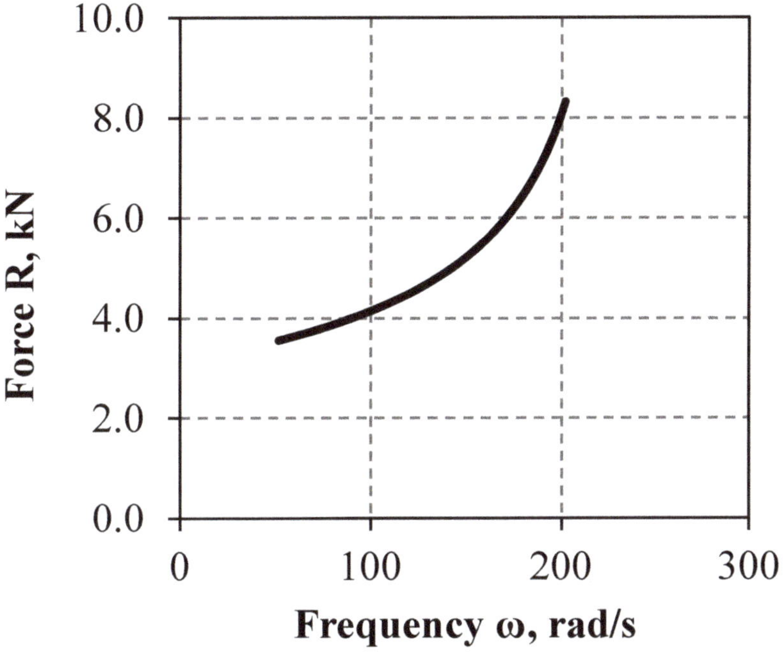

Figure 6. Nonlinear relation of resulting inertial force and rotation frequency.

4.2. FEM Analysis of the Vibrator Units

Taking into account that sieving screens are usually above resonance machines, there is no threat of higher harmonics matching with machine main natural modes of vibration, which may cause overloading of its structural elements. Besides the technological and energetic advantages, the proposed inertial drive will spread out the supplied electrical motor energy among several frequencies of vibration, and in such a way, the dangerous amplitudes of machine transient resonances will be significantly reduced during start-up and slow down. Nevertheless, the reliability of vibrating machines greatly depends on the right choice of bearings and their load capacity under the action of significant periodical radial forces. Therefore, the main units of the developed drive are checked by

FEM simulations. A detailed FEM analysis of strength capacity is carried out based on values of reactions in the bearings supports of coaxially rotating masses.

The main elements of a dual-frequency inertial vibrator (see Figure 7) are the central unbalanced shaft and unbalanced frame, on the edges of which two unbalanced masses are installed. Inertial force F_1 of the unbalanced shaft produces the reaction forces R_A and R_B in the supports. They act as the additional impacts on unbalanced frame loaded by their inertial force F_2. The values of the reactions in the supports of the unbalanced frame R_C and R_D are accounted for together with reaction forces R_A and R_B for the selection of bearings in the supports of the vibrator shaft.

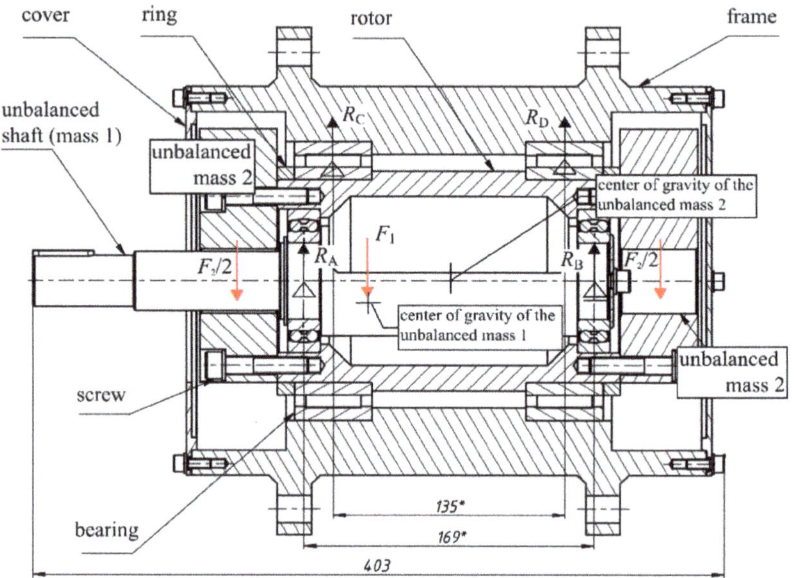

Figure 7. The scheme of reactions forces in the supports of a dual-frequency inertial vibrator.

The resulting periodic force of the vibrator $R(t)$ acting on the vibrating machine served to check its strength by FEM simulation. According to the results of the model analysis (see Figure 8), the values of the maximum von Mises equivalent stresses 40.96 MPa in the inner unbalanced shaft and 5.46 MPa in outer unbalanced housing do not exceed the endurance limits of structural steel under cyclic fatigue, which is the condition of durable and stable drive operation.

Besides, based on the reaction forces used in calculations of strength capacity, the power losses for rolling friction are also estimated in bearings of shafts supports. Results are summarized in Table 2.

In general, the presented design of a dual-frequency vibrating drive with variable kinematic and force characteristics can work as an autonomous unit on the industrial vibration machines. It can also be equipped with an internal combustion engine. In manual vibrating machines, poly-frequency vibration is better for perception by operators.

(a)   (b)

Figure 8. The results of the calculation of equivalent stresses in unbalanced masses: (**a**) inner unbalanced shaft, (**b**) outer unbalanced housing.

Table 2. Power losses on rolling friction in bearings.

Pos.	Rotation Velocity ω, [rad/s]	Diameter d, [mm]	Reaction Force R, [kN]	Coefficient of Friction	Moment of Friction M, [Nm]	Power Loss P, [W]
A	314	45	2.004	0.0015	0.068	21.35
B	314	45	0.534	0.0015	0.018	5.65
C	157	120	5.083	0.0040	1.22	191.54
D	157	120	3.236	0.0040	0.78	122.46

5. Discussion and Conclusions

Although different types of treated media activation principles have been proposed in many published studies and declared in patents, the main amount of bulk materials processing in industrial plants is still provided by the rotating unbalanced vibrators. The efficiency of bulk media sieving, compaction and transportation machines is made low enough by the energy consumed in electrical inertial actuators. Also, the well-known Sommerfeld effect requires additional power of electric motors to pass resonance ranges of rotation speeds which correspond to the main natural modes of machines' structural vibrations.

The implementation of electromagnetic and other types of excitation in resonant types of vibrating machines, which consume much less energy than conventional above-resonant machines, is a rare case for industrial plants. The standard approach to activate the treatment process of prone to aggregation fine fractions is to increase the amplitude and acceleration of oscillated media. This approach significantly (power function of rotation speed) reduces service time of bearings and increases the overall cost of vibrating machines.

To solve these problems, recent developments of vibratory units and corresponding technologies are directed to dual and multi-frequency excitation by different methods. High-frequency, up to ultrasound range, activators are used including devices generating impact forces with a wideband spectrum of random oscillations. However, tuning of such devices for the required elliptical trajectory of particles motion is difficult for realization.

The proposed design of a dual-frequency vibrating drive includes only a single motor and allows the realization of complex trajectories of working surfaces with a variable force both in amplitude value (3.55–8.32 kN) and in the frequency of oscillations (51.7–202.0 rad/s).

The investigated exemplary design of inertial drive showed that the maximum value $R_{\max} = 8.32$ kN does not depend on the difference in the angle of phase shift between two unbalanced masses, but depends on their inertial characteristics and rotational frequencies. This design can be easily scaled to implement in large-size vibrating machines.

Due to a created special waveform, the amplitude and corresponding energy ratio of high harmonics excited in the resulting force are about 14.1% for dual-frequency vibration and about 25.4% if compared with a single-frequency vibrator. Such redistribution of supplied to electrical motor energy allows less dangerous passing of transient modes for the above resonance industrial machines. Besides, the poly-frequency vibration with moderate amplitudes of harmonics is better for perception by operators and allows them to achieve better energy efficiency.

The finite element model simulations of a dual-frequency inertial drive allowed us to check the strength capacity of the most loaded units to provide their durable operation. Power losses due to rolling friction in the bearing supports correspond to the generally assumed values in machine design and do not cause additional energy consumption. Instead, only one electrical motor implementation for dual-frequency force generation significantly reduces the power losses in the inertial drive.

Such variable periodic characteristics allow the implementation of technological systems that will be more efficient at energy consumption and processing materials with uncertain physical and mechanical characteristics in contrast to systems with constant kinematic and force generation parameters.

Author Contributions: Conceptualization, V.G. and I.K.; methodology, V.G.; software, P.K. and V.G.; validation, P.K., and V.G.; resources, R.Z.; writing—original draft preparation, P.K.; writing—review and editing, V.G.; project administration, funding acquisition, R.Z. All authors have read and agreed to the published version of the manuscript.

Funding: This activity has received partial funding from the European Institute of Innovation and Technology (EIT), a body of the European Union, under the Horizon 2020, the EU Framework Programme for Research and Innovation. This work is supported by EIT RawMaterials GmbH under Framework Partnership Agreements No. 18253 (OPMO. Operation monitoring of mineral crushing machinery).

Data Availability Statement: The data presented in this study are available on request from the corresponding author.

References

1. Palacios, J.L.; Balthazar, J.M.; Brasil, R.M.L.R.F. A short note on a nonlinear system vibrations under two non-ideal excitations. *J. Braz. Soc. Mech. Sci. Eng.* **2003**, *25*, 391–395. [CrossRef]
2. Shah, K.P. Construction, Working and Maintenance of Electric Vibrators and Vibrating Screens. Available online: www.practicalmaintenance.net (accessed on 5 November 2020).
3. Gasior, K.; Urbanska, H.; Grzesiek, A.; Zimroz, R.; Wyłomańska, A. Identification, decomposition and segmentation of impulsive vibration signals with deterministic components—A sieving screen case study. *Sensors* **2020**, *20*, 5648. [CrossRef] [PubMed]
4. Krot, P.; Zimroz, R. Methods of springs failures diagnostics in ore processing vibrating screens. *IOP Conf. Ser. Earth Environ. Sci.* **2019**, *362*, 1–9. [CrossRef]
5. Lyan, I.P.; Panovko, G.Y.; Shokhin, A.E. On the issue of energy consumption of vibration technological machines. *IOP Conf. Ser. Mater. Sci. Eng.* **2020**, *747*, 012055. [CrossRef]
6. Gursky, V.; Kuzio, I.; Korendiy, V. Optimal synthesis and implementation of resonant vibratory systems. *Univ. J. Mech. Eng.* **2018**, *6*, 38–46. [CrossRef]
7. Kuzio, I.V.; Lanets, O.V.; Gursky, V.M. Substantiation of technological efficiency of two-frequency resonant vibration machines with pulse electromagnetic disturbance. *Nauk. Visnyk Natsionalnoho Hirnychoho Universytetu* **2013**, *3*, 71–77.
8. Chen, B.; Yan, J.; Yin, Z.; Tamma, K.K. A new study on dynamic adjustment of vibration direction angle for dual-motor-driven vibrating screen. *Proc. Inst. Mech. Eng. Part E J. Process Mech. Eng.* **2020**. [CrossRef]
9. Zou, M.; Fang, P.; Peng, H.; Hou, D.; Du, M.; Hou, Y. Study on synchronization characteristics for self-synchronous vibration system with dual-frequency and dual-motor excitation. *J. Mech. Sci. Technol.* **2019**, *33*, 1065–1078. [CrossRef]
10. Zou, M.; Fang, P.; Hou, Y.; Wang, Y.; Hou, D.; Peng, H. Synchronization analysis of two eccentric rotors with double-frequency excitation considering sliding mode control. *Commun. Nonlinear Sci. Numer. Simul.* **2021**, *92*, 105458. [CrossRef]
11. Krot, P.; Zimroz, R.; Michalak, A.; Wodecki, J.; Ogonowski, S.; Drozda, M.; Jach, M. Development and verification of the diagnostic model of the sieving screen. *Shock Vib.* **2020**, *2020*, e8015465. [CrossRef]

12. Blekhman, I.I.; Semenov, Y.A.; Yaroshevych, M.P. On the possibility of designing adaptive vibration machinery using self-synchronizing exciters. In *Advanced Technologies in Robotics and Intelligent Systems*; Misyurin, S.Y., Arakelian, V., Avetisyan, A.I., Eds.; Springer: Cham, Switzerland, 2020; pp. 231–236. [CrossRef]

13. Hou, Y.J.; Liu, Q.Y. The electromechanical dynamics simulation of complex frequency vibrating screen with two-motor-driving. *Appl. Mech. Mater.* **2013**, *271–272*, 1211–1217. [CrossRef]

14. Yang, D. The Self-Synchronization theory of a vibrating system by two-motor driven with isolation frame. *Appl. Mech. Mater.* **2014**, *602–605*, 1490–1494. [CrossRef]

15. Hou, Y.J.; Gao, J.; Liang, J. Dynamics Simulation for Solid Conveyance of Dual-Frequency Vibrating Screen. *Adv. Sci. Lett.* **2012**, *15*, 391–395. [CrossRef]

16. Hou, Y.J.; Fang, P.; Liu, Q.Y. Motion simulation of dual-frequency vibrating screen. *Appl. Mech. Mater.* **2012**, *204–208*, 4916–4921. [CrossRef]

17. Gursky, V.M.; Kuzio, I.V.; Lanets, O.S.; Kisala, P.; Tolegenova, A.; Syzdykpayeva, A. Implementation of dual-frequency resonant vibratory machines with pulsed electromagnetic drive. *Przegląd Elektrotechniczny* **2019**, *1*, 41–46. [CrossRef]

18. Zou, M.; Fang, P.; Hou, Y.; Chai, G.; Chen, J. Self-synchronization theory of tri-motor excitation with double-frequency in far resonance system. *Proc. Inst. Mech. Eng. Part C J. Mech. Eng. Sci.* **2020**, *234*, 3166–3184. [CrossRef]

19. Kong, X.; Zhang, X.; Chen, X.; Wen, B.; Wang, B. Phase and speed synchronization control of four eccentric rotors driven by induction motors in a linear vibratory feeder with unknown time-varying load torques using adaptive sliding mode control algorithm. *J. Sound Vib.* **2016**, *370*, 23–42. [CrossRef]

20. Filimonikhin, G.; Yatsun, V.V.; Dumenko, K. Research into excitation of dual frequency vibrational-rotational vibrations of screen duct by ball-type auto-balancer. *East.-Eur. J. Enterp. Technol.* **2016**, *3*, 47. [CrossRef]

21. Yatsun, V.V.; Filimonikhin, G.; Dumenko, K.; Nevdakha, A. Search for dual-frequency motion modes of dual-mass vibratory machine with vibration exciter in the form of passive auto-balancer. *East.-Eur. J. Enterp. Technol.* **2018**, *1*, 47–54. [CrossRef]

22. Lawinska, K.; Modrzewski, R.; Wodzinski, P. Results of mineral aggregates tests on double-frequency screen. *Min. Sci.* **2014**, *21*, 129–138.

23. Modrzewski, R.; Wodzinski, P. Grained material classification on a double frequency screen. *Physicochem. Probl. Miner. Process.* **2011**, *46*, 5–12.

24. Malewski, J. Control of the screening process in multi-deck screens. *Min. Sci.* **2016**, *23*, 95–106. [CrossRef]

25. Luo, T.; Gan, X.; Luo, W. Nonlinear dynamics simulation of compacting mechanism with double-eccentric vibrator of asphalt-paver. In Proceedings of the International Conference on Intelligent Computation Technology and Automation, Changsha, China, 11–12 May 2010; pp. 800–803. [CrossRef]

26. Wang, N.; Jiang, D.; Xu, H. Dynamic characteristics analysis of a dual-rotor system with inter-shaft bearing. *Proc. Inst. Mech. Eng. Part G J. Aerosp. Eng.* **2017**, *233*, 1147–1158. [CrossRef]

27. Wang, N.; Jiang, D.; Behdinan, K. Vibration response analysis of rubbing faults on a dual-rotor bearing system. *Arch. Appl. Mech.* **2017**, *87*, 1–17. [CrossRef]

28. Bukin, S.L.; Kondrakhin, V.P.; Belovodsky, V.N.; Khomenko, V.N. Excitation of polyharmonic vibrations in single-body vibration machine with inertia drive and elastic clutch. *J. Min. Sci.* **2014**, *50*, 101–107. [CrossRef]

29. Belovodskiy, V.N.; Bukin, S.L.; Sukhorukov, M.Y. Nonlinear antiresonance vibrating screen. In *Advances in Mechanisms Design. Mechanisms and Machine Science*; Beran, J., Bílek, M., Hejnova, M., Zabka, P., Eds.; Springer: Dordrecht, The Netherlands, 2012; Volume 8, pp. 167–173. [CrossRef]

30. Maslov, A.; Batsaikhan, J. The research of the operating mode of the concrete mixture plane depth compactor with a circular vibration exciter. *Acad. J. Ser. Ind. Mach. Build. Civ. Eng.* **2019**, *1*, 5–12. [CrossRef]

31. Nazarenko, I.; Svidersky, A.; Kostenyuk, A.; Dedov, O.; Kyzminec, N.; Slipetskyi, V. Determination of the workflow of energy-saving vibration unit with polyphase spectrum of vibrations. *East.-Eur. J Enterp. Technol.* **2020**, *1*, 43–49. [CrossRef]

32. KROOSH Screening Technology. Available online: http://kroosh.com/technology (accessed on 5 November 2020).

33. Kroosh, I.; Obodan, Y. Vibrating Separator for Dimensional Separation of Bulk Substances (Options). Patent of Ukraine UA105212 C2, Priority, 17 October 2011.

34. Hou, Y.J.; Fang, P.; Zeng, L. Finite element analysis of dual-frequency vibrating screen. *Adv. Mater. Res.* **2012**, *479–481*, 2124–2128. [CrossRef]

Article

Characteristics of Waste Generated in Dimension Stone Processing

Paweł Strzałkowski 

Department of Mining, Faculty of Geoengineering, Mining and Geology, Wroclaw University of Science and Technology, Wybrzeże Wyspiańskiego 27, 50-370 Wrocław, Poland; pawel.strzalkowski@pwr.edu.pl

Abstract: Natural dimension stone processing generates large volumes of stone waste, which have a significant impact on the environment, as well as on the efficiency and profitability of the stone-processing plant. The article presents the characteristics of waste produced as a result of natural dimension stone processing and the structure of the waste production process. Solid stone scraps and sludge were distinguished. On the basis of the performed analyses, it was shown that stone waste constitutes 10–35% in relation to the quantity of the processed stone material, with the quantity of sludge being even threefold greater than the volume of solid scraps. According to the circular economy principles, the aim should be to reduce the amount of waste generated by reducing primary resources in favour of secondary material. Reducing the volume of stone waste is possible through rational planning of stone production while at the same time maximising the efficiency of stone material usage and introducing the most modern processing machines. This significant volume of stone waste encourages efforts to find solutions for both its management and reduction. This paper reviews the utility potential of stone waste. Sensible use of waste is important to increase the profitability and productivity of processing plants while incentivising environmental protection.

Keywords: stone waste; waste generation; waste recycling; industrial waste treatment; sustainable manufacturing; dimension natural stone processing

Citation: Strzałkowski, P. Characteristics of Waste Generated in Dimension Stone Processing. *Energies* **2021**, *14*, 7232. https://doi.org/10.3390/en14217232

Academic Editors: Abdulnaser Sayma, Robert Król, Witold Kawalec, Izabela Sówka and Krzysztof M. Czajka

Received: 23 September 2021
Accepted: 26 October 2021
Published: 2 November 2021

Publisher's Note: MDPI stays neutral with regard to jurisdictional claims in published maps and institutional affiliations.

1. Introduction

Natural stone owes its popularity to its availability, performance, and decorative qualities. An accelerating trend towards the use of natural stone is also related to a wide variety of stones which may serve various purposes, for example, for use in window sills, work surfaces, cladding, or floor tiles. As a natural material, stone has unique physical and mechanical properties, and therefore, it is widely appreciated in construction. Some construction products made of natural stone and their parameters are standardised, e.g., in EN 1341 or EN 1469 [1]. The growing interest in natural stone entails increased production, which requires a number of processing machines and tools. It is important to select and use stone-processing machinery which would be adjusted to the manufacturing of particular natural stone products having defined properties and parameters of a final element.

Stones are processed with various devices, which give them proper shape and dimensions, as well as surface texture. Natural stone processing technology comprises a number of actions aimed at delivering stone products for various applications. However, apart from practical products, stone processing generates significant quantities of stone waste. This waste, as well as its efficient management, represents a great environmental problem in many countries [2–7]. The volume of generated stone waste largely depends on the amount of the processed material (and the efficiency of the processing plant), on the type and size of the generated waste, the type and geological properties of the stone, the type of machinery used for stone processing [4], as well as on the applied technology of dimension natural stone processing, the degree to which the block of stone is used in order to produce the final product and the needs of the clients. In order to identify its potential applications, it is important to identify the volume of stone waste generated, as well as its character [7]. However, this task is sometimes difficult, particularly when different types of natural stone are supplied to processing plants.

Issues related to stone waste have already been discussed in the literature [2,4,7]. However, these works discuss the problem from a general perspective or with specific examples and do not directly indicate to what extent the volume of generated stone waste depends on the technology used and type of waste. Meanwhile, it is important to know and understand the production processes of natural stone products, which, consequently, lead to the generation of different types of stone waste. Knowledge of the share of such waste in the total volume of processed stone material indicates the scale of the problem and the need to find a method of reusing it. This article characterises the process of generating stone wastes and indicates their types and quantities, as well as their potential application.

2. Stone Wastes as a by-Product of Natural Dimension Stone Processing

2.1. Definition of Stone Wastes

Waste is generated in any company and should be understood as any substance or object which its owner disposes of, intends to dispose of, or is obliged to dispose of [8]. Depending on the specific nature of a company, the type and volume of waste differ. According to the European Parliament and European Commission Directive 2008/98/EC of 19 November 2008 on waste [8], each economic entity is obliged to have an adequate waste management policy. Economic activity should result in a possibly limited waste generation. If this is impossible, waste should be recovered, then recycled, or if no other option is available, stored in dedicated sites. Stone wastes that are recovered and reused should be understood as scraps.

Mining wastes are a characteristic type of waste. They comprise by-products generated during the exploration and mining stages, as well as in physical and chemical processing and treatment of minerals. Careddu [2], Kaźmierczak et al. [5,9], Yurdakul [7], Tayebi-Khorami et al. [10], and Woźniak and Pactwa [11] note that these wastes pose significant problems despite their vast potential for further use. Moreover, they seem impossible to be completely eliminated. Therefore, it is important to explore different solutions to reduce and manage them [12]. A reduction in waste is possible by following the circular economy rule, which promotes a drastic reduction in primary resources in favour of secondary material flowing through internal cycles. Lèbre et al. [13] emphasise that it is unreasonable to believe that mining is becoming an unnecessary economic sector. The growing demand for mineral resources will continue to render mining processes indispensable. However, proper management of Earth's resources and adoption of the circular economy rule is the basis for reducing the amount of waste generated in mineral mining and processing.

Stone wastes generated as a result of natural stone processing in stonemasonry companies represent a special type of mining waste. Stone wastes are typically large and medium-size fragments, as well as small parts of stone produced in stone processing, or ready stone products which do not meet the quality standards. According to the European Commission Decision of 18 December 2014 amending Decision 2000/532/EC on waste registers in accordance with European Parliament and EC Directive 2008/98/EC [14], wastes generated in natural stone processing are classified as wastes from stone cutting and sawing (code 01 04 13). Importantly, these wastes are not classified as dangerous wastes.

The waste from natural dimension stone processing becomes a serious problem because the amount of waste generated reaches enormous volumes, which makes it practically impossible to neutralise this waste properly [15]. Additionally, the stone processing technologies and types of stone products make it difficult to limit the volume of waste produced in stonemasonry companies. Production may be more effective and economic and may result in a smaller quantity of waste, if some part of the waste is reused into other stone products or if stones are processed with the use of innovative technologies. However, the most significant reductions in natural stone wastes may be achieved through proper organisation of the processing plants and increased awareness of the management staff [6].

Mitchell et al. [16] and Shamrai et al. [17] observe numerous benefits of rational stone waste management, the most significant of which are the potential revenue from the sale of stone waste, thus becoming an additional income for the company, and also a rational use of natural resources. The benefits also include a decreased amount of material lost in extraction and processing, lower costs of waste storage, transportation, and disposal, as well as increased social responsibility of the company.

2.2. Structure of Stone Waste Production in Processing Plants

Stonemasonry companies process natural stone supplied in the form of raw blocks or pre-processed elements requiring further treatment. The fundamental stage in producing a stone element is its processing, which consists of cutting, milling, or providing the desired texture to its surface. Usually, various types of stone (e.g., granite, sandstone, marble, limestone, etc.) are processed in stonemasonry companies, and various stone products are manufactured. Much less often, stonemasonry companies specialise in the manufacturing of one product (serial production) or in processing one type of natural stone. Stonemasonry companies that process one type of stone most often are plants located near mines extracting this particular stone type.

Except for the finished stone product, the processing of natural dimension stone results in solid scraps (Figure 1a) and sludge (Figure 1b). Depending on the particular process, technology involved and type of natural stone (e.g., granite, sandstone), the size and type of waste are different (Figure 2).

(**a**)	(**b**)

Figure 1. Stone waste: (**a**) crushed slabs (different waste size); (**b**) sludge.

Solid scraps are produced in the process of cutting off larger parts of natural stone or giving the texture of stone surfaces without the use of water. Solid scraps have various sizes, from a few millimetres to several hundred centimetres [4]. The most common type of solid scrap includes offcuts, i.e., rough-edged stone cut off the stone block in order to achieve regular surfaces of the stone block (Figure 3). The length and width of the offcuts are usually equal to the size of the stone block being worked on (e.g., 2.5 × 1.5 m). However, the thickness of this waste is from several to several dozen cm. It should be noted that the sizes of these scraps depend on the quality of the processed stone block (regular and equal shape of the stone block), which is influenced by the adopted mining technology. Therefore, a properly selected dimension stone mining technology may significantly reduce the size of the produced stone waste.

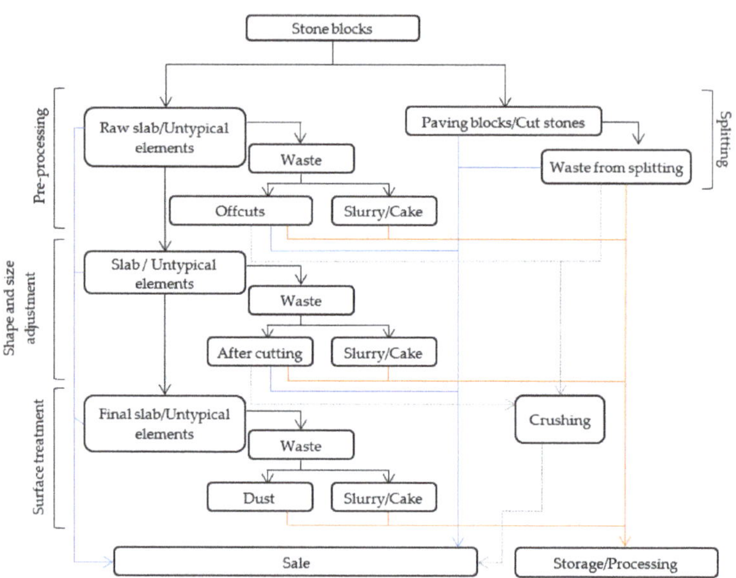

Figure 2. General diagram representing waste production in the processing of natural stone.

Figure 3. Offcuts (2.5 m long, 1.5 m wide, 0–0.1 m thick).

Depending on the pre-processing technology employed, a valvestone can be formed in this process. Valvestone is the lower part of the cut stone block that remains in the block cutting machines for safety reasons. The size of this type of scrap is similar to the size of the offcuts.

Other scraps include stone fragments cut or split (in the process of giving the stone element its desired size and shape, and when split stone elements are produced) or dust, generated as a result of texturising the stone by an impact action (e.g., pointing, chiselling) or by flame treatment. The dimensions of scraps produced after cutting or splitting are from several to several dozen centimetres, while dust waste is usually up to 1 mm.

Stone sludge is produced by using water as a medium for cooling and removing fine stone particles from underneath the processing tool. They are a mixture of ground rock mass with water and may additionally contain some small quantities of abrasive material (particles of the synthetic diamond). The majority of grains in sludge are smaller than 100 μm and rarely larger than 150 μm [4]. Importantly, stone sludge represents the greatest volume of processing-related wastes. While the volume of solid scraps may be reduced by rational stone material handling, a reduction in the volume of sludge is difficult without alternating the employed stone-processing technology.

The above-mentioned wastes are accompanied by wastes from damaged stone blocks or final stone products. These include materials with inherent defects (fractures, voids, etc.) and secondary defects (e.g., defects resulting from the manufacturing process). Such materials are frequently reprocessed or sold at discount prices. Table 1 contains a detailed characteristic of wastes generated in the processing of natural stone.

Table 1. Characteristics of waste from the processing of natural stone.

Type of Waste	Definition of Waste	Description of Waste Source
SOLID STONE SCRAPS		
Damaged stone blocks	Stone blocks that have significant defects or have been damaged and are characterised by different sizes and irregular shapes.	Stone blocks that have insufficient material quality or have been damaged during transport or unsuccessfully divided into smaller parts.
Damaged final stone products	Final stone products with inherent and secondary defects.	During the processing operation, fractures or defects in the final stone products may occur (secondary defects) or primary defects are revealed.
Offcuts	The first and last slab of a stone block cut in a head saw, having one surface smooth and one surface raw/split.	Offcuts occur when a stone block with uneven surfaces is cut. The basic operation behind cutting a stone block is to approximate its shape to a cube. Offcuts are due to the technology employed in natural stone processing.
Valvestones	Lower part of the cut stone block	Waste is generated after cutting a stone block due to safety reasons and in order not to damage the cut raw slabs. This waste is generated as a result of using certain stone block cutting technologies (e.g., disc saw) and is less and less frequent.
Waste from splitting	Parts of the split material outside the size standard of the product (e.g., paving stone).	When splitting natural stone into smaller-size elements (e.g., in the production of paving stone), oversize rock parts of the desired element are split off.
Waste from cutting	Parts of rock material which are smaller in volume than offcuts and have a minimum of 3 smooth surfaces.	Waste is produced as part of the size and shape adjustment. Rock material is produced as a result of cutting off the oversize parts of natural stone. The quantity of this waste depends on the volume of the cutoff stone parts and on the planned cutting locations.
Dust	Fine fraction rock and abrasive material.	As stone is processed, fine fractions of rock material and spalls are split/chipped off the rock. In addition, depending on the surface treatment technology used, this type of waste may include abrasive material (e.g., sand being the product of sandblasting).
STONE SLUDGE		
Slurry/Cake	A mixture of water, ground fine rock, and particles of the cutting tool.	This type of waste is generated at each stage of stone processing. Stone is abraded by the processing elements and subsequently mixed with water. This type of waste additionally comprises small amounts of particles from the working tools.

2.3. Analysis of Stone Waste Production

Analysing the volume and type of the produced stone materials is an important parameter in selecting proper machinery for a processing plant, as well as the basis for the adoption of a production management strategy with a view to reducing the volume of waste produced. The machinery is selected on the basis of numerous other factors, the most important of which is the type of stone products. Table 2 shows types of processed natural stone, types of stone products, and machinery used in 10 plants in which the scale and type of processing-related stone waste were analysed. The plants process various types of rocks (granites, sandstones, marbles, etc.) and produce various elements (stone slabs, paving blocks, untypical and formed elements, etc.).

The volumes of natural stone waste produced in stonemasonry companies vary. Table 3 shows the analysis of the volume and type of stone wastes (scraps). The volume of stone wastes (scraps) was analysed based on data obtained from processing plants on the flow of semi-finished products in the individual stages of stone processing, as well as on the author's own observations. The volume of stone wastes generated at individual stages was estimated from the differences in the volume of stone elements before and after the processing stage. The same basis was used to identify the masses of processed stone and semi-finished products. The volume and mass of sludge were determined from measurements (weighing) conducted by the processing plants. The volume of solid stone scraps was calculated from the difference in initial weight and volume of processed dimension natural stone, finished stone products, and sludge.

The analyses demonstrated that the volumes of stone wastes (scraps) are between 10% and 35% in relation to the quantity of the processed stone material. Interestingly, the greatest volume of waste is produced at the pre-processing stage. This phenomenon is related to the greatest amount of work involved and to the amount of processed natural stone, as well as to the type of machinery employed. In the complete stone processing cycle, the volume of sludge is in all analysed plants more than the volume of solid wastes (solid scraps) (up to three times more). Solid wastes (solid scraps) are generated in the pre-processing and, to a lower extent, in the shape and size adjustment processes. They are characterised by larger fragments of natural stone, which can be processed again, thus significantly reducing the volume of stone waste produced. Stone sludge is a much more significant problem because it is a fine material and cannot be reprocessed. Therefore, it is necessary to use it in other industrial sectors.

Table 2. Types of stone products and machinery used in plants in which the scale and type of stone waste were analysed.

	Processing Plant 1	Processing Plant 2	Processing Plant 3	Processing Plant 4	Processing Plant 5	Processing Plant 6	Processing Plant 7	Processing Plant 8	Processing Plant 9	Processing Plant 10
Type of natural stone	granite, limestone, marble	granite, marble	granite, sandstone	granite, gneiss, marble, sandstone, limestone, onyx	sandstone	granite	sandstone, limestone, marble	granite, sandstone, limestone, marble	sandstone	granite, sandstone
Production	cladding slabs, stairs, window sills, paving blocks	raw slabs stairs, window sills, paving blocks, curbs	cladding slabs, stairs, window sills, curbs	raw slabs	raw slabs, cladding slabs, formed elements, untypical elements	raw slabs, paving blocks, curbs	raw slabs, cladding slabs, stairs, window sills, formed elements	raw slabs, cladding slabs, stairs, window sills, formed elements, curbs	raw slabs, cladding slabs, formed elements	raw slabs
Pre-processing	gang saw, disc saw, multi-wire diamond saw	gang saw, disc saw	gang saw	multi-wire diamond saw	gang saw	gang saw, multi-wire diamond saw, stone splitting machine	gang saw, multi-wire diamond saw	gang saw, multi-wire diamond saw	multi-wire diamond saw	gang saw
Shape and size adjustment	table saw	table saw	table saw	-	table saw	-	table saw	table saw	table saw	-
Surface treatment	abrasive-polishing line, side grinder, thermal burner, graining machine, sandblaster	abrasive-polishing line, side grinder, thermal burner	abrasive-polishing line, side grinder, graining machine, sandblaster	abrasive-polishing line, thermal burner	abrasive-polishing line	-	abrasive-polishing line	abrasive-polishing line, graining machine	abrasive-polishing line	abrasive-polishing line, sandblaster

Table 3. Average volume of stone waste produced per month in each of the processing stages.

		Processing Plant 1			Processing Plant 2			Processing Plant 3			Processing Plant 4			Processing Plant 5		
Quantity of rock material to be processed, m^3		268.00			244.88			175.25			291.67			179.79		
		Quantity of stone waste		Loss of stone material	Quantity of stone waste		Loss of stone material	Quantity of stone waste		Loss of stone material	Quantity of stone waste		Loss of stone material	Quantity of stone waste		Loss of stone material
		m^3	Mg	%	m^3	Mg	%	m^3	Mg	%	m^3	Mg	%	m^3	Mg	%
Pre-processing	Stone sludge *	22.35	211.50	8.34	19.15	181.72	7.82	38.51	182.00	21.98	42.00	112.14	14.40	8.55	23.00	4.76
	Solid stone scraps	8.45	82.30	3.15	8.76	71.18	3.58	14.39	68.00	8.21	16.80	44.86	5.76	9.18	24.70	5.11
Shape and size adjustment	Stone sludge *	1.91	6.00	0.71	0.08	0.30	0.03	1.54	5.00	0.88	-	-	0.00	1.49	4.00	0.83
	Solid stone scraps	5.09	16.00	1.90	0.42	1.50	0.17	2.46	8.00	1.40	-	-	0.00	3.79	10.20	2.11
Surface treatment	Stone sludge *	5.10	15.60	1.90	0.60	10.70	0.25	4.10	17.20	2.34	18.90	50.46	6.48	2.96	8.50	1.65
	Solid stone scraps	2.80	11.40	1.04	0.70	11.00	0.29	1.10	10.60	0.63	2.10	5.61	0.72	-	-	0.00
Total	Stone sludge *	29.36	233.10	10.96	19.83	192.72	8.10	44.15	204.20	25.19	60.90	162.60	20.88	13.00	35.50	7.23
	Solid stone scraps	16.34	109.70	6.10	9.88	83.68	4.03	17.95	86.60	10.24	18.90	50.46	6.48	12.97	34.90	7.22
	Total volume	45.70	342.80	17.06	29.71	276.40	12.13	62.10	290.80	35.43	79.80	213.06	27.36	25.97	70.40	14.45

		Processing plant 6			Processing plant 7			Processing plant 8			Processing plant 9			Processing plant 10		
Quantity of rock material to be processed, m^3		330.00			259.75			360.00			86.48			250.60		
		Quantity of stone waste		Loss of stone material	Quantity of stone waste		Loss of stone material	Quantity of stone waste		Loss of stone material	Quantity of stone waste		Loss of stone material	Quantity of stone waste		Loss of stone material
		m^3	Mg	%	m^3	Mg	%	m^3	Mg	%	m^3	Mg	%	m^3	Mg	%
Pre-processing	Stone sludge *	18.74	51.54	5.68	28.71	73.21	11.05	41.49	112.02	11.53	5.79	15.63	6.70	15.82	43.82	6.31
	Solid stone scraps	18.70	51.43	5.67	16.34	41.67	6.29	21.44	57.89	5.96	3.10	8.37	3.58	8.50	23.55	3.39
Shape and size adjustment	Stone sludge *	-	-	0.00	2.12	5.41	0.82	5.67	15.31	1.58	2.70	7.29	3.12	-	-	0.00
	Solid stone scraps	-	-	0.00	5.40	13.77	2.08	7.60	20.52	2.11	3.44	9.29	3.98	-	-	0.00
Surface treatment	Stone sludge *	-	-	0.00	4.59	11.70	1.77	8.01	21.62	2.22	2.90	7.84	3.36	9.33	25.86	3.72
	Solid stone scraps	-	-	0.00	-	-	0.00	1.33	3.60	0.37	-	-	0.00	0.57	1.57	0.23
Total	Stone sludge *	18.74	51.54	5.68	35.42	90.32	13.64	55.17	148.95	15.32	11.39	30.77	13.18	25.15	69.68	10.04
	Solid stone scraps	18.70	51.43	5.67	21.74	55.44	8.37	30.37	82.01	8.44	6.54	17.66	7.56	9.07	25.11	3.62
	Total volume	37.44	102.97	11.35	57.16	145.76	22.01	85.54	230.97	23.76	17.93	48.42	20.74	34.22	94.79	13.66

* The volume of stone sludge was determined on the basis of the dried volume of hydrated grated stone material.

3. The Use of Waste from the Processing of Natural Stone

Natural stone wastes from stonemasonry plants constitute a substantial part of all waste produced there and are a significant environmental problem, as they are not biodegradable [18]. The efficient production management in processing plants and the resulting rational stone waste management require the planning and designing of stone elements to be performed in such a way that the processed material is fully used as various stone products. It is also important to introduce technical solutions to reduce waste. Such an approach will affect a reduction in stone scrap production. Still, the generated stone scraps should continue to be properly processed. Although reusing waste is socially and environmentally important, it should also be economically profitable and technically feasible [16].

Generally, stone wastes from processing plants are environmentally neutral. However, according to Simsek et al. [19], Rizzo et al. [20], Nasserdine et al. [21], and Luodes et al. [22], they may also have a negative environmental impact. Therefore, the European Parliament and EC Directive 2008/98/EC of 19 November 2008 on waste [8] encourages other applications of stone wastes. Although stone wastes are generally stored in waste disposal facilities or in post-mining excavations, the literature on the subject mentions numerous proposals for their application (Table 4).

Table 4. Literature research on the possibilities of application of stone scraps.

Application of Stone Scrap	Type of Stone Scrap	Material	References
Building materials (mortar/concrete/brick)	powder/fine aggregate	granite	[23–33]
		marble	[34–55]
		limestone	[56–61]
	powder/fine aggregate	basalt	[62–65]
	coarse/fine aggregate	sandstone	[66–71]
	powder	mix/unidentified	[72–77]
	fine grained	granite	[78–80]
Ceramic materials	waste/powder	marble	[81–85]
		gneiss	[86–88]
	powder	serpentinite	[89,90]
		mix/unidentified	[91,92]
Stabilised clay soil	powder	marble	[93]
		limestone	[94–97]
		mix/unidentified	[98,99]
Fertilisation	unidentified	marble	[100]
		basalt	[101,102]
	powder	gneiss	[103–106]
		granite	[107]
Various composite materials	powder	marble	[108–111]
		sandstone	[112]
		basalt	[113,114]
Other applications	powder	granite	[115]
		marble	[116–119]

In processing plants, various types and volumes of natural stone (e.g., granite, sandstone, limestone, marble) are often processed, which indicates the variety of stone waste and the different physical and chemical properties of these materials. The literature has repeatedly described both the possibilities of using stone scrap for the production of construction materials and their physicomechanical parameters. The largest number of studies showing the possibilities of using stone waste concerns granite and marble because of their widespread use in architecture and civil engineering. In addition, the largest number of studies were related to the use of stone powder.

The analysis of the literature on the subject indicates the possibilities of using stone scrap in the construction industry, which is concerned with the selection of various mixtures with the use of stone waste (most often from the processing of granite and marble) to produce mortar, concrete, or bricks. These works reveal that the use of up to 35% of the volume of stone waste material for the production of Portland cement does not affect the quality parameters of this product [26]. Ghannam et al. [24] and Prošek et al. [46] indicate that using stone waste can increase the strength parameters of concrete. Additionally, the use of fine stone waste can be a good solution for the production of ceramic products. Luiz et al. [92] and Munir et al. [83,84] emphasise that the inclusion of 15% of ornamental stone waste in the production of ceramic products does not affect the properties of these products.

Saygili [93], Ibrahim et al. [94], Ogila [95], Pastor et al. [96], Sabat and Muni [97], Igwe and Adepehin [98], and Sivrikaya et al. [99] have shown that the application of stone scrap to stabilise clay soils is beneficial and, at the same time, improves geotechnical properties and reduces soil swelling. The use of fine rock material for fertiliser production, used in agriculture, as well as in the reclamation process, can have a positive impact on the growth of vegetation and improve soil properties [100,102–105].

Stone scraps can also be used to produce various composite materials. Karimi et al. [107] propose to use granite waste for the production of ecological stone composite based on acrylonitrile–butadiene–styrene (ABS), while Conde-Vázquez et al. [112] indicate the possibility of using sandstone waste for the production of artificial arenite using cement polymerisation. Kurańska et al. [113] point to the possibility of applying stone waste to the production of highly efficient porous polyurethane composites. Basalt waste can also be used as admixtures for gypsum composites [114]. The use of fine marble waste is possible to produce various other composite materials: geopolymer hybrid composite materials [108,111], composite materials with the structure of unsaturated polyester [110], or composite materials produced from waste PET [109].

Other applications of stone scrap do not represent a large-scale use. However, more research should be conducted into the possibilities of their usefulness. Alves et al. [115] point to the use of granite waste for rock wool production. Agrawal et al. [116], Marras, and Careddu [117] propose to use marble waste in the rubber industry, whereas Özkaya et al. [119] have conducted a study on the possibility of applying waste marble powder as an adhesive filler in the manufacturing of laminated veneer lumber (LVL). Navar et al. [118] have tested the possibility of using waste marble powder as a potential alternative to current commercial calcium carbonate sorbents for capturing CO_2.

The reprocessing and reusing of stone scraps increase productivity and profitability while reducing the final production costs. In addition, it simultaneously limits the threat to the environment, reduces the number of non-biodegradable waste deposition sites, and offers alternative raw materials for various industrial activities. Table 5 shows general methods for using scraps generated in the processing of natural stone.

Table 5. Different applications of scrap from dimension natural stone processing (own work based on Shamrai et al. [17] and Shirazi [120]).

Type of Stone Scrap	Different Applications of Scrap from Dimension Natural Stone Processing
Small stone waste (including sludge)	Asphalt and concrete production Brick manufacturing Construction fill Production of synthetic aggregate Media for biofiltration systems or soil remediation Mineral content for soil Tire mixtures production
Waste in the form of aggregates	Construction fill Construction mixture ingredient For road filling For reclamation in landscaping and decorative use
Larger stone pieces and paving	Media for biofiltration systems Fill for gabion retaining walls and foundations For reclamation in landscaping and decorative use Use as a foundation filler
Damaged blocks or slabs	Production of aggregates For cutting tiles of small size Production of paving stones or tiles

4. Conclusions

Production processes should be planned to maximise resource usage and environmental protection while balancing costs. One of the cost-intensive factors in the production process is energy. Energy efficiency is an important optimisation issue and should be understood as the amount of energy used to obtain a product. This issue has been addressed in many works on mining [121,122]. In the context of natural stone processing, this issue is also important because such processes require high energy input. The generated stone waste, which generally represents a loss of raw material, also affects the number of production costs. In addition, stone wastes produced in stone processing pose significant problems, which considerably affect the environment, as well as the production efficiency and profitability.

The produced stone waste, i.e., solid scraps and sludge, comprises 10–35% of the quantity of the processed natural stone, solid scraps accounting for almost threefold greater quantities. Most of this waste is generated in the first stage of stone processing. A reduction in the quantity of stone waste is possible if the production of stone elements is rationally planned in order to use the material with maximum efficiency and if companies decide to introduce modern machinery, which is now designed with a view to reducing stone waste. In addition to the need to reduce stone scrap, it is necessary to search for its usage. The use of scrap, once its potential has been discovered, is now considered to be an activity that can contribute to product diversification, reduce final costs, and provide alternative raw materials for a variety of industrial sectors [79].

Based on the analysis conducted in the article, the following two important research areas can be identified that should be developed with a view to waste reduction or reprocessing:

- Research efforts to find or improve a technology that reduces the volume of waste produced or the development of a waste-free technology;
- Research efforts to find possibilities of stone waste application.

Author Contributions: Conceptualization, P.S.; methodology, P.S.; software, P.S.; validation, P.S.; writing—original draft preparation, P.S.; writing—review and editing, P.S. All authors have read and agreed to the published version of the manuscript.

Funding: This research was funded by the Polish Ministry of Education and Science Subsidy 2021 for the Department of Mining WUST, Grant Number 8211104160.

Institutional Review Board Statement: Not applicable.

Informed Consent Statement: Not applicable.

Data Availability Statement: Not Applicable.

Conflicts of Interest: The author declares no conflict of interest.

References

1. Strzałkowski, P. Requirements and test methods for selected natural stone products. *Gór. Odkryw.* **2018**, *59*, 34–40. (In Polish)
2. Careddu, N. Dimension stones in the circular economy world. *Resour. Policy* **2019**, *60*, 243–245. [CrossRef]
3. Careddu, N.; Dino, G.A. Reuse of residual sludge from stone processing: Differences and similarities between sludge coming from carbonate and silicate stones—Italian experiences. *Environ. Earth Sci.* **2016**, *75*, 1–9. [CrossRef]
4. Karaca, Z.; Pekin, A.; Deliormanlı, A.H.; Deliormanli, A.H. Classification of dimension stone wastes. *Environ. Sci. Pollut. Res.* **2012**, *19*, 2354–2362. [CrossRef]
5. Kaźmierczak, U.; Blachowski, J.; Górniak-Zimroz, J.; Wirth, H. Quantitative and Qualitative Research on the Waste from the Mining of Rock Raw Materials in Lower Silesia. *Minerals* **2018**, *8*, 375. [CrossRef]
6. Mosaferi, M.; Dianat, I.; Khatibi, M.S.; Mansour, S.N.; Fahiminia, M.; Hashemi, A.A. Review of environmental aspects and waste management of stone cutting and fabrication industries. *J. Mater. Cycles Waste Manag.* **2014**, *16*, 721–730. [CrossRef]
7. Yurdakul, M. Natural stone waste generation from the perspective of natural stone processing plants: An industrial-scale case study in the province of Bilecik, Turkey. *J. Clean. Prod.* **2020**, *276*, 123339. [CrossRef]
8. *Directive 2008/98/EC of the European Parliament and of the Council of 19 November 2008 on Waste and Repealing Certain Directives, 2008/98/EC*; EC: Brussels, Belgium, 2008.
9. Kaźmierczak, U.; Blachowski, J.; Górniak-Zimroz, J. Multi-Criteria Analysis of Potential Applications of Waste from Rock Minerals Mining. *Appl. Sci.* **2019**, *9*, 441. [CrossRef]
10. Tayebi-Khorami, M.; Edraki, M.; Corder, G.; Golev, A. Re-Thinking Mining Waste through an Integrative Approach Led by Circular Economy Aspirations. *Minerals* **2019**, *9*, 286. [CrossRef]
11. Woźniak, J.; Pactwa, K. Overview of Polish Mining Wastes with Circular Economy Model and Its Comparison with Other Wastes. *Sustainbility* **2018**, *10*, 3994. [CrossRef]
12. Yilmaz, E. Advances in reducing large volumes of environmentally harmful mine waste rocks and tailings. *Gospod. Surowcami Miner.–Miner. Resour. Manag.* **2011**, *27*, 89–112.
13. Lèbre, É.; Corder, G.; Golev, A. The Role of the Mining Industry in a Circular Economy: A Framework for Resource Management at the Mine Site Level. *J. Ind. Ecol.* **2017**, *21*, 662–672. [CrossRef]
14. *Commission Decision of 18 December 2014 Amending Decision 2000/532/EC on the List of Waste Pursuant to Directive 2008/98/EC of the European Parliament and of the Council, 2014/955/UE*; EC: Brussels, Belgium, 2014.
15. Singh, S.; Nagar, R.; Agrawal, V. Performance of granite cutting waste concrete under adverse exposure conditions. *J. Clean. Prod.* **2016**, *127*, 172–182. [CrossRef]
16. Mitchell, C.; Harrison, D.; Robinson, H.; Ghazireh, N. Minerals from Waste: Recent BGS and Tarmac experience in finding uses for mine and quarry waste. *Miner. Eng.* **2004**, *17*, 279–284. [CrossRef]
17. Shamrai, V.I.; Korobiichuk, V.V.; Sobolevskyi, R.V. Management of waste of stone processing in the framework of Euro integration of Ukraine. *J. Zhytomyr State Technol. Univ. Ser. Eng.* **2017**, *1*, 234–239. [CrossRef]
18. Lakhani, R.; Kumar, R.; Tomar, P. Utilization of Stone Waste in the Development of Value Added Products: A State of the Art Review. *J. Eng. Sci. Technol. Rev.* **2014**, *7*, 180–187. [CrossRef]
19. Simsek, C.; Karaca, Z.; Gemici, U.; Gunduz, O. The assessment of the impacts of a marble waste site on the water and sedi-ment quality in a river system. *Fresenius Environ. Bull.* **2005**, *14*, 1013–1023.
20. Rizzo, G.; D'Agostino, F.; Ercoli, L. Problems of soil and groundwater pollution in the disposal of "marble" slurries in NW Sicily. *Environ. Earth Sci.* **2008**, *55*, 929–935. [CrossRef]
21. Nasserdine, K.; Mimi, Z.; Bevan, B.; Elian, B. Environmental management of the stone cutting industry. *J. Environ. Manag.* **2009**, *90*, 466–470. [CrossRef] [PubMed]
22. Luodes, H.; Kauppila, P.M.; Luodes, N.; Aatos, S.; Kallioinen, J.; Luukkanen, S.; Aalto, J. Characteristics and the environmental acceptability of the natural stone quarrying waste rocks. *Bull. Int. Assoc. Eng. Geol.* **2012**, *71*, 257–261. [CrossRef]
23. Da Silva, J.L.; Campos, D.B.D.C.; Lordsleem, A.C.; Povoas, Y.V. Influence of the partial substitution of fine aggregate by granite powder in mortar on the process of natural carbonation. *Waste Manag. Res.* **2020**, *38*, 254–262. [CrossRef] [PubMed]
24. Ghannam, S.; Najm, H.; Vasconez, R. Experimental study of concrete made with granite and iron powders as partial replacement of sand. *Sustain. Mater. Technol.* **2016**, *9*, 1–9. [CrossRef]
25. Gupta, L.K.; Vyas, A.K. Impact on mechanical properties of cement sand mortar containing waste granite powder. *Constr. Build. Mater.* **2018**, *191*, 155–164. [CrossRef]

26. Hussain, S.A.; Shirisha, P.; Kumar, T.; Krishna, S.V.; Satyanarayana, S.V. The Use of Granite Industry Waste as a Cement Substitute. *Int. J. Eng. Res. Technol.* **2013**, *2*, 894–898.
27. Li, L.; Wang, Y.; Tan, Y.; Kwan, A. Adding granite dust as paste replacement to improve durability and dimensional stability of mortar. *Powder Technol.* **2018**, *333*, 269–276. [CrossRef]
28. Lokeshwari, M.; Jagadish, K. Eco-friendly Use of Granite Fines Waste in Building Blocks. *Procedia Environ. Sci.* **2016**, *35*, 618–623. [CrossRef]
29. Boadella, Í.L.; Gayarre, F.L.; González, J.S.; Gómez-Soberón, J.M.; Pérez, C.L.-C.; López, M.S.; De Brito, J. The Influence of Granite Cutting Waste on the Properties of Ultra-High Performance Concrete. *Materials* **2019**, *12*, 634. [CrossRef]
30. Ramos, T.; Matos, A.M.; Schmidt, B.; Rio, J.; Sousa-Coutinho, J. Granitic quarry sludge waste in mortar: Effect on strength and durability. *Constr. Build. Mater.* **2013**, *47*, 1001–1009. [CrossRef]
31. Shamsabadi, E.A.; Ghalehnovi, M.; De Brito, J.; Khodabakhshian, A. Performance of Concrete with Waste Granite Powder: The Effect of Superplasticizers. *Appl. Sci.* **2018**, *8*, 1808. [CrossRef]
32. Singh, S.; Nagar, R.; Agrawal, V.; Rana, A.; Tiwari, A. Sustainable utilization of granite cutting waste in high strength concrete. *J. Clean. Prod.* **2016**, *116*, 223–235. [CrossRef]
33. Vijayalakshmi, M.; Sekar, A.; Sivabharathy, M.; Prabhu, G.G. Utilization of Granite Powder Waste in Concrete Production. *Defect Diffus. Forum* **2012**, *330*, 49–61. [CrossRef]
34. Alyamaç, K.E.; Ghafari, E.; Ince, R. Development of eco-efficient self-compacting concrete with waste marble powder using the response surface method. *J. Clean. Prod.* **2017**, *144*, 192–202. [CrossRef]
35. Alyamaç, K.E.; Ince, R. A preliminary concrete mix design for SCC with marble powders. *Constr. Build. Mater.* **2009**, *23*, 1201–1210. [CrossRef]
36. Bilgin, N.; Yeprem, H.; Arslan, S.; Bilgin, A.; Günay, E.; Marşoglu, M. Use of waste marble powder in brick industry. *Constr. Build. Mater.* **2012**, *29*, 449–457. [CrossRef]
37. Corinaldesi, V.; Moriconi, G.; Naik, T.R. Characterization of marble powder for its use in mortar and concrete. *Constr. Build. Mater.* **2010**, *24*, 113–117. [CrossRef]
38. Eliche-Quesada, D.; Iglesias, F.J.; Pérez-Villarejo, L.; Iglesias-Godino, F. Recycling of sawdust, spent earth from oil filtration, compost and marble residues for brick manufacturing. *Constr. Build. Mater.* **2012**, *34*, 275–284. [CrossRef]
39. Grilo, M.J.; Pereira, J.; Costa, C. Waste Marble Dust Blended Cement. *Mater. Sci. Forum* **2012**, *730-732*, 671–676. [CrossRef]
40. Kabeer, K.S.A.; Vyas, A.K. Utilization of marble powder as fine aggregate in mortar mixes. *Constr. Build. Mater.* **2018**, *165*, 321–332. [CrossRef]
41. Khyaliya, R.K.; Kabeer, K.S.A.; Vyas, A.K. Evaluation of strength and durability of lean mortar mixes containing marble waste. *Constr. Build. Mater.* **2017**, *147*, 598–607. [CrossRef]
42. Kore, S.D.; Vyas, A.K.; Syed, A.K.K.I. A brief review on sustainable utilisation of marble waste in concrete. *Int. J. Sustain. Eng.* **2020**, *13*, 264–279. [CrossRef]
43. Li, L.; Huang, Z.; Tan, Y.; Kwan, A.; Chen, H. Recycling of marble dust as paste replacement for improving strength, microstructure and eco-friendliness of mortar. *J. Clean. Prod.* **2019**, *210*, 55–65. [CrossRef]
44. Li, L.; Huang, Z.; Tan, Y.; Kwan, A.; Liu, F. Use of marble dust as paste replacement for recycling waste and improving durability and dimensional stability of mortar. *Constr. Build. Mater.* **2018**, *166*, 423–432. [CrossRef]
45. Mashaly, O.; El-Kaliouby, B.A.; Shalaby, B.; El-Gohary, A.; Rashwan, M. Effects of marble sludge incorporation on the properties of cement composites and concrete paving blocks. *J. Clean. Prod.* **2016**, *112*, 731–741. [CrossRef]
46. Prošek, Z.; Nežerka, V.; Tesárek, P. Enhancing cementitious pastes with waste marble sludge. *Constr. Build. Mater.* **2020**, *255*, 119372. [CrossRef]
47. Prošek, Z.; Trejbal, J.; Topič, J.; Plachý, T.; Tesárek, P. Utilization of the waste from the marble industry for application in transport infrastructure: Mechanical properties of cement pastes. *IOP Conf. Ser. Mater. Sci. Eng.* **2017**, *236*, 012092. [CrossRef]
48. Rodrigues, R.; de Brito, J.; Sardinha, M. Mechanical properties of structural concrete containing very fine aggregates from marble cutting sludge. *Constr. Build. Mater.* **2015**, *77*, 349–356. [CrossRef]
49. Seghir, N.T.; Mellas, M.; Sadowski, Ł.; Krolicka, A.; Żak, A.; Ostrowski, K. The Utilization of Waste Marble Dust as a Cement Replacement in Air-Cured Mortar. *Sustainbility* **2019**, *11*, 2215. [CrossRef]
50. Seghir, N.T.; Mellas, M.; Sadowski, Ł.; Żak, A. Effects of marble powder on the properties of the air-cured blended cement paste. *J. Clean. Prod.* **2018**, *183*, 858–868. [CrossRef]
51. Siddique, Z.; Bhargava, R.; Ansari, M.; Khan, W. Experimental study for the utilization of marble powder and in construction industry. In Proceedings of the International Conference on Sustainable Materials and Structures for Civil Infrastructures (SMSCI2019); AIP Publishing: Melville, NY, USA, 2019; Volume 2158, p. 020025.
52. Tunc, E.T. Recycling of marble waste: A review based on strength of concrete containing marble waste. *J. Environ. Manag.* **2019**, *231*, 86–97. [CrossRef]
53. Ulubeyli, G.C.; Bilir, T.; Artir, R. Durability Properties of Concrete Produced by Marble Waste as Aggregate or Mineral Additives. *Procedia Eng.* **2016**, *161*, 543–548. [CrossRef]
54. Vardhan, K.; Goyal, S.; Siddique, R.; Singh, M. Mechanical properties and microstructural analysis of cement mortar incorporating marble powder as partial replacement of cement. *Constr. Build. Mater.* **2015**, *96*, 615–621. [CrossRef]

55. Vardhan, K.; Siddique, R.; Goyal, S. Influence of marble waste as partial replacement of fine aggregates on strength and drying shrinkage of concrete. *Constr. Build. Mater.* **2019**, *228*, 116730. [CrossRef]
56. Cavaleri, L.; Borg, R.P.; La Mantia, F.P.; Liguori, V. Quarry limestone dust as fine aggregate for concrete. *IOP Conf. Ser. Mater. Sci. Eng.* **2018**, *442*. [CrossRef]
57. Chouhan, H.S.; Kalla, P.; Nagar, R.; Gautam, P.K. Gainful utilization of dimensional limestone waste as fine aggregate in cement mortar mixes. *Constr. Build. Mater.* **2019**, *221*, 363–374. [CrossRef]
58. Chouhan, H.S.; Kalla, P.; Nagar, R.; Gautam, P.K.; Arora, A.N. Investigating use of dimensional limestone slurry waste as fine aggregate in mortar. *Environ. Dev. Sustain.* **2018**, *22*, 2223–2245. [CrossRef]
59. Kibriya, T.; Tahir, L. Sustainable Construction—High Performance Concrete Containing Limestone Dust as Filler. *World J. Eng. Technol.* **2017**, *5*, 404–411. [CrossRef]
60. Rana, A.; Kalla, P.; Csetenyi, L. Recycling of dimension limestone industry waste in concrete. *Int. J. Min. Reclam. Environ.* **2017**, *31*, 231–250. [CrossRef]
61. Turgut, P.; Algin, H.M. Limestone dust and wood sawdust as brick material. *Build. Environ.* **2007**, *42*, 3399–3403. [CrossRef]
62. Andrade, F.R.D.; Pecchio, M.; Bendoraitis, D.P.; Montanheiro, T.J.; Kihara, Y. Basalt mine-tailings as raw-materials for Portland clinker. *Cerâmica* **2010**, *56*, 39–43. [CrossRef]
63. Dobiszewska, M. Zastosowanie pyłu bazaltowego, jako substytutu piasku w zaprawie i betonie cementowym. *Bud. Arch.* **2016**, *15*, 075–085. [CrossRef]
64. Dobiszewska, M.; Barnes, R.W. Properties of Mortar Made with Basalt Powder as Sand Replacement. *Mater. J.* **2020**, *117*, 3–9. [CrossRef]
65. Dobiszewska, M.; Beycioğlu, A. Physical Properties and Microstructure of Concrete with Waste Basalt Powder Addition. *Materials* **2020**, *13*, 3503. [CrossRef] [PubMed]
66. Chandar, K.R.; Gayana, B.; Sainath, V. Experimental investigation for partial replacement of fine aggregates in concrete with sandstone. *Adv. Concr. Constr.* **2016**, *4*, 243–261. [CrossRef]
67. Gupta, R.C.; Basu, P.; Agrawal, S. Mechanical properties of self-compacting concrete containing sandstone slurry. *IOP Conf. Ser. Mater. Sci. Eng.* **2019**, *689*, 012002. [CrossRef]
68. Kumar, S.; Gupta, R.C.; Shrivastava, S. Long term studies on the utilisation of quartz sandstone wastes in cement concrete. *J. Clean. Prod.* **2017**, *143*, 634–642. [CrossRef]
69. Kumar, S.; Gupta, R.C.; Shrivastava, S.; Csetenyi, L.; Thomas, B.S. Preliminary study on the use of quartz sandstone as a partial replacement of coarse aggregate in concrete based on clay content, morphology and compressive strength of combined gradation. *Constr. Build. Mater.* **2016**, *107*, 103–108. [CrossRef]
70. Kumar, S.; Thomas, B.S.; Gupta, V.; Basu, P.; Shrivastava, S. Sandstone wastes as aggregate and its usefulness in cement concrete—A comprehensive review. *Renew. Sustain. Energy Rev.* **2018**, *81*, 1147–1153. [CrossRef]
71. Singhal, A.; Goel, S.; Sengupta, D. Physicochemical and elemental analyses of sandstone quarrying wastes to assess their impact on soil properties. *J. Environ. Manag.* **2020**, *271*, 111011. [CrossRef] [PubMed]
72. Almada, B.S.; Melo, L.D.S.; Dutra, J.B.; Bubani, L.C.; Silva, G.J.B.; dos Santos, W.J.; Aguilar, M.T.P. Influence of the heterogeneity of waste from wet processing of ornamental stones on the performance of Portland cement composites. *Constr. Build. Mater.* **2020**, *262*, 120036. [CrossRef]
73. Amin, S.K.; Allam, M.E.; Garas, G.L.; Ezz, H. A study of the chemical effect of marble and granite slurry on green mortar compressive strength. *Bull. Natl. Res. Cent.* **2020**, *44*, 1–5. [CrossRef]
74. Bacarji, E.; Filho, R.T.; Koenders, E.; Figueiredo, E.; Lopes, J. Sustainability perspective of marble and granite residues as concrete fillers. *Constr. Build. Mater.* **2013**, *45*, 1–10. [CrossRef]
75. Chouhan, H.S.; Kalla, P.; Nagar, R.; Gautam, P.K. Influence of dimensional stone waste on mechanical and durability properties of mortar: A review. *Constr. Build. Mater.* **2019**, *227*, 116662. [CrossRef]
76. Galetakis, M.; Soultana, A. A review on the utilisation of quarry and ornamental stone industry fine by-products in the construction sector. *Constr. Build. Mater.* **2016**, *102*, 769–781. [CrossRef]
77. Rana, A.; Kalla, P.; Verma, H.; Mohnot, J. Recycling of dimensional stone waste in concrete: A review. *J. Clean. Prod.* **2016**, *135*, 312–331. [CrossRef]
78. Araújo, A.J.M.; O Sousa, A.R.; Macedo, D.A.; Dutra, R.P.S.; Campos, L.F.A.; De Sousa, A.R.O. Effects of granite waste addition on the technological properties of industrial silicate based-ceramics. *Mater. Res. Express* **2019**, *6*, 125205. [CrossRef]
79. Menezes, R.R.; Ferreira, H.S.; Neves, G.A.; Lira, H. Use of granite sawing wastes in the production of ceramic bricks and tiles. *J. Eur. Ceram. Soc.* **2005**, *25*, 1149–1158. [CrossRef]
80. Panna, W.; Wyszomirski, P.; Gajek, M. Characteristics of the fine-grained fractions of the crushed Strzegom granites as possi-ble materials in manufacture of ceramic tiles. *Gospod. Surowcami Miner.–Miner. Resour. Manag.* **2015**, *31*, 59–75. [CrossRef]
81. El-Mahllawy, M.S.; Kandeel, A.M.; Latif, M.L.A.; El Nagar, A.M. The Feasibility of Using Marble Cutting Waste in a Sustainable Building Clay Industry. *Recycling* **2018**, *3*, 39.
82. Montero, M.A.; Jordán, M.; Crespo, M.S.H.; Sanfeliu, T. The use of sewage sludge and marble residues in the manufacture of ceramic tile bodies. *Appl. Clay Sci.* **2009**, *46*, 404–408. [CrossRef]
83. Munir, M.J.; Abbas, S.; Nehdi, M.L.; Kazmi, S.M.S.; Khitab, A. Development of Eco-Friendly Fired Clay Bricks Incorporating Recycled Marble Powder. *J. Mater. Civ. Eng.* **2018**, *30*, 04018069. [CrossRef]

84. Munir, M.J.; Kazmi, S.M.S.; Wu, Y.-F.; Hanif, A.; Khan, M.U.A. Thermally efficient fired clay bricks incorporating waste marble sludge: An industrial-scale study. *J. Clean. Prod.* **2018**, *174*, 1122–1135. [CrossRef]
85. Yeşilay, S.; Çakı, M.; Ergun, H. Usage of marble wastes in traditional artistic stoneware clay body. *Ceram. Int.* **2017**, *43*, 8912–8921. [CrossRef]
86. El-Mehalawy, N.; El-Omla, M.; Naga, S.M. Stoneware Tiles Based on Gneiss Rocks. *Interceram—Int. Ceram. Rev.* **2018**, *67*, 26–33. [CrossRef]
87. Moreira, J.; Manhães, J.; Holanda, J. Processing of red ceramic using ornamental rock powder waste. *J. Mater. Process. Technol.* **2008**, *196*, 88–93. [CrossRef]
88. Souza, A.; Pinheiro, B.; Holanda, J. Recycling of gneiss rock waste in the manufacture of vitrified floor tiles. *J. Environ. Manag.* **2010**, *91*, 685–689. [CrossRef] [PubMed]
89. Cavallo, A. Serpentinitic waste materials from the dimension stone industry: Characterization, possible reuses and critical issues. *Resour. Policy* **2018**, *59*, 17–23. [CrossRef]
90. Díaz, L.; Torrecillas, R. Porcelain stoneware obtained from the residual muds of serpentinite raw materials. *J. Eur. Ceram. Soc.* **2007**, *27*, 2341–2345. [CrossRef]
91. Amaral, L.F.; de Carvalho, J.P.R.G.; da Silva, B.M.; Delaqua, G.C.G.; Monteiro, S.N.; Vieira, C.M.F. Development of ceramic paver with ornamental rock waste. *J. Mater. Res. Technol.* **2019**, *8*, 599–608. [CrossRef]
92. Luiz, N.F.; Cecchin, D.; Azevedo, A.R.G.; Alexandre, J.; Marvila, M.T.; Da Silva, F.C.; Paes, A.L.C.; Pinheiro, V.D.; Do Car-mo, D.F.; Ferraz, P.F.P.; et al. Characterization of materials used in the manufacture of ceramic tile with incorporation of or-namental rock waste. *Agron. Res.* **2020**, *18*, 904–914. [CrossRef]
93. Saygili, A. Use of Waste Marble Dust for Stabilization of Clayey Soil. *Mater. Sci.* **2015**, *21*, 601–606. [CrossRef]
94. Ibrahim, H.H.; Alshkane, Y.M.; Mawlood, Y.I.; Noori, K.M.G.; Hasan, A.M. Improving the geotechnical properties of high expansive clay using limestone powder. *Innov. Infrastruct. Solutions* **2020**, *5*, 1–11. [CrossRef]
95. Ogila, W.A.M. The impact of natural ornamental limestone dust on swelling characteristics of high expansive soils. *Environ. Earth Sci.* **2016**, *75*, 1493. [CrossRef]
96. Pastor, J.L.; Tomás, R.; Cano, M.; Riquelme, A.; Gutiérrez, E. Evaluation of the Improvement Effect of Limestone Powder Waste in the Stabilization of Swelling Clayey Soil. *Sustainbility* **2019**, *11*, 679. [CrossRef]
97. Sabat, A.K.; Muni, P.K. Effects of limestone dust on geotechnical properties of an expansive soil. *Int. J. Appl. Eng. Res.* **2015**, *10*, 37724–37730.
98. Igwe, O.; Adepehin, E.J. Alternative Approach to Clay Stabilization Using Granite and Dolerite Dusts. *Geotech. Geol. Eng.* **2017**, *35*, 1657–1664. [CrossRef]
99. Sivrikaya, O.; Kıyıldı, K.R.; Karaca, Z. Recycling waste from natural stone processing plants to stabilise clayey soil. *Environ. Earth Sci.* **2014**, *71*, 4397–4407. [CrossRef]
100. Tozsin, G.; Arol, A.I.; Oztas, T.; Kalkan, E. Using marble wastes as a soil amendment for acidic soil neutralization. *J. Environ. Manag.* **2014**, *133*, 374–377. [CrossRef]
101. Artico, M.; Firpo, B.A.; Artico, L.L.; Tubino, R.M.C. Integrated use of sewage sludge and basalt mine waste as soil substitute for environmental restoration. *REM-Int. Eng. J.* **2020**, *73*, 225–232. [CrossRef]
102. Zagożdżon, P.P. Basalt Powder in Agricultural Use. *Min. Sci.* **2008**, *10*, 133–142. (In Polish)
103. De Souza, M.E.P.; Cardoso, I.M.; De Carvalho, A.M.X.; Lopes, A.P.; Jucksch, I. Gneiss and steatite vermicomposted with organic residues: Release of nutrients and heavy metals. *Int. J. Recycl. Org. Waste Agric.* **2019**, *8*, 233–240. [CrossRef]
104. De Souza, M.E.P.; Cardoso, I.M.; De Carvalho, A.M.X.; Lopes, A.P.; Jucksch, I.; Janssen, A. Rock Powder Can Improve Ver-micompost Chemical Properties and Plant Nutrition: An On-farm Experiment. *Commun. Soil Sci. Plant Anal.* **2018**, *49*, 1–12. [CrossRef]
105. Dino, G.A.; Passarella, I.; Ajmone-Marsan, F. Quarry rehabilitation employing treated residual sludge from dimension stone working plant. *Environ. Earth Sci.* **2014**, *73*, 7157–7164. [CrossRef]
106. Kufka, D.; Cichoń, T.; Pomorski, A. Use of Gneiss Waste for Cultivation of Maize under Natural Sunflow Conditions. *Gór. Odkryw.* **2019**, *60*, 25–29. (In Polish)
107. Karimi, D.; Crawford, B.; Milani, A.S. Manufacturing process and mechanical properties of a novel acrylonitrile butadiene styrene-based composite, with recycled natural granite micro-particles. *Manuf. Lett.* **2020**, *23*, 79–84. [CrossRef]
108. Bakshi, P.; Pappu, A.; Patidar, R.; Gupta, M.K.; Thakur, V.K. Transforming Marble Waste into High-Performance, Water-Resistant, and Thermally Insulative Hybrid Polymer Composites for Environmental Sustainability. *Polymers* **2020**, *12*, 1781. [CrossRef] [PubMed]
109. Çınar, M.E.; Kar, F. Characterization of composite produced from waste PET and marble dust. *Constr. Build. Mater.* **2018**, *163*, 734–741. [CrossRef]
110. Doan, T.T.L.; Brodowsky, H.M.; Gohs, U.; Mäder, E. Re-Use of Marble Stone Powders in Producing Unsaturated Polyester Composites. *Adv. Eng. Mater.* **2018**, *20*, 1701061. [CrossRef]
111. Thakur, A.K.; Pappu, A.; Thakur, V.K. Synthesis and characterization of new class of geopolymer hybrid composite materials from industrial wastes. *J. Clean. Prod.* **2019**, *230*, 11–20. [CrossRef]
112. Conde-Vázquez, C.; Martín, O.D.M.-S.; García-Herbosa, G. Artificial arenite from wastes of natural sandstone industry. *Mater. Constr.* **2019**, *69*, 178. [CrossRef]

113. Kurańska, M.; Barczewski, M.; Uram, K.; Lewandowski, K.; Prociak, A.; Michałowski, S. Basalt waste management in the production of highly effective porous polyurethane composites for thermal insulating applications. *Polym. Test.* **2019**, *76*, 90–100. [CrossRef]

114. Petropavlovskaya, V.B.; Novichenkova, T.B.; Zavadko, M.Y.; Petropavlovskii, K.S. On the potential use of basalt waste as mineral fillers. *IOP Conf. Ser. Mater. Sci. Eng.* **2020**, *791*, 012070. [CrossRef]

115. Alves, J.O.; Espinosa, D.C.R.; Tenório, J.A.S. Recovery of Steelmaking Slag and Granite Waste in the Production of Rock Wool. *Mater. Res.* **2015**, *18*, 204–211. [CrossRef]

116. Agrawal, S.; Mandot, S.; Bandyopadhyay, S.; Mukhopadhyay, R.; Dasgupta, M.; De, P.; Deuri, A. Use of Marble Waste in Rubber Industry: Part I (in NR Compound). *Prog. Rubber Plast. Recycl. Technol.* **2004**, *20*, 229–246. [CrossRef]

117. Marras, G.; Careddu, N. Sustainable reuse of marble sludge in tyre mixtures. *Resour. Policy* **2018**, *59*, 77–84. [CrossRef]

118. Nawar, A.; Ghaedi, H.; Ali, M.; Zhao, M.; Iqbal, N.; Khan, R. Recycling waste-derived marble powder for CO_2 capture. *Process. Saf. Environ. Prot.* **2019**, *132*, 214–225. [CrossRef]

119. Özkaya, K.; Ayrilmis, N.; Özdemir, S. Potential use of waste marble powder as adhesive filler in the manufacture of laminated veneer lumber. *BioRes* **2015**, *10*, 1686–1695. [CrossRef]

120. Shirazi, E.K. Reusing of stone waste in various industrial activities. In Proceedings of the 2nd International Conference on Environmental Science and Development, Singapore, 26–28 February 2011; Volume 4, pp. 217–219.

121. Kawalec, W.; Suchorab, N.; Konieczna-Fuławka, M.; Król, R. Specific Energy Consumption of a Belt Conveyor System in a Continuous Surface Mine. *Energies* **2020**, *13*, 5214. [CrossRef]

122. Bortnowski, P.; Gładysiewicz, L.; Król, R.; Ozdoba, M. Energy Efficiency Analysis of Copper Ore Ball Mill Drive Systems. *Energies* **2021**, *14*, 1786. [CrossRef]

Article

Assessment of GHG Interactions in the Vicinity of the Municipal Waste Landfill Site—Case Study

Maciej Górka [1], Yaroslav Bezyk [2] and Izabela Sówka [2,*]

1   Faculty of Earth Science and Environmental Management, University of Wrocław, Cybulskiego 32, 50-205 Wroclaw, Poland; maciej.gorka@uwr.edu.pl
2   Faculty of Environmental Engineering, Wroclaw University of Science and Technology, Plac Grunwaldzki 13, 50-377 Wroclaw, Poland; jaroslaw.bezyk@pwr.edu.pl
*   Correspondence: izabela.sowka@pwr.edu.pl

Abstract: Landfills have been identified as one of the major sources of greenhouse gas (GHG) emissions and as a contributor to climate change. Landfill facilities exhibit considerable spatial and temporal variability of both methane (CH_4) and carbon dioxide (CO_2) rates. The present work aimed to evaluate the spatial distribution of CH_4 and CO_2 and their $\delta^{13}C$ isotopic composition originating from a municipal landfill site, to identify its contribution to the local GHG budget and the potential impact on the air quality of the immediate surroundings in a short-term response to environmental conditions. The objective was met by performing direct measurements of atmospheric CO_2 and CH_4 at the selected monitoring points on the surface and applying a binary mixing model for the determination of carbon isotopic ratios in the vicinity of the municipal waste landfill site. Air samples were collected and analysed for isotopic composition using flask sampling with a Picarro G2201-I Cavity Ring-Down Spectroscopy (CRDS) technique. Kriging and Inverse distance weighting (IDW) methods were used to evaluate the values at unsampled locations and to map the excess of GHGs emitted from the landfill surface. The large off-site dispersion of methane from the landfill site at a 500 m distance was identified during field measurements using isotopic data. The mean $\delta^{13}C$ of the landfill biogas emitted to the surrounded atmosphere was $-53.9 \pm 2.2‰$, which corresponded well to the microbial degradation processes during acetate fermentation in the waste deposits. The calculated isotopic compositions of CO_2 ($\delta^{13}C = -18.64 \pm 1.75‰$) indicate the domination of biogenic carbon reduction by vegetation surrounding the landfill. Finally, amounts of methane escaping into the air can be limited by the appropriate landfill management practices (faster covers active quarter through separation layer), and CH_4 reduction can be achieved by sealing the cover on the leachate tank.

Keywords: GHG emissions; stable isotopes; waste management; energy recovery

Citation: Górka, M.; Bezyk, Y.; Sówka, I. Assessment of GHG Interactions in the Vicinity of the Municipal Waste Landfill Site—Case Study. *Energies* **2021**, *14*, 8259. https://doi.org/10.3390/en14248259

Academic Editor: Rajender Gupta

Received: 7 November 2021
Accepted: 7 December 2021
Published: 8 December 2021

Publisher's Note: MDPI stays neutral with regard to jurisdictional claims in published maps and institutional affiliations.

1. Introduction

Carbon dioxide (CO_2) and methane (CH_4) derive from a majority of natural and anthropogenic sources and play a critical role in regulating the Earth's climate by trapping heat and contributing to overall global warming [1–3]. There are limitations of the accuracy of emissions data derived from the official inventories, especially fugitive emissions from landfills, and estimates obtained from direct atmospheric measurements. Methane and carbon dioxide are major components of landfill gas (biogas), accounting for ~45–60% and ~40–60% of the total gas emissions [4], respectively; their variations depend on the composition and age of the waste [5], as well as landfill operating procedures [6]. Biogas also includes other gases, such as volatile organic compounds (VOCs), non-methane organic compounds (NMOCs), hydrogen sulfide (H_2S), and ammonia (NH_3), as well as trace amounts of harmful compounds [7].

Landfill gas is produced by the degradation of organic matter in the waste mix in a series of biological and physicochemical mechanisms under anaerobic conditions [5,8,9]. The production of landfill gas consists of the following five phases: (1) hydrolysis (in the presence of oxygen complex organic polymers, become converted into simple compounds, sugar, fatty acids, and amino acids, further O_2 becomes depleted while CO_2 is generated), (2) acidogenesis (anaerobic acidic phase after the oxygen in the landfill is exhausted and anaerobic bacteria convert compounds produced by hydrolysis into acetic, lactic and formic acids and alcohols; CO_2 and H_2 are then generated as end products), (3) acetogenesis (the methanogenesis process starts and acids from acidogenesis are decomposed into acetate, decrease in pH; the production of CH_4 increases up to a level of around 70%, and the production of CO_2 decreases from 70 to 40%), (4) methanogenesis (the equilibrium is reached and acetophilic methanogens directly produce CH_4 and CO_2 from acetate or hydrogenophilic methanogenic bacteria using hydrogen to yield CH_4 and CO_2), and (5) maturation (most organic waste is already degraded and most of the landfill gas emissions occur, then the volume of gas generation decreases with respect to the landfill stabilization) [10–12].

Emissions of landfill gases (LFG) and the presence of contaminants in wastewater (leachate) are the two major contributors to environmental impacts in the immediate surroundings [13–15]. Moreover, inappropriate waste maintenance and landfilling constitute a serious threat to public health and the welfare of the ecosystem [16,17]. The negative environmental impact can be limited by extracting the captured biogas, which can be flared or collected and purified, and used as renewable fuel whenever environmentally and economically convenient (e.g., electricity generation or vehicle fueling). Landfill gas collection begins in the extraction wells (network composed of slotted plastic pipes) installed and operated vertically and/or horizontally inside the waste mass, depending on site-specific factors [5]. It is necessary to reduce the abrasive and corrosive nature of raw landfill gas streams by treating and further removing moisture with a moisture separator and mist eliminator, and removing particulates and other impurities through the use of filtration [5,18]. Collected LFG condensate is commonly combined with the formation of landfill leachate that requires optimal system operations to conduct and store the leachate for its further treatment or disposal [5,18].

It is still difficult to assess the scale of CO_2 and CH_4 emissions from landfill facilities due to the large temporal and spatial variations in the landfill source strength [7,19,20]. The combination of high-resolution atmospheric and precise stable isotope measurements acts as an effective tool for monitoring the strength of major CO_2 and CH_4 sources and understanding GHGs biogeochemistry [21–25]. The ranges of isotopic signatures ($\delta^{13}C$) of carbon dioxide and methane emitted from the major source categories are very large. With respect to carbon isotopic fractionation during methanogenesis, three types of methane sources have distinct $\delta^{13}C$ signatures, whereby biogenic CH_4 is heavily depleted in ^{13}C (from −75 to −55‰) [26,27] whereas thermogenic methane typically has $\delta^{13}C$ values (from −60 to −20‰) [28,29] and pyrogenic methane is assigned more enriched values (−50 to −40‰) [21,30]. The isotopic range of $\delta^{13}C(CO_2)$ for terrestrial vegetation varies from −29 to −12‰ [31,32] whereas the $\delta^{13}C$ composition of fossil fuel combustion ranges from −44‰ to −22‰ [22,33].

Methane released by biogas production processes, leachate characterization, and the maturity stage of the landfill site, can also be detected by measuring the stable isotopic composition of methane $\delta^{13}C(CH_4)$ [8,34–37]. Carbon sources at landfill sites access a narrow range of isotopic methane signatures, however, the oxidation of CH_4 processes alongside alteration and isotopic exchange can influence the raw CH_4 biogenic genesis signal [37]. According to Fischer et al. [38], isotopic signatures of methane emission from gas-collection systems in landfills located in Germany and the Netherlands were in the range of −60.3 to −57.4‰, with an average of −57.4 ± 2.2‰; samples derived from the atmospheric upwind and downwind measurements values of $\delta^{13}C(CH_4)$ varied within the range typical for terrestrial biogenic CH_4 sources (from −58.0 to −54.2‰, with an average of −55.4 ± 1.4‰). In the study conducted in Kuwait [39], reported methane isotopic signatures for samples obtained at the vicinity of a landfill site were in a range from −59.4 to −51.9‰, with an average $\delta^{13}C(CH_4)$ value of −56.6 ± 3.1‰. A study on the isotopic composition of methane emissions from active landfill sites in the United Kingdom [40] reveal the isotopic signature of $\delta^{13}C(CH_4)$ to be in the range -60 to $-58‰$. The $\delta^{13}C(CH_4)$ values measured at two Danish landfills were in the range of -59.94 and $-48.45‰$ for downwind samples and ranged between -49.43 and $-46.68‰$ for upwind gas samples from the source area [41].

An isotopic analysis is important in studies on the metabolic pathways of methane production. For biogenic methane formation from the microbial decomposition of organic matter (OM), two main methanogenic pathways can be isotopically differentiated, namely, acetate fermentation (AF) and hydrogen-utilizing CO_2-reduction (CR). The third major methanogenic pathway in OM degradation is through the utilization of methyl (amines and mercaptans) and methanol, yet the generated CH_4 is less isotopically distinct and can resemble CH_4 acetate fermentation. Methanogenesis leads to CH_4 depletion in ^{13}C compared to the organic substrate. Typical $\delta^{13}C(CH_4)$ values, in the range -65 to $-50‰$, $\Delta^{13}C(CO_2\text{-}CH_4)$ between 30 and $55‰$, and $\delta^2H < -300‰$ are indicative of an AF pathway. The CR pathway generates more ^{13}C-depleted CH_4, typically in the range -60 to $-80‰$, $\Delta^{13}C(CO_2\text{-}CH_4)$ between 55 and $100‰$, although less ^2H-depleted in the range -250 to $-150‰$. The reason for the large $\Delta^{13}C(CO_2\text{-}CH_4)$ is that the fractionation during CO_2-reduction (multiple enzymatic steps) is higher than in cases of acetate disproportionation to CO_2 and CH_4, and the residual CO_2 becomes progressively ^{13}C-enriched, depending on how open/closed the system is [8,37,38,41–44].

There have been few studies that attempted to estimate mechanisms and the relative contribution of emissions of CH_4 from landfill sources [45–48]. An investigation of spatial and temporal variations of CH_4 fluxes from the landfill facilities with the use of interpolation methods was conducted by Haro et al. [48], Gonzalez-Valencia et al. [49] and Zhang et al. [50]. In general, landfills are known to be a source of significant fugitive methane emissions, however, there is also a large spatial variability in CH_4 rates from similar sources in various locations across countries. Moreover, most of the investigations reported in the literature focused on: (1) the biogeochemical transformations between CH_4/CO_2 from municipal waste landfill (MWL) and dissolved inorganic carbon (DIC) in leachate, so the process is quite well known; (2) investigations of geochemical and isotopic maturation in CH_4/CO_2 in MWL, although mostly in reclaimed MWL and well-drilled/exploited biogas systems. Therefore, the objective of this study was to assess the spatial distribution of GHGs (CH_4 and CO_2 and their $\delta^{13}C$ isotopic composition) originating from the municipal waste landfill (MWL) site, and to elucidate the interactions and effects of biogenic CO_2 components (assimilation by neighboring vegetation) to examine the MWL effect on the local atmosphere around the dump area and short-term responses to environmental factors. The analyzed MWL is quite unique because the half area is a reclaimed quarter (with biogas drilling well system and co-generator use), whereas the second half is still an active location, to which a huge amount of municipal waste (segregated and not) is supplied, which can act as a completely uncontrolled bio-reactor and a source of a significant amount of GHGs. The research tasks included field measurements of concentrations and isotopic analyses of atmospheric CO_2 and CH_4 in combination with landfill-site investigations and a review of existing reported emission reports, in order to provide a comprehensive approach for distinguishing the signal and confirming the off-site migration of landfill gases in the surrounding atmosphere. The closest location to the MWL is a forest as well as an arable land area, both of which can work as a possible buffer from the MWL genesis of GHGs.

2. Materials and Methods

2.1. Site Description

The research site (51°38′03.3″ N 16°41′56.8″ E) is located in the Rudna Wielka village in the north-central part of the Lover Silesia Voivodeship (Figure 1), Góra County, Wąsosz commune. Rudna Wielka is located 80 km north-west from Wrocław (4th biggest polish city and main supplier of waste to landfill). The climate of Lower Silesian Voivodship is characterized by variability due to the topography. The northern area is on the border of the temperate zone of the oceanic and continental climate. The climatic conditions throughout the entire area of the Góra County are practically uniform and is characterized as a temperate climate—dry and warm. Temperature fluctuations here are smaller than the average fluctuations in Poland, the springs and summers are early and warm, and winters are mild. The average annual temperature is 8 °C, which is classified as high, the average temperature in July is 18 °C, and in January it is −2 °C. The average monthly atmospheric pressure ranges from 1014.4 hPa in April to 1018.2 hPa in January. The average annual sums of atmospheric precipitation over many years (1951–1980) range from 500 mm in the west to 600 mm in the east of the county. The winds prevail from the west and their average speed at a height of 10 m above the ground is 3.0–3.5 m·s^{-1} [51].

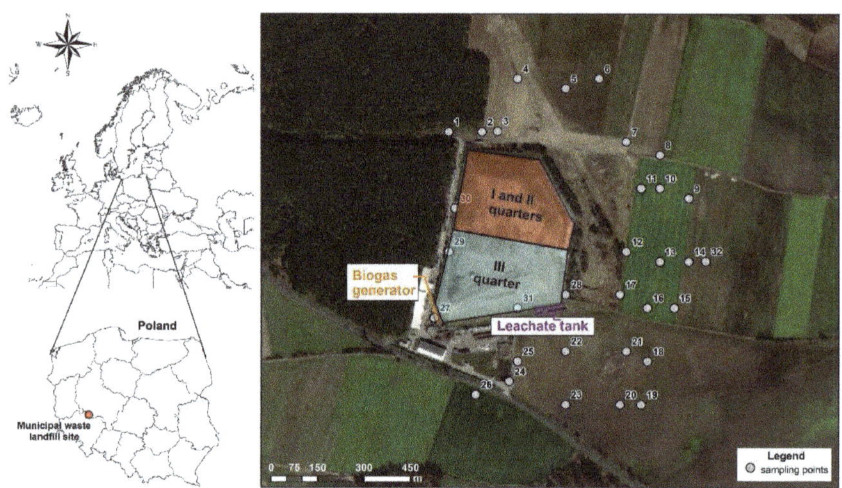

Figure 1. Location of the municipal waste landfill site (51°38′03.3″ N 16°41′56.8″ E) and 32 atmospheric air sampling points on 22 August 2017. Base map source [52].

The municipal waste landfill (MWL) in Rudna Wielka was opened in 2004. The area where the waste is stored is secured by an artificial geological sealing barrier (clays) and a synthetic geomembrane so that the leachate does not penetrate into the ground and, in the next stage, into groundwater. The leachate that arises in the landfill is discharged through a drainage system to the leachate tank. Biogas in the Rudna Wielka landfill (which is produced by the decomposition of biochemical processes) is absorbed by degassing wells and then converted into energy and heat. This process allows for the reduction of biogas emissions (including methane) to the atmosphere.

The landfill in Rudna Wielka meets all the conditions stipulated by Polish law and European Union directives. The owner of the MWL aims to minimize the negative impact on the environment. The company oversees waste collection, segregation, recovery and the management of waste that is no longer suitable for use. According to the obtained data, the most often segregated waste types (typical composition) in the Rudna Wielka municipal waste landfill site received from the Voivodeship city of Wroclaw in 2017 [53] are shown in Table 1. Based on the amount of waste collected in MWL, the main component was non-combustible mixed waste (73.95% mass), which contained unsorted (mixed) municipal waste and bulky waste. The waste composition also has a high organic/biodegradable content, which constitutes 20.53% of the mass of the waste material deposited on the landfill. The waste contained a 3.46% mass of packaging fractions, (mainly glass) and residual minerals (1.55% mass), which mainly includes other non-biodegradable waste, soil and stones, a small percentage (0.51% mass) of construction and demolition debris.

Table 1. Amount and composition of waste in the municipal waste landfill site collected in 2017, according to the survey in [53].

Waste Component	Classification Codes for Waste in EU	Assumed ($Mg \cdot Year^{-1}$)	% Mass in Wet Basis
Organic, biodegradable	20 02 01	23,797.33	20.53
Non-combustible mixed waste	20 03 01, 20 03 07	85,701.55	73.95
Minerals	20 02 02; 20 02 03	1792.27	1.55
Textile	20 01 11	2.63	0.00
Mixed construction and demolition debris	17 01 07; 17 09 04	587.70	0.51
Packaging fractions:			
- glass	15 01 07	3814.43	3.29
- multilayer packages	15 01 06	202.60	0.17
* Total		115,898.51	100

* Information on waste composition received from Wrocław urban area.

The landfill for non-hazardous and inert waste consists of three quarters with a total area of 8.07 ha. The capacity of quarters 1 and 2 is approximately 960,000 m^3, while the capacity of quarter 3 is 700,000 m^3. Currently, headquarters No. 3 is in operation, and plots 1 and 2 have been reclaimed (Figure 1). There is an installation for mechanical and biological waste treatment in the landfill. The mechanical part consists of a modular waste segregation station (140,000 Mg·year^{-1}) and an installation for the mechanical treatment of waste (151,200 Mg·year^{-1}) [54].

In the biological part, the installation allows for the replacement or parallel performance of biological drying processes (100,000 Mg·year^{-1} for the biodegradable fraction at least 0–80 mm or 21,765 Mg·year^{-1} for biological drying of mixed waste), biological waste processing under aerobic conditions (55,000 Mg·year^{-1}) and waste composting in the recovery process (13,000 Mg·year^{-1}). In addition, there was an installation for composing alternative fuels (110,000 Mg·year^{-1} and a maximum of 20 Mg·h^{-1}) in the landfill [54,55]. At the landfill (in the reclaimed quarters 1 and 2), there is an installation for the disposal/recovery of landfill gas (accounting for 46.85% of CH$_4$) by burning in a flare (about 447,620 m^3 in 2017) or for energy use in a cogeneration system [54]. The leachate water is collected with a drainage system located at the bottom of the quarters and collected in the leachate tank, and then transported to the sewage treatment plant (approximately 10,174 m^3 in 2017) [54] or recirculated to the top of the landfill.

2.2. Sampling Methodology

The ambient air samples (Figure 1) were collected in 32 locations across the study site on 22 August 2017, in 1-L PTFE bags (SKC, Bag, Tedlar®, 1 L, Single PP Fitting) by using a vacuum pump with a miniature in-line magnesium perchlorate trap. All sites were sampled windward between 12.00–3.00 PM local time c.a. 1.5–2 m above ground level. Filled Tedlar bags were subsequently analyzed within 24 h at the laboratory for $\delta^{13}C(CO_2)$ and $\delta^{13}C(CH_4)$, CO$_2$ and CH$_4$ mole fraction by CRDS laser spectroscopy (G2201-i, Picarro Inc., Santa Clara, CA, USA). The sample air during measurements had a relatively constant water vapour mole fraction, typically of around 0.5–0.8%. As the background of atmospheric methane and carbon dioxide mole fraction and as $\delta^{13}C(CH_4)$ and $\delta^{13}C(CO_2)$ values are not affected by local biogenic and anthropogenic processes, the data for Mace Head (MHD) Ireland NOAA station (August 2017) were used [56]. The main air masses arrive in Poland from the North Atlantic area, hence MHD station (53.3260° N, 9.899° W) was deemed to be the most appropriate as a clean background sample.

2.3. CO$_2$ and CH$_4$ Mole Fraction and Carbon Isotope Analysis

The CO$_2$ and CH$_4$ mole fractions and $^{13}C/^{12}C$ isotope ratios in ambient air (transferred into Tedlar bags) were measured by cavity ring-down spectroscopy (CRDS—Picarro G2201-i isotopic analyzer). Measurements were verified using fixed working reference gas in synthetic air, with a CO_2 concentration of 408 ± 2 ppm, a $\delta^{13}C(CO_2)$ value around $-30.2 \pm 0.3‰$ and 1.88 ± 0.04 ppm for CH_4 with a $\delta^{13}C(CH_4)$ value of $-42.6 \pm 0.4‰$. In order to confirm the quality of methane measurements and the long-term stability of the analyzer, three standard gases with high and low concentrations of CH_4 were applied. The cylinders of the certified (VPDB) CH_4 mole fraction and the isotopic composition ranged between (standard 1) 9.5 ppm, $\delta^{13}C(CH_4)$: $-69.8‰$; (standard 2) 1.88 ppm, $\delta^{13}C(CH_4)$: $-42.6‰$, (standard 3) 3.3 ppm, $\delta^{13}C(CH_4)$: $-55.1‰$.

2.4. Meteorological Data

Weather parameters (barometric pressure, relative humidity, temperature, wind speed and direction) were recorded during the field campaign using an on-site weather station (Kestrel 5500 Weather Meter, Kestrel Instruments, Boothwyn, PA, USA). The meteorological station was placed at the level of 1.5 m a.g.l. on a portable tripod at each ambient air sampling point.

2.5. Data Analysis

The statistical analysis (using Statistica 13.0 Software) was carried out on samples collected over a sampling campaign to test CO_2 and CH_4 mole fractions, and $\delta^{13}C$ analysed both GHG gases as well as measured meteorological parameters. Normality tests were performed using Shapiro–Wilk estimates. Due to a lack of normality, Spearman's rank correlation coefficient was used to test possible relationships between the analysed parameters.

The Keeling plot method [57] was used to determine the isotopic composition of the atmospheric carbon dioxide and methane mixing ratio, respectively. A two-end-component mixing model, for estimating the intercept and standard error of the intercept of the Keeling plot, is represented by the carbon-isotope ($\delta^{13}C$) signatures as a function of the inverse ($1/CO_2$) of atmospheric gas mixing ratios, derived from a geometric mean regression as follows (Equation (1)) [58]:

$$\delta_{mix} = \frac{C_{atm} \times (\delta_{atm} - \delta_{src})}{C_{mix}} + \delta_{src}, \tag{1}$$

where:

δ_{mix}—background C isotope mixing ratio;
δ_{src}—mean source C isotope mixing ratio;
C_{mix}—background atmospheric C concentration;
C_{atm}—C concentrations of the mean source.

The graphical relations between measured parameters were prepared using Grapher® (from Golden Software LLC, Golden, CO, USA, www.goldensoftware.com (accessed on 27 November 2021)), whereas a spatial map distribution of analysed parameters was produced using Surfer® (from Golden Software, LLC, Golden, CO, USA, www.goldensoftware.com (accessed on 27 November 2021)). The maps of CO_2 mole fraction, as well as $\delta^{13}C(CO_2)$ and $\delta^{13}C(CH_4)$, were extrapolated using the Kriging method, whereas the CH_4 mole fraction map, due to the large data differences, was extrapolated using an inverse-distance to a power method [59].

3. Results

3.1. Weather Conditions

During the investigation on 22 August 2017, the weather conditions were quite stable with occasionally small wind velocity and direction fluctuations (Table 2 and Figure 2).

Table 2. Meteorological parameters and geochemical data of ambient air samples gathered in the vicinity of the municipal waste landfill site during the campaign on 22 August 2017.

No Sample	Wind Direction [o]	Wind Velocity [m/s]	Temperature [°C]	RH [%]	Pressure [hPa]	CO_2 [ppm]	$1/CO_2$ [ppm^{-1}]	$\delta^{13}C(CO_2)$ [‰]	CH_4 [ppm]	$1/CH_4$ [ppm^{-1}]	$\delta^{13}C(CH_4)$ [‰]
1 *	249	1.6	18.6	52.8	1005.2	394.3	2.536	−7.8	2.21	0.452	−50.9
2 *	253	3.8	18.3	55.1	1005.4	388.4	2.575	−7.4	2.02	0.495	−51.6
3 *	264	3.3	17.9	55.4	1005.7	386.6	2.587	−7.3	2.01	0.497	−52.2
4 *	220	0.0	20.2	54.2	1005.5	390.5	2.561	−7.5	1.92	0.521	−51.4
5 *	272	3.7	17.5	51.3	1005.9	385.9	2.591	−7.3	1.90	0.528	−51.5
6 *	270	5.5	18.1	53.6	1006.2	386.4	2.588	−7.3	1.90	0.528	−52.1
7 *	283	5.2	17.7	53.1	1006.4	388.8	2.572	−7.5	1.90	0.527	−52.0
8 *	330	3.0	17.7	55.6	1006.2	385.1	2.596	−7.2	1.90	0.525	−51.8
9 *	255	3.2	18.8	54.5	1006.4	385.5	2.594	−7.4	2.16	0.463	−53.0
10 *	252	3.9	18.2	54.6	1006.5	387.2	2.583	−7.3	2.07	0.484	−52.9
11 *	345	4.4	18.4	54.0	1006.2	386.7	2.586	−7.4	1.92	0.520	−53.1
12 *	297	0.8	19.5	49.6	1006.2	385.9	2.591	−7.3	2.43	0.411	−54.1
13 *	300	6.9	17.1	50.5	1006.5	390.8	2.559	−7.4	2.12	0.471	−53.2
14 *	267	5.1	17.6	53.5	1006.5	386.2	2.590	−7.2	2.17	0.461	−53.8
15 *	248	9	18.2	52.8	1006.2	393.6	2.540	−7.3	129.48	0.008	−57.9
16 *	264	2.4	19.6	54.1	1006.2	392.4	2.548	−7.5	18.85	0.053	−56.9
17 *	285	3.7	20.2	52.6	1006.0	390.1	2.564	−7.6	4.49	0.223	−56.3
18 *	305	2.2	19.1	52.8	1006.0	391.0	2.558	−7.4	4.71	0.212	−57.7
19 *	240	6.9	18.9	48.8	1006.2	386.7	2.586	−7.1	2.37	0.421	−53.8
20 *	274	4.1	19.3	49.6	1005.9	392.8	2.546	−7.5	1.92	0.522	−52.2
21 *	276	1.7	19.2	49.2	1005.9	390.8	2.559	−7.4	2.41	0.415	−54.4
22 *	286	2.6	19.8	50.3	1005.5	390.5	2.561	−7.3	3.03	0.331	−55.9
23 *	273	3.3	18.9	50.6	1005.5	384.9	2.598	−7.4	1.89	0.528	−52.1
24 *	273	4.0	18.5	47.9	1005.0	384.9	2.598	−7.2	2.01	0.499	−52.5
25 *	300	3.3	19.9	48.8	1005.2	389.9	2.565	−7.4	3.04	0.329	−55.7
26 *	259	5.0	18.4	47.4	1005.5	388.2	2.576	−7.3	1.90	0.525	−51.6
27 *	203	0.5	18.4	55.6	1005.2	396.4	2.523	−7.6	2.50	0.400	−54.0
28 *	248	3.5	18.7	56.2	1005.4	406.4	2.461	−8.4	11.48	0.087	−56.2
29 *	27	0.6	19.3	56.5	1005.0	417.1	2.398	−12.4	23.72	0.042	−56.2
30 *	111	1.9	17.8	52.2	1004.7	395.7	2.527	−7.5	5.63	0.178	−59.1
31 *	261	4.6	18.9	55.9	1005.0	419.0	2.387	−7.8	21.12	0.047	−57.0
32 *	277	4.7	17.7	53.7	1006.5	385.0	2.598	−7.1	2.17	0.462	−52.9
Min.	27	0.0	17.1	47.4	1004.7	384.9	2.387	−12.4	1.89	0.008	−59.1
Max.	345	6.9	20.2	56.5	1006.5	419.0	2.598	−7.1	129.48	0.528	−50.9
AVG	258	3.4	18.6	52.6	1005.8	391.4	2.556	−7.6	8.48	0.380	−53.9
Median	269	3.4	18.6	53.0	1005.9	389.3	2.569	−7.4	2.17	0.461	−53.2
SD	57	1.7	0.8	2.6	0.5	8.2	0.051	0.9	22.46	0.171	2.2

* data from [60].

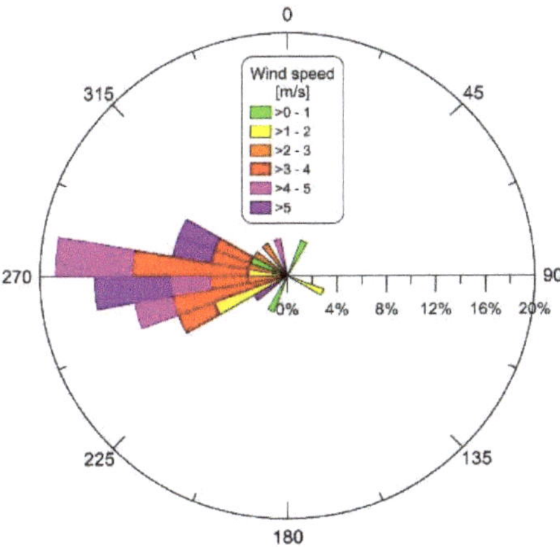

Figure 2. Wind rose measured on 22 August 2017 in the vicinity of the municipal waste landfill.

Weather parameters were as follows: the air temperature varied between 17.1 and 20.2 °C (average of 18.6 ± 0.8 °C), wind speed varied between 0.0 and 6.9 m·s^{-1} (average of 3.4 ± 1.7 m·s^{-1}), wind direction fluctuated between 27 and 345° (average of $258 \pm 57°$ which indicated prevailing westerly winds Figure 2), relative humidity varied between 47.4 and 56.5% (average of 52.6 ± 2.6%) and the atmospheric pressure increased from 1004.6 to 1006.5 hPa (average of 1005.8 ± 0.5 hPa). The notably weak statistical relations (Table 3) between the different meteorological parameters due to very low parameter fluctuations are accidental rather than representative of environmental dependence.

Table 3. Spearman's rank correlation coefficient between analyzed parameters ($n = 32$). Statistically significant coefficients for $p < 0.05$ are bolded.

	Wind Direction [°]	Wind Velocity [m/s]	Temperature [°C]	RH [%]	Pressure [hPa]	CO_2 [ppm]	$1/CO_2$ [ppm^{-1}]	$\delta^{13}C(CO_2)$ [‰]	CH_4 [ppm]	$1/CH_4$ [ppm^{-1}]
Wind velocity [m/s]	0.27									
Temperature [°C]	0.00	**−0.45**								
RH [%]	**−0.35**	−0.19	−0.17							
Pressure [hPa]	**0.38**	**0.42**	**−0.36**	−0.01						
CO_2 [ppm]	**−0.40**	**−0.35**	0.30	0.24	**−0.45**					
$1/CO_2$ [ppm^{-1}]	**0.40**	**0.35**	−0.30	−0.24	**0.45**	**−1.00**				
$\delta^{13}C(CO_2)$ [‰]	0.29	**0.39**	**−0.37**	−0.31	**0.48**	**−0.80**	**0.80**			
CH_4 [ppm]	−0.27	**−0.42**	**0.39**	0.17	−0.25	**0.67**	**−0.67**	**−0.41**		
$1/CH_4$ [ppm^{-1}]	0.27	**0.42**	**−0.39**	−0.17	0.25	**−0.67**	**0.67**	**0.41**	**−1.00**	
$\delta^{13}C(CH_4)$ [‰]	0.03	0.24	−0.33	−0.03	0.10	**−0.52**	**0.52**	0.28	**−0.88**	**0.88**

3.2. Mole Fraction and Stable Carbon Isotopic Composition of CH_4 in the Vicinity of the Municipal Waste Landfill

The CH_4 mole fraction measured during the 22 August 2017 sampling investigation varied from 1.89 to 129.48 ppm with an average value of 8.48 ± 22.46 ppm and a median of 2.17 ppm (Table 2). At some sampling points (e.g., No. 15), the mole fraction of ambient atmospheric CH_4 in the vicinity of the municipal waste landfill showed significant enrichment (up to 65 times) compared to the global methane Mace Head Ireland NOAA (1.92 ppm, August 2017) background [56]. The significant differences between the average and median values indicate that some CH_4 hot spots were observed rather than an enrichment of the entire analyzed area (Figure 3A).

The $\delta^{13}C(CH_4)$ in the ambient atmosphere follows the variability with ^{13}C depletion in samples in the direction of the wind from landfill relation to samples with other wind directions. The $\delta^{13}C(CH_4)$ values vary between −59.1 and −50.9‰ with an average value of -53.9 ± 2.2‰ and median of −53.2‰; ^{13}C was clearly depleted when compared to the Mace Head Ireland NOAA (−47.6‰, August 2017) background [56]. The carbon isotope composition clearly indicates hot spots (Figure 3B), however, similar values of the average and median suggest more mixed local air than the CH_4 mole fraction data. The Spearman's rank correlation coefficient between CH_4 mole fraction data and $\delta^{13}C(CH_4)$ values reached −0.88 (Table 3) and indicate a significant negative relationship between both parameters.

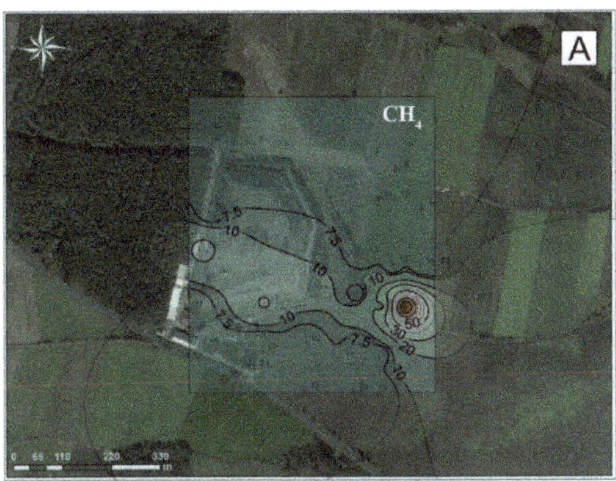

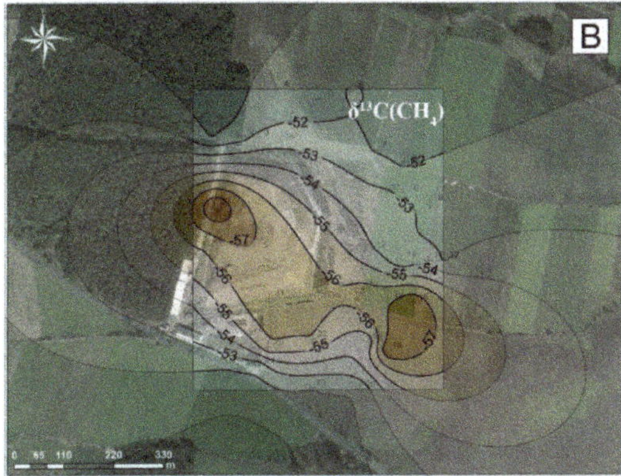

Figure 3. Spatial distribution of CH_4 mole fraction (**A**) and the $\delta^{13}C(CH_4)$ values (**B**) in the vicinity of the municipal waste landfill sampled on 22 August 2017 campaign. Brightness rectangle covered sampling points, whereas the area outside has been extrapolated.

3.3. Mole Fraction and Stable Carbon Isotopic Composition of CO_2 in the Vicinity of the Municipal Waste Landfill

The mole fraction of ambient atmospheric CO_2 measured in the vicinity of the municipal waste landfill showed similar values to the global carbon dioxide Mace Head Ireland NOAA (402.5 ppm, August 2017) background [56], slightly changed due to local assimilation during the vegetation season (Table 2). The CO_2 mole fraction during the 22 August 2017 sampling campaign varied from 384.9 to 419.0 ppm, with an average value of 391.4 ± 8.2 ppm and median value of 389.3 ppm. The lack of significant differences between the average and median as well as the similarity to the background NOAA CO_2 mole fraction indicate that rather moderate CO_2 hot spots existed in the analyzed area (Figure 4A).

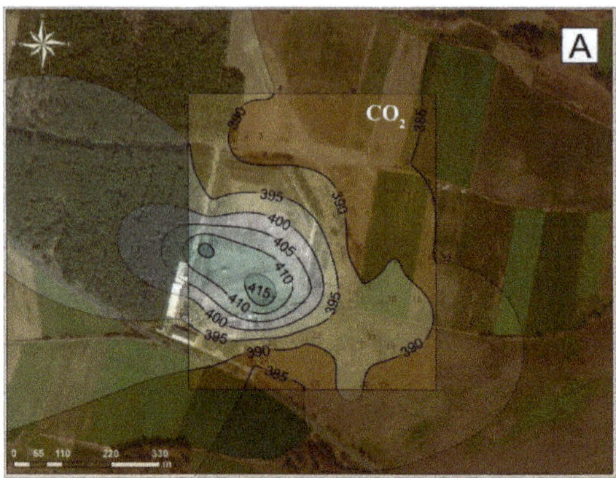

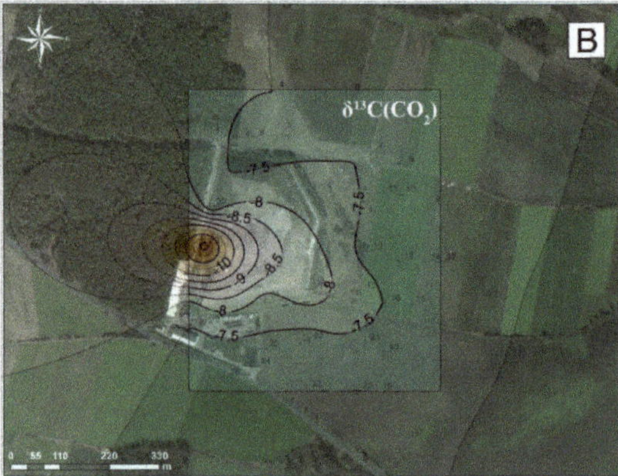

Figure 4. Spatial distribution of CO_2 mole fraction (**A**) and the $\delta^{13}C(CO_2)$ values (**B**) in the vicinity of the municipal waste landfill sampled on 22 August 2017 campaign. Brightness rectangle covered sampling points, whereas the area outside has been extrapolated.

The $\delta^{13}C(CO_2)$ in the ambient atmosphere follows the variability with ^{13}C depletion in samples in the direction of the wind in relation to the landfill, observed for samples with other wind directions. However, the range of ^{13}C depletion compared to the background values indicates a negligible but notable influence on local air. The $\delta^{13}C(CO_2)$ values vary between -12.4 and $-7.1‰$ with an average value of $-7.6 \pm 0.9‰$ and median of $-7.4‰$. The carbon isotope composition clearly indicates hot spots (Figure 4B), however, similar values of average and median suggest a mixed local-air similarly to the CO_2 mole fraction data. Moreover, most samples indicate ^{13}C enrichment compared to with the NOAA background ($-8.2‰$, August 2017), especially on arable areas [56]. The Spearman's rank correlation coefficient between CO_2 mole fraction data and $\delta^{13}C(CO_2)$ values reaches -0.80 (Table 3) and indicated significant negative relationships between both parameters.

4. Discussion

4.1. General Divagation about GHGs on Investigated Landfill

The present work reveals the adverse impact of an active municipal waste landfill on GHGs levels in the neighboring atmosphere. Some studies also provided evidence that biogeochemical processes concerning bio-waste degradation in landfills ultimately lead to increased CO_2 and CH_4 emissions [4,50]. The rates of landfill-gas generation depend on the composition (organic content), age (or time since emplacement), moisture content, particle size and compaction and methods of landfilling, climate variables, etc. [7]. In general, standard gas production rates vary from 0.0007 to 0.0080 m^3 per kg·year^{-1} [61].

The average mole fraction of CH_4 and CO_2 reported by the automatic measurement system in the biogas installation of our investigated landfill in 2017 was 46.85% and 37.73%, respectively and the amount of biogas (methane) collected and neutralized during burning in the flare and cogeneration processes (for electricity and heat production in generator) was 447,620 m^3·year^{-1} [54]. Our investigations focused on two main gases (CH_4 and CO_2), due to the fact that official emissions of this municipal waste landfill reported in the KOBIZE (Poland's National database on greenhouse gases and other substances emissions) database from the 2017 year are: 1073.920 Mg·year^{-1} CH4 from landfill quarters and 9.963 Mg·year^{-1} CO_2 biogas used in electricity and heat generator in landfills [62]. The present results, compared to other reported data, are quite unique, due to analyzing an MWL for which a section has been reclaimed with a well-working biogas collection system, whereas the other section is still active with a completely uncontrolled emission of GHGs to the surrounding atmosphere.

The amount of biogas generated by fugitive emissions noted in [54] was 191,837 m^3·year^{-1}. In spite of effective and environmentally friendly procedures and installations such as reclaimed quarters I and II, with a biogas collection/burning system and alternative fuel production line, etc., the negative impact of GHGs on the local atmosphere should exist. The Keeling plots revealed the contribution from landfill methane and leachate treatment methane, with an average end-member $\delta^{13}C_{source}$ of $-58.36 \pm 0.44‰$ (Figures 5 and 6). The intercept of CH_4 contributed to by landfilling was also confirmed by a highly significant linear regression model ($R^2 = 0.80$, $p < 0.05$). In contrast, the obtained $\delta^{13}C_{source}$ of carbon dioxide end-member in the mixing model (Figures 5 and 6) has an average value of $-18.64 \pm 1.75‰$, which is in the range of photosynthetic CO_2 assimilation in C3 plants. The significance of biogenic CO_2 sources during the vegetative season was previously reported for the Wroclaw urban area [63]. In general, the ranges of $\delta^{13}C$ values of CO_2 and CH_4 observed in our study are comparable to those from other European landfills, with a mean $\delta^{13}C(CO_2) = -16.8 \pm 0.4‰$, $\delta^{13}C(CH_4) = -52.9 \pm 5.4‰$. The obtained values of $\delta^{13}C$ are consistent with other isotopic studies and indicate the main pathway of CH_4 production in the MWL, resulting from acetate fermentation [37,38,44].

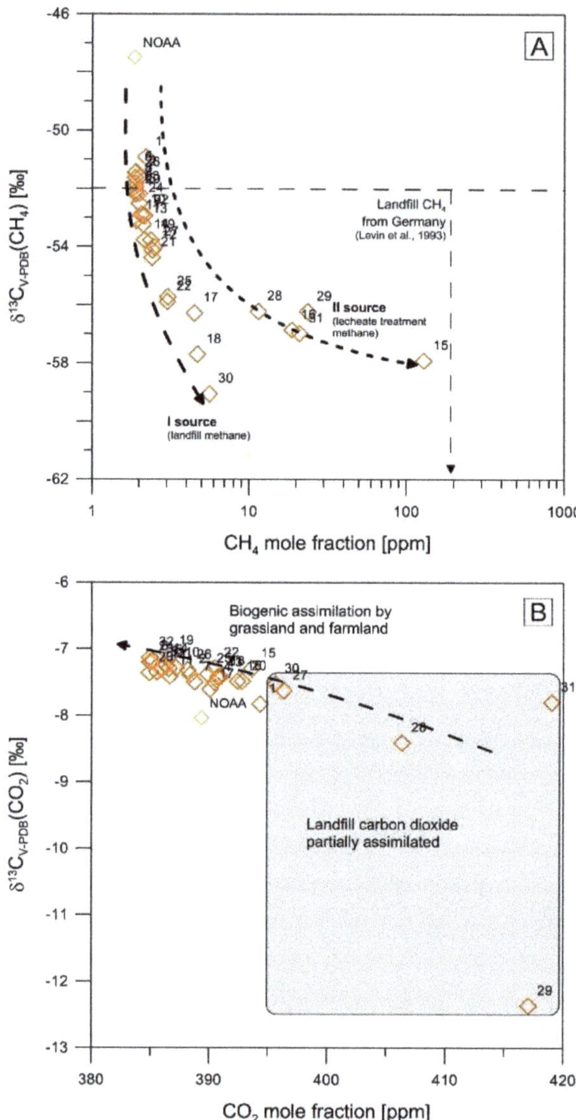

Figure 5. Relations between mole fraction and carbon isotopes composition of methane (**A**) and carbon dioxide (**B**) from ambient air gathered in the vicinity of municipal waste landfill sampled during the 22 August 2017 campaign. Dashed lines show the linear regression fit for the data sampled. The green dots represent average background values according Mace Head Research Station, Ireland, NOAA. Black solid/dashed lines represent expected source contribution range.

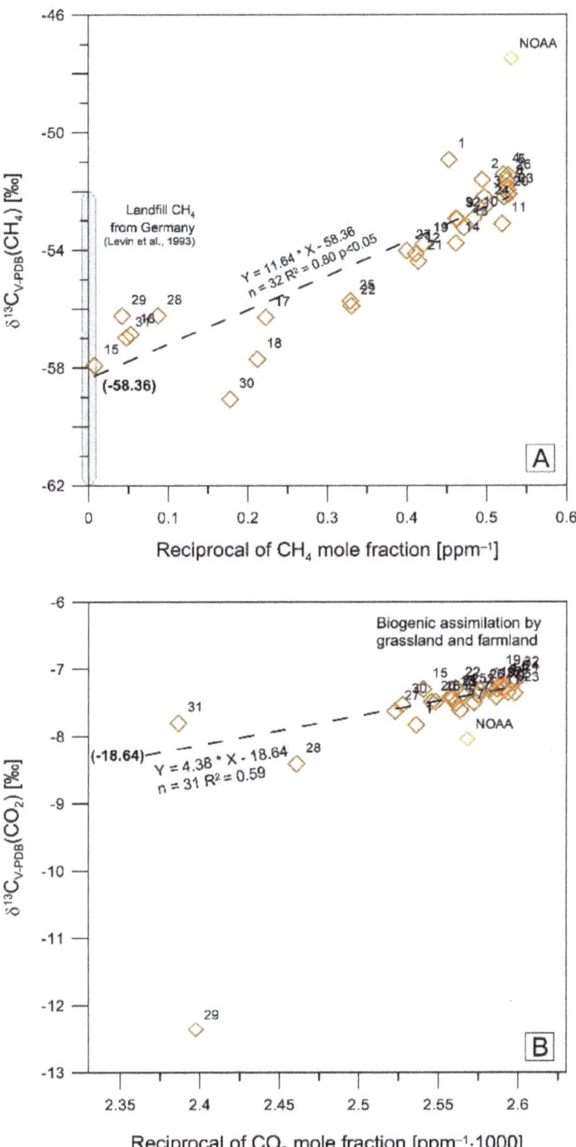

Figure 6. Relations between reciprocal of mole fraction and carbon isotopes composition of methane (**A**) and carbon dioxide (**B**) from ambient air gathered in the vicinity of municipal waste landfill sampled on 22 August 2017 campaign. Dashed lines show the linear regression fit of the data sampled. The green dots represent average background values according to Mace Head Research Station, Ireland, NOAA.

The largest sources of methane emissions from the analyzed landfill site were: (I) active landfill quarter III; (II) landfill leachate tank; (III) biogas collecting well. According to carbon dioxide, the most probable hotspots (Figure 1) were (I) the active landfill quarter III; (II) the biogas heat and power generator. The highest mole fractions of CH_4 and CO_2 were expected in the nearest eastward points from the landfill (Figure 1) due to the dominant wind existing westward (Figure 2) during the sampling campaign. Finally, the field sampling campaign revealed that substantial amounts of methane escaped from the leachate tank and quarter No. 3 (in operation) in the MWL. The plume of the gases released from the mentioned sources towards the east direction had a potential negative impact on the air quality of the immediate surroundings. The enhanced methane–mole fraction and strongly diluted $\delta^{13}C(CH_4)$ in ambient air were measured at a few points on the east side of the landfill (e.g., point No. 15—129.48 ppm; point No. 16—18.85 ppm; point No. 28—11.48 ppm) (Table 2, Figure 3).

4.2. CH_4 Balance in the Vicinity of Municipal Waste Landfill

During the sampling campaign, we observed notable CH_4 mole fractions not only in the nearest landfill points (Figures 1 and 3A) but also at a distance of 300–500 m eastward/southeastward. The spatial distribution (Figure 3A) found a high methane–mole fraction in the air, even 25 times above the background (e.g., No. 15 allocated on the leeward side), compared to other analyzed points in the vicinity of landfill. Moreover, samples with higher CH_4 mole fraction reported notable depletion of ^{13}C, confirmed by the negative -0.88 coefficient of Spearman's rank (Table 3). Graphical representation (Figure 5A) allowed us to hypothesize that: (i) the carbon isotope composition of many samples indicate biogenic landfill methane according to Levin et al. [30] data; (ii) two main CH_4 hotspot sources probably exist, connected with landfill methane derived from active quarter III as well as generated in the open leachate tank (Figures 1 and 3A,B). Landfill methane [30,45,64] has a similar range of $\delta^{13}C(CH_4)$ values (between -60 and $-50‰$) for methane generated from leachates [65] and from sewage [24].

The Keeling plot (Figure 6A) clearly indicates a biogenic origin (landfill + leachates) of methane with an intercept $\delta^{13}C(CH_4)$ value $-58.36 \pm 0.44‰$ as a dominant CH_4 source in the analyzed area. Biogenic CO_2 derived from decomposition processes in landfills/or soil/root respiration, as well as from anthropogenic CO_2 from biogas/fossil fuels burning was only connected with sample No. 29. In spite of the lack of $\delta D(CH_4)$ values, according to the C/H isotopes plot in Coleman et al. [66], the range of our $\delta^{13}C(CH_4)$ values indicate that the acetate-fermentation pathway dominates the CO_2-reduction pathway. The $\delta^{13}C(CO_2)$ values did not show a genetic relation with the CO_2-reduction pathway, both at its preliminary stage (when ^{13}C depleted CO_2 exists), as well as on its final stage, where it appears to be mostly reduced and enriched in $^{13}C(CO_2)$.

We did not discount the fact that intensive CO_2/CH_4 exchange can exist in covered and reclaimed landfill part in quarters I and II, however, almost the entire biogas from these quarters was found to be captured in wells and burned in a generator. Hence, methane observed in the vicinity is probably generated mostly in situ from "fresh" municipal waste and from landfill leachates. The lack or the very negligible increase of the CH_4 mole fraction, as well as the fact that it was generally not isotopically different from the background in north and northeast sampling points (near reclaimed fragment of landfill) confirmed our hypothesis. The rose-type diagram for the CH_4 mole fraction (Figure 7A) and $\delta^{13}C(CH_4)$ values (Figure 7B) revealed the dominant wind direction and velocity to be factors that controlled methane spread in the local atmosphere, in spite of pure or a lack of official statistical relations (Table 3).

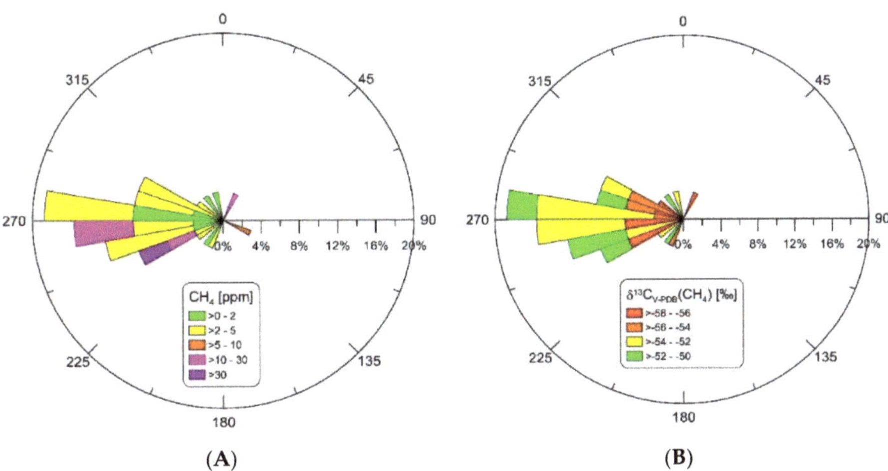

Figure 7. The circular distribution of directional data of CH_4 mole fraction (**A**) and $\delta^{13}C(CH_4)$ values (**B**) from ambient air gathered in the vicinity of municipal waste landfill sampled during 22 August 2017 campaign.

4.3. CO_2 Balance in the Vicinity of Municipal Waste Landfill

Surprisingly, for a given study period, the above-atmospheric levels of CO_2 (>390 ppm) were possible to observe only in the nearest landfill points (No. 27–31 Figures 1 and 4A). The other sampling points reached similar or slightly lower (up to 5 ppm) CO_2 mole fractions compared to the background level [56]. Similarly to the CO_2 mole fraction, the $\delta^{13}C(CO_2)$ values of most air samples collected further from landfills showed a slight ^{13}C enrichment compared to the background, and only the nearest (No. 28 and 29) points indicate a notable depletion in ^{13}C (up to 4.4‰). The relationship between the CO_2 mole fraction and $\delta^{13}C(CO_2)$ values (Figure 5B) clearly indicated a biogenic input of the surrounding landfill agriculture fields and grassland. An assimilation process of CO_2 by vegetation or by soil microorganisms [67–69] caused a lower CO_2 mole fraction of residue carbon dioxide in ambient air, as well as the enrichment of ^{13}C due to preferential ^{12}C consumption.

The slightly marked CO_2 hotspot near generator burned biogas (methane) (Figure 1) and leachate tank influence the local landfill atmosphere (Table 2 and Figure 4A,B). As a result, the very intensive aerobic bacterial reactions on active quarter III are not spatially observed in $CO_2/\delta^{13}C(CO_2)$ values plum. A total of 447,600 m^3 of biogas was burned in the flare and generator (0.294 MW of electric power) in 2017. However, a CO_2 signal was not detected, probably due to the hot exhaust gases being conventionally raised and mixed in the upper layers of the atmosphere, although not affecting the local atmosphere.

Many authors noted extremely enriched in $\delta^{13}C(CO_2)$ values, reaching up to +20‰ for landfill CO_2 and CO_2/DIC in landfill leachates [8,45,64]. Therefore, in the case of the existing bacterial CO_2 reduction process, extremely ^{13}C enriched residual ambient CO_2 in landfill and neighboring sampling points have been expected. Instead of this, the slightly depleted in $\delta^{13}C(CO_2)$, observed in the vicinity of the landfill, was probably connected with exhaust gasses from the exploitation of heavy machinery (garbage trucks, landfill compactors) and from the biogas generator (marked on Figure 5B as landfill CO_2). Conversely, the oxidation of biogas methane in landfills results in a significant depletion in $^{13}C(CO_2)$ in the landfill atmosphere [66]. However, in this situation, both the CO_2 mole fraction, as well as the carbon isotopic signal of carbon dioxide have not been observed in the analysed landfill.

We hypothesize that intensive CO_2 reduction on this landfill does not occur, especially in active open quarter III, and only local grassland/agriculture fields were able to buffer and assimilate CO_2 that potentially derived from the landfill. This thesis is confirmed by the positive ($\rho = 0.80$) coefficients of Spearman's rank between CO_2 and $\delta^{13}C(CO_2)$ and negative ($\rho = -0.41$) between the CH_4 mole fraction and $\delta^{13}C(CO_2)$ (Table 3). The Keeling plot (Figure 6B) for $n = 31$ points' mixing ratio source ($\delta^{13}C_{source} = -18.64 \pm 1.75‰$) indicates assimilation by vegetation surrounding the landfill, rather than (i) biogenic CO_2 derived from decomposition processes on landfills/or soil/root respiration, as well as (ii) anthropogenic CO_2 from biogas/fossil fuels burning (only connected with sample No. 29).

5. Conclusions

This study presented the spatial characteristic of atmospheric CO_2 and CH_4 levels in short-term responses to environmental conditions observed in the vicinity of the municipal waste landfill for our August 2017 sampling campaign. Our investigation confirmed the negative influence of the municipal waste landfill site, especially at the hot-spot zones, on the local GHG balance in the surrounding atmosphere, but less than assumed. We observe a negligible CO_2 mole fraction difference in relation to that of the background, as well as its carbon isotopes signal in samples at a distance of 100 m from the landfill. A very significant CO_2 assimilation process in the surrounded agriculture/grassland area completely compensated for the possible negative CO_2 input. Most of the ambient CO_2 samples collected around the landfill had prevailing background CO_2 levels (varied from 384.9 to 419.0 ppm), with an average value of 391.4 ± 8.2 ppm, which were calculated by the binary mixing model isotopic signature and had a mean source value of $-18.64 \pm 1.75‰$.

The obtained results suggest that the isotopic content of methane in the immediate surroundings (average value of $\delta^{13}C = -53.9 \pm 2.2‰$) is strongly influenced by microbial origins (acetate -fermentation pathway) in the studied landfill site. Methane contribution was likely connected with two of the landfill CH_4 hotspots, including (1) active quarter III and (2) the leachate system/tank. The highest excess of CH_4 and notable methane depletion in ^{13}C was observed c.a. 300 m eastward/northeastward from the landfill. The CH_4 mole fraction in the emitted gases close to the hotspots reached up to 129.48 ppm, with an average $\delta^{13}C$ value of $-57.3 \pm 1.0‰$. The reclaimed quarters I and II showed negligible methane emission to the local atmosphere, due to the existing well-working biogas collection and processing system (with heat and power co-generator).

Biogas emissions showed a good correlation with air temperature and wind velocity, however, no significant relationships between relative humidity, atmospheric pressure and methane levels were observed. The effect of moderate to strong westward wind (average of 3.4 ± 1.7 m·s^{-1}) on the dispersion of landfill origin GHG gases (CO_2 and CH_4) in the local atmosphere dominated up to 500 m from the landfill. The dispersion of GHG emissions from landfill areas should be further investigated using in situ surveys, as their contribution to the local carbon budget vary significantly in different spatial and temporal settings. Finally, to maximize the potential of CH_4 mitigation, emissions from landfill sites should be reduced by utilizing waste composition as a source. Therefore, further research should focus on assessing the level of biogas production from active quarter III and the leachate tank, using chamber methods and experimental well drilling, followed by dispersion modeling, which should help to minimize the negative impact of GHGs derived from MWLs on the environment.

Author Contributions: Supervision, writing original draft, investigation, methodology, visualization: M.G.; writing original draft, methodology, investigation: Y.B.; writing, review and editing, funding acquisition: I.S. All authors have read and agreed to the published version of the manuscript.

Funding: This research received no external funding.

Institutional Review Board Statement: Not applicable.

Informed Consent Statement: Not applicable.

Data Availability Statement: Not applicable.

Acknowledgments: The authors thank Nikola Nowak for assistance with field sample collection. We would like to thank Dariusz Strąpoć for his helpful advice on principal methane production processes during the preparation of this manuscript. We are also grateful to the National Center for Emissions Management, Poland (KOBiZE) for access to emission data for particular source categories in the selected years.

Conflicts of Interest: The authors declare no conflict of interest. The funders had no role in the design of the study; in the collection, analyses, or interpretation of data; in the writing of the manuscript, or in the decision to publish the results.

References

1. Friedlingstein, P.; O'Sullivan, M.; Jones, M.W.; Andrew, R.M.; Hauck, J.; Olsen, A.; Peters, G.P.; Peters, W.; Pongratz, J.; Sitch, S.; et al. Global Carbon Budget 2020. *Earth Syst. Sci. Data* **2020**, *12*, 3269–3340. [CrossRef]
2. Smith, K.R.; Desai, M.A.; Rogers, J.V.; Houghton, R.A. Joint CO_2 and CH_4 accountability for global warming. *Proc. Natl. Acad. Sci. USA* **2013**, *110*, E2865–E2874. [CrossRef] [PubMed]
3. Saunois, M.; Stavert, A.R.; Poulter, B.; Bousquet, P.; Canadell, J.G.; Jackson, R.B.; Raymond, P.A.; Dlugokencky, E.J.; Houweling, S.; Patra, P.K.; et al. The Global Methane Budget 2000–2017. *Earth Syst. Sci. Data* **2020**, *12*, 1561–1623. [CrossRef]
4. Pehme, K.-M.; Orupõld, K.; Kuusemets, V.; Tamm, O.; Jani, Y.; Tamm, T.; Kriipsalu, M. Field Study on the Efficiency of a Methane Degradation Layer Composed of Fine Fraction Soil from Landfill Mining. *Sustainability* **2020**, *12*, 6209. [CrossRef]
5. Speight, J. *Natural Gas*; Elsevier: Amsterdam, The Netherlands, 2019; ISBN 9780128095706.
6. Yang, L.; Chen, Z.; Zhang, X.; Liu, Y.; Xie, Y. Comparison study of landfill gas emissions from subtropical landfill with various phases: A case study in Wuhan, China. *J. Air Waste Manag. Assoc.* **2015**, *65*, 980–986. [CrossRef]
7. Robertson, T.; Dunbar, J. Guidance for evaluating landfill gas emissions from closed or abandoned facilities. In *EPA Report*; United States Environmental Protection Agency: Washington, DC, USA, 2005.
8. de Medeiros Engelmann, P.; dos Santos, V.H.J.M.; Barbieri, C.B.; Augustin, A.H.; Ketzer, J.M.M.; Rodrigues, L.F. Environmental monitoring of a landfill area through the application of carbon stable isotopes, chemical parameters and multivariate analysis. *Waste Manag.* **2018**, *76*, 591–605. [CrossRef]
9. Kjeldsen, P.; Barlaz, M.A.; Rooker, A.P.; Baun, A.; Ledin, A.; Christensen, T.H. Present and Long-Term Composition of MSW Landfill Leachate: A Review. *Crit. Rev. Environ. Sci. Technol.* **2002**, *32*, 297–336. [CrossRef]
10. Białowiec, A. Some Aspects of Environmental Impact of Waste Dumps. In *Contemporary Problems of Management and Environmental Protection*; University of Warmia and Mazury in Olsztyn: Olsztyn, Poland, 2011; ISBN 978-83-929462-7-4.
11. Khalil, D.M.J.; Gupta, R.; Sharma, K. Microbiological Degradation of Municipal Solid Waste in Landfills for LFG Generation. In *National Conference on Synergetic Trends in engineering and Technology (STET-2014) International Journal of Engineering and Technical Research*; ISSN 2321-0869. Special Issue; Available online: https://www.erpublication.org/published_paper/IJETR_APRIL_2014_STET_03.pdf (accessed on 2 November 2021).
12. Njoku, P.O.; Odiyo, J.O.; Durowoju, O.S.; Edokpayi, J.N. A Review of Landfill Gas Generation and Utilisation in Africa. *Open Environ. Sci.* **2018**, *10*, 1–15. [CrossRef]
13. Vaverková, M.D. Landfill Impacts on the Environment—Review. *Geosciences* **2019**, *9*, 431. [CrossRef]
14. Brennan, R.B.; Healy, M.G.; Morrison, L.; Hynes, S.; Norton, D.; Clifford, E. Management of landfill leachate: The legacy of European Union Directives. *Waste Manag.* **2016**, *55*, 355–363. [CrossRef] [PubMed]
15. Ozbay, G.; Jones, M.; Gadde, M.; Isah, S.; Attarwala, T. Design and Operation of Effective Landfills with Minimal Effects on the Environment and Human Health. *J. Environ. Public Health* **2021**, *2021*, 1–13. [CrossRef] [PubMed]
16. Toro, R.; Morales, L. Landfill fire and airborne aerosols in a large city: Lessons learned and future needs. *Air Qual. Atmos. Health* **2018**, *11*, 111–121. [CrossRef]
17. Rao, M.N.; Sultana, R.; Kota, S. *Solid and Hazardous Waste Management*; Elsevier: Amsterdam, The Netherlands, 2016; ISBN 9780128098769.
18. U.S. Environmental Protection Agency. *International Best Practices Guide for Landfill Gas Energy Projects*; United States Environmental Protection Agency: Washington, DC, USA, 2012.
19. Xu, L.; Lin, X.; Amen, J.; Welding, K.; McDermitt, D. Impact of changes in barometric pressure on landfill methane emission. *Glob. Biogeochem. Cycles* **2014**, *28*, 679–695. [CrossRef]
20. Goldsmith, C.D.; Chanton, J.; Abichou, T.; Swan, N.; Green, R.; Hater, G. Methane emissions from 20 landfills across the United States using vertical radial plume mapping. *J. Air Waste Manag. Assoc.* **2012**, *62*, 183–197. [CrossRef]
21. Menoud, M.; van der Veen, C.; Necki, J.; Bartyzel, J.; Szénási, B.; Stanisavljević, M.; Pison, I.; Bousquet, P.; Röckmann, T. Methane (CH_4) sources in Krakow, Poland: Insights from isotope analysis. *Atmos. Chem. Phys.* **2021**, *21*, 13167–13185. [CrossRef]
22. Zimnoch, M.; Necki, J.; Chmura, L.; Jasek, A.; Jelen, D.; Galkowski, M.; Kuc, T.; Gorczyca, Z.; Bartyzel, J.; Rozanski, K. Quantification of carbon dioxide and methane emissions in urban areas: Source apportionment based on atmospheric observations. *Mitig. Adapt. Strateg. Glob. Chang.* **2019**, *24*, 1051–1071. [CrossRef]
23. Pang, J.; Wen, X.; Sun, X. Mixing ratio and carbon isotopic composition investigation of atmospheric CO_2 in Beijing, China. *Sci. Total Environ.* **2016**, *539*, 322–330. [CrossRef] [PubMed]

24. Górka, M.; Lewicka-Szczebak, D.; Fuß, R.; Jakubiak, M.; Jędrysek, M.O. Dynamics and origin of atmospheric CH4 in a Polish metropolitan area characterized by wetlands. *Appl. Geochem.* **2014**, *45*, 72–81. [CrossRef]

25. Stolper, D.A.; Martini, A.M.; Clog, M.; Douglas, P.M.; Shusta, S.S.; Valentine, D.L.; Sessions, A.L.; Eiler, J.M. Distinguishing and understanding thermogenic and biogenic sources of methane using multiply substituted isotopologues. *Geochim. Et Cosmochim. Acta* **2015**, *161*, 219–247. [CrossRef]

26. Douglas, P.M.J.; Stolper, D.A.; Eiler, J.M.; Sessions, A.L.; Lawson, M.; Shuai, Y.; Bishop, A.; Podlaha, O.G.; Ferreira, A.A.; Santos Neto, E.V.; et al. Methane clumped isotopes: Progress and potential for a new isotopic tracer. *Org. Geochem.* **2017**, *113*, 262–282. [CrossRef]

27. Schaefer, H.; Fletcher, S.E.M.; Veidt, C.; Lassey, K.R.; Brailsford, G.W.; Bromley, T.M.; Dlugokencky, E.J.; Michel, S.E.; Miller, J.B.; Levin, I.; et al. A 21st-century shift from fossil-fuel to biogenic methane emissions indicated by 13CH4. *Science* **2016**, *352*, 80–84. [CrossRef]

28. Zazzeri, G.; Lowry, D.; Fisher, R.E.; France, J.L.; Lanoisellé, M.; Kelly, B.F.J.; Necki, J.M.; Iverach, C.P.; Ginty, E.; Zimnoch, M.; et al. Carbon isotopic signature of coal-derived methane emissions to the atmosphere: From coalification to alteration. *Atmos. Chem. Phys.* **2016**, *16*, 13669–13680. [CrossRef]

29. Tait, D.R.; Maher, D.T.; Santos, I.R. Seasonal and Diurnal Dynamics of Atmospheric Radon, Carbon Dioxide, Methane, $\delta 13C\text{-}CO_2$ and $\delta 13C\text{-}CH_4$ in a Proposed Australian Coal Seam Gas Field. *Water Air Soil Pollut.* **2015**, *226*, 350. [CrossRef]

30. Levin, I.; Bergamaschi, P.; Dörr, H.; Trapp, D. Stable isotopic signature of methane from major sources in Germany. *Chemosphere* **1993**, *26*, 161–177. [CrossRef]

31. Hare, V.J.; Loftus, E.; Jeffrey, A.; Ramsey, C.B. Atmospheric CO_2 effect on stable carbon isotope composition of terrestrial fossil archives. *Nat. Commun.* **2018**, *9*, 252. [CrossRef] [PubMed]

32. Jasek, A.; Zimnoch, M.; Gorczyca, Z.; Smula, E.; Rozanski, K. Seasonal variability of soil CO_2 flux and its carbon isotope composition in Krakow urban area, Southern Poland. *Isot. Environ. Health Stud.* **2014**, *50*, 143–155. [CrossRef] [PubMed]

33. Graven, H.; Keeling, R.F.; Rogelj, J. Changes to Carbon Isotopes in Atmospheric CO_2 Over the Industrial Era and Into the Future. *Glob. Biogeochem. Cycles* **2020**, *34*, e2019GB006170. [CrossRef]

34. North, J.C.; Frew, R.D.; Van Hale, R. Can stable isotopes be used to monitor landfill leachate impact on surface waters? *J. Geochem. Explor.* **2006**, *88*, 49–53. [CrossRef]

35. Raco, B.; Dotsika, E.; Battaglini, R.; Bulleri, E.; Doveri, M.; Papakostantinou, K. A Quick and Reliable Method to Detect and Quantify Contamination from MSW Landfills: A Case Study. *Water Air Soil Pollut.* **2013**, *224*, 1380. [CrossRef]

36. Wimmer, B.; Hrad, M.; Huber-Humer, M.; Watzinger, A.; Wyhlidal, S.; Reichenauer, T.G. Stable isotope signatures for characterising the biological stability of landfilled municipal solid waste. *Waste Manag.* **2013**, *33*, 2083–2090. [CrossRef] [PubMed]

37. Widory, D.; Proust, E.; Bellenfant, G.; Bour, O. Assessing methane oxidation under landfill covers and its contribution to the above atmospheric CO_2 levels: The added value of the isotope ($\delta 13C$ and $\delta 18O$ CO_2; $\delta 13C$ and δD CH_4) approach. *Waste Manag.* **2012**, *32*, 1685–1692. [CrossRef] [PubMed]

38. Bergamaschi, P.; Lubina, C.; Königstedt, R.; Fischer, H.; Veltkamp, A.C.; Zwaagstra, O. Stable isotopic signatures (●13C, gD) of methane from European landfill sites. *J. Geophys. Res.* **1998**, *103*, 8251–8265. [CrossRef]

39. Al-Shalan, A.; Lowry, D.; Fisher, R.E.; Nisbet, E.G.; Zazzeri, G.; Al-Sarawi, M.; France, J.L. Methane emissions in Kuwait: Plume identification, isotopic characterisation and inventory verification. *Atmos. Environ.* **2021**, *268*, 118763. [CrossRef]

40. Bakkaloglu, S.; Lowry, D.; Fisher, R.; France, J.; Lanoiselle, M.; Fernandez, J. Characterization and Quantification of Methane Emissions from Waste in the UK. In Proceedings of the EGU General Assembly 2020, Online, 4–8 May 2020. [CrossRef]

41. Aghdam, E.F.; Fredenslund, A.M.; Chanton, J.; Kjeldsen, P.; Scheutz, C. Determination of gas recovery efficiency at two Danish landfills by performing downwind methane measurements and stable carbon isotopic analysis. *Waste Manag.* **2018**, *73*, 220–229. [CrossRef] [PubMed]

42. Sanci, R.; Panarello, H.O. Carbon stable isotopes as indicators of the origin and evolution of CO_2 and CH_4 in urban solid waste disposal sites and nearby areas. *Environ. Earth Sci.* **2016**, *75*, 294. [CrossRef]

43. Whiticar, M.J. Stable isotope geochemistry of coals, humic kerogens and related natural gases. *Int. J. Coal Geol.* **1996**, *32*, 191–215. [CrossRef]

44. Whiticar, M.; Faber, E.; Schoell, M. Biogenic methane formation in marine and freshwater environments: CO_2 reduction vs. acetate fermentation—Isotope evidence. *Geochim. Et Cosmochim. Acta* **1986**, *50*, 693–709. [CrossRef]

45. Baldassare, F.J.; Laughrey, C.D. Identifying the sources of stray methane by using geochemical and isotopic fingerprinting. *Environ. Geosci.* **1997**, *4*, 85–94.

46. Tecle, D.; Lee, J.; Hasan, S. Quantitative analysis of physical and geotechnical factors affecting methane emission in municipal solid waste landfill. *Environ. Geol.* **2009**, *56*, 1135–1143. [CrossRef]

47. Zhang, C.; Xu, T.; Feng, H.; Chen, S. Greenhouse Gas Emissions from Landfills: A Review and Bibliometric Analysis. *Sustainability* **2019**, *11*, 2282. [CrossRef]

48. Haro, K.; Ouarma, I.; Nana, B.; Bere, A.; Tubreoumya, G.C.; Kam, S.Z.; Laville, P.; Loubet, B.; Koulidiati, J. Assessment of CH_4 and CO_2 surface emissions from Polesgo's landfill (Ouagadougou, Burkina Faso) based on static chamber method. *Adv. Clim. Chang. Res.* **2019**, *10*, 181–191. [CrossRef]

49. Gonzalez-Valencia, R.; Magana-Rodriguez, F.; Martinez-Cruz, K.; Fochesatto, G.J.; Thalasso, F. Spatial and temporal distribution of methane emissions from a covered landfill equipped with a gas recollection system. *Waste Manag.* **2021**, *121*, 373–382. [CrossRef] [PubMed]
50. Zhang, C.; Guo, Y.; Wang, X.; Chen, S. Temporal and spatial variation of greenhouse gas emissions from a limited-controlled landfill site. *Environ. Int.* **2019**, *127*, 387–394. [CrossRef]
51. Climate-Data.org. Available online: https://en.climate-data.org/europe/poland/lower-silesian-voivodeship-456/ (accessed on 9 October 2021).
52. OpenStreetMap. Available online: OpenStreetMap.org (accessed on 11 November 2020).
53. Information on Individual Types of Municipal Waste Collected from the Wroclaw Urban Area. Available online: https://www.google.com/url?sa=t&rct=j&q=&esrc=s&source=web&cd=&ved=2ahUKEwjXt-vzwpr0AhUEt4sKHdw9CDcQFnoECAkQAQ&url=https%3A%2F%2Fbip.um.wroc.pl%2Fattachments%2Fdownload%2F57300&usg=AOvVaw3O4-ZvhtKb5GEwKlTRZ1yG (accessed on 2 November 2021).
54. GIOŚ. *Chief Inspectorate of Environmental Protection (GIOS, Poland). Inspection Report no. WIOŚ-Wroc 301/2018*; GIOŚ: Wrocław, Poland, 2018.
55. Chybiński, S.; Gaworecka, M.; Olearnik, M.; Krzyśków, A. *Analysis of the State of Municipal Precipitation Management in the Kobierzyce Commune*; Gmina Kobierzyce: Wrocław, Poland, 2015.
56. NOAA. Carbon Cycle Gases Mace Head, County Galway, Ireland. Available online: https://gml.noaa.gov/dv/iadv/graph.php?code=MHD&program=ccgg&type=lg (accessed on 28 April 2021).
57. Keeling, C.D. The concentration and isotopic abundances of atmospheric carbon dioxide in rural areas. *Geochim. Et Cosmochim. Acta* **1958**, *13*, 322–334. [CrossRef]
58. Pataki, D.E. Seasonal cycle of carbon dioxide and its isotopic composition in an urban atmosphere: Anthropogenic and biogenic effects. *J. Geophys. Res.* **2003**, *108*, 4735. [CrossRef]
59. Bezyk, Y.; Sówka, I.; Górka, M.; Blachowski, J. GIS-Based Approach to Spatio-Temporal Interpolation of Atmospheric CO_2 Concentrations in Limited Monitoring Dataset. *Atmosphere* **2021**, *12*, 384. [CrossRef]
60. Nowak, N. Assessment of Greenhouse Gas Emissions from the Municipal Waste Landfill in Rudna Wielka on the Basis of Concentrations and Carbon Isotopic Composition of Atmospheric CO_2 and CH_4. Master's thesis, University of Wroclaw, Wroclaw, Poland, 2018.
61. Ezekwe, I.C.; Arokoyu, S.B. Landfill Emissions and their Urban Planning and Environmental Health Implications in Port Harcourt, South-South Nigeria. *Desenvolv. E Meio Ambiente* **2017**, *42*, 42–224. [CrossRef]
62. KOBiZE. *Poland's National Database on Greenhouse Gases and other Substances Emissions. Data on the Waste Management Plant. Installations and Sources, Emission Levels of Selected Substances for 2016–2020*; IOŚ-PIB: Warsaw, Poland, 2021.
63. Bezyk, Y.; Sówka, I.; Górka, M. Assessment of urban CO_2 budget: Anthropogenic and biogenic inputs. *Urban Clim.* **2021**, *39*, 100949. [CrossRef]
64. Hackley, K.C.; Liu, C.L.; Coleman, D.D. Environmental Isotope Characteristics of Landfill Leachates and Gases. *Ground Water* **1996**, *34*, 827–836. [CrossRef]
65. Bréas, O.; Guillou, C.; Reniero, F.; Wada, E. The Global Methane Cycle: Isotopes and Mixing Ratios, Sources and Sinks. *Isot. Environ. Health Stud.* **2001**, *37*, 257–379. [CrossRef] [PubMed]
66. Coleman, D.D.; Liu, C.L.; Hackley, K.C.; Pelphrey, S.R. Isotopic Identification of Landfill Methane. *AAPG Div. Environ. Geosci. J.* **1995**, *2*, 95–103.
67. Schubert, B.A.; Jahren, A.H. The effect of atmospheric CO_2 concentration on carbon isotope fractionation in C3 land plants. *Geochim. Et Cosmochim. Acta* **2012**, *96*, 29–43. [CrossRef]
68. Bezyk, Y.; Dorodnikov, M.; Sówka, I. Effects of climate factors and vegetation on the CO_2 fluxes and $\delta^{13}C$ from re-established grassland. *E3S Web Conf.* **2017**, *22*, 00017. [CrossRef]
69. Werth, M.; Kuzyakov, Y. 13C fractionation at the root–microorganisms–soil interface: A review and outlook for partitioning studies. *Soil Biol. Biochem.* **2010**, *42*, 1372–1384. [CrossRef]

Article

3D Spatial Analysis of Particulate Matter (PM_{10}, $PM_{2.5}$ and $PM_{1.0}$) and Gaseous Pollutants (H_2S, SO_2 and VOC) in Urban Areas Surrounding a Large Heat and Power Plant

Robert Cichowicz * and Maciej Dobrzański *

Faculty of Architecture, Civil and Environmental Engineering, Lodz University of Technology, Al. Politechniki 6, 90-924 Lodz, Poland
* Correspondence: robert.cichowicz@p.lodz.pl (R.C.); maciej.dobrzanski@p.lodz.pl (M.D.)

Citation: Cichowicz, R.; Dobrzański, M. 3D Spatial Analysis of Particulate Matter (PM_{10}, $PM_{2.5}$ and $PM_{1.0}$) and Gaseous Pollutants (H_2S, SO_2 and VOC) in Urban Areas Surrounding a Large Heat and Power Plant. *Energies* **2021**, *14*, 4070. https://doi.org/10.3390/en14144070

Academic Editors: Robert Król, Witold Kawalec, Izabela Sówka and Krzysztof M. Czajka

Received: 29 May 2021
Accepted: 1 July 2021
Published: 6 July 2021

Abstract: In many regions of the world, the winter period is a time of poor air quality, due primarily to the increased use of individual and district heating systems. As a consequence, the atmospheric air contains increased concentrations of both particulate matter and gaseous pollutants (as a result of "low" emissions at altitudes of up to 40 m and "high" emissions more than 40 m above ground level). In winter, the increased pollution is very often exacerbated by meteorological conditions, including air temperature, pressure, air speed, wind direction, and thermal inversion. Here, we analyze the concentrations of particulate matter (PM_{10}, $PM_{2.5}$, and $PM_{1.0}$) and gaseous pollutants (H_2S, SO_2, and VOC) in the immediate vicinity of a large solid fuel-fired heat and power plant located in an urban agglomeration. Two locations were selected for analysis. The first was close to an air quality measurement station in the center of a multi-family housing estate. The second was the intersection of two main communication routes. To determine the impact of "low" and "high" emissions on air quality, the selected pollutants were measured at heights of between 2 and 50 m using an unmanned aerial vehicle. The results were compared with permissible standards for the concentration of pollutants. Temperature inversion was found to have a strong influence on the level of pollutants at various heights, with higher concentrations of particulate matter registered at altitudes above 40 m. The source of PM, H_2S, and SO_2 pollutants was confirmed to be "low emission" from local transport, industrial plant areas, and the housing estate comprising detached houses located in the vicinity of the measuring points. "High emission" was found to be responsible for the high concentrations of VOC at altitudes of more than 40 m above the intersection and in the area of the housing estate.

Keywords: air quality monitoring; SO_2; VOC; H_2S; PM_{10}; $PM_{2.5}$; $PM_{1.0}$; outdoor air quality; air flow aerodynamics; street canyon

1. Introduction

Air quality depends on the volume of pollutant emissions, the intensity and type of physico-chemical changes occurring in the atmosphere, and the large-scale movement of air pollutants. Pollutants occur in the air for natural reasons, independent of humans, and from anthropogenic sources. Anthropogenic pollutants occur at high concentrations in highly urbanized areas, where there is often high population density, and, thus, have a particularly negative impact on human health and quality of life. Increased levels of pollutants are mainly caused by "low emissions" [1], mainly from road transport and household and municipal waste, including from heating individual homes [2,3].

Among the most dangerous pollutants emitted into the atmosphere are particulate matter pollutants (PM_{10}, $PM_{2.5}$, $PM_{1.0}$) and gaseous pollutants: (sulphur dioxide, carbon dioxide, carbon monoxide, hydrogen sulphide, or odors) [4–7]. Particulate matter pollutants have a negative impact on human health, both directly by penetrating the body (causing allergies and lung diseases), and indirectly by acting as carriers for heavy metals, microorganisms, and bacteria [8,9]. In the European Union, the permissible dust concentration has only been defined for the PM_{10} and $PM_{2.5}$ fractions (in accordance with Directive 2008/50/EC). There are no standards for the $PM_{1.0}$ fraction, which is increasingly considered the most dangerous type of dust pollution. The permissible level of PM_{10} is 50 $\mu g/m^3$ for the daily average and 40 $\mu g/m^3$ for the annual average. For $PM_{2.5}$, the maximum limit is 25 $\mu g/m^3$ (annual average). According to WHO recommendations, the daily average concentration of $PM_{2.5}$ should not exceed 25 $\mu g/m^3$.

Gaseous pollutants are another dangerous category of emissions. For example, hydrogen sulphide is released into the atmosphere as a result of the activities of crude oil refineries, during natural gas purification, as well as by food processing and animal husbandry plants [10]. Other sources include deaeration, drainage and sewage systems in buildings [11]. Small concentrations of hydrogen sulphide in the air may cause only eye irritation, but at higher concentrations of 1400–2800 mg/m^3 it can cause respiratory damage and, ultimately, respiratory arrest. At concentrations above 7000 mg/m^3, hydrogen sulphide can cause death in seconds [12,13]. For this reason, the permissible level of hydrogen sulphide has been defined as 0.02 mg/m^3 (daily average) [14]. According to the WHO [15], the concentration of hydrogen sulphide in the air should not exceed 0.15 mg/m^3 (daily average), and exposure to average concentrations of 7 $\mu g/m^3$ for more than 30 min should be avoided. Due to the ability of gaseous pollutants to more easily penetrate buildings, they present a significantly greater challenge than dust pollutants [16].

Volatile organic compounds (VOC) are a diverse group of organic chemicals that are ubiquitous in most urban environments. According to the EPA (The US Environmental Protection Agency) [17], the most common VOC emissions come from chemicals used in new furniture, paints, aromatic hydrocarbons, cleaners, wood, various fuels, copying and printing machines, perfumes, varnishes, and tobacco products [18]. Common VOCs include ethanol, formaldehydes, benzene, and acetone. Due to the variety of VOCs, there are no unified guidelines for the permissible levels of VOC in the outside air. For example, benzene is a Group 1 carcinogen (as classified by the International Agency for Research on Cancer) [19].

Another important pollutant is sulphur dioxide, which is highly toxic with a suffocating smell. It has a high specific gravity and high relative density, which causes it to slowly spread through the atmosphere. Sulphur compounds contribute to acidification of the environment. Sulphur dioxide emissions can be described as "seasonal", since higher concentrations are observed in the heating/winter seasons, whereas in the vegetation/summer seasons the concentrations decrease. Sulphur dioxide is produced mainly by the combustion of solid and liquid fuels (e.g., coal, crude oil) contaminated with sulphur, in internal combustion engines, power plants, and combined heat and power plants. The largest proportion of sulphur dioxide comes from the fuel and energy industry—i.e., combustion processes in industry and combustion processes during energy production [20]. The amount of SO_2 introduced into the environment depends largely on the quality of the fuel. According to Directive 2008/50/EC [21], the permissible average daily concentration of SO_2 is 125 $\mu g/m^3$ and the average hourly concentration is 350 $\mu g/m^3$. These levels are the acceptable values for the protection of human health. On the other hand, the WHO [22] has set a much lower value for the permissible average daily concentration of SO_2 at 20 $\mu g/m^3$.

Air can transport pollutants over long distances. Rough terrain and dense agglomerations of buildings constitute a barrier that causes the accumulation of pollutants. This is combined in some cities with the development partial or complete urban ventilation corridors, which additionally increase levels of local pollution. As a consequence, dense and tall buildings are sometimes responsible both for increased levels of harmful substances in the air (especially in city centers) and for strong local air turbulence. There are also complex phenomena described by the theory of aerodynamics [23,24]. Models of pollution flow and spread in urban areas [25] have been created to reflect real phenomena and present information on how building affect the accumulation of pollutants. So-called street canyons where roads are surrounded by relatively tall buildings have been associated with the movement of pollutants [26,27]. The transport of pollutants is influenced by the layout and coefficient of the shape of streets, the geometry of the buildings, the source of pollution (height and location), and the direction and speed of the wind [25]. Due to the complexity of the phenomena and the number of parameters that should be taken into account, field studies are a very important source of information [28–31]. The purpose of this study was to determine spatial changes in the state of air pollution in a city in relation to sources of "low" and "high" emissions.

Currently, unmanned aerial vehicles are increasingly used for various purposes, such as mapping, transport, monitoring. The development and miniaturization of air quality measuring devices allowed the use of UAVs for their transport and measurement of pollutant concentrations in the air table [32–34]. Currently, research on this subject is carried out in many research centers. Most of them concern the stage of UAV and measuring equipment construction [35,36]. In the article, the authors present the previously unpublished use of UAV with measuring equipment, as well as the method of analyzing and interpreting the obtained results concerning 3D spatial air quality analysis.

2. Methodology

Measurement Area

Measurements were made in the city of Lodz, which is the third largest settlement unit in Poland (Central-Eastern Europe) in terms of the number of inhabitants (population 677,286, population density 2309.6 people/km^2). The area of analysis covers the area of a municipal communication junction, which is located next to a large housing estate in the immediate vicinity of a solid fuel-fired heat and power plant. The critical points in this area are a housing estate with an air quality measurement station owned and run by the Voivodship Inspectors of Environmental Protection (VIEP) (Figure 1 point 1), the intersection of Puszkina street and Przybyszewskiego street (Figure 1, point 2), and the EC4 heat and power plant (Figure 1, point 3). The VIEP station (station ID: PL0096A) is a container station that automatically measures the parameters for PM_{10}, $PM_{2.5}$, and O_3. Around the air quality measurement station is an area of multi-family housing, characterized by a predominance of five-story residential buildings (15–20 m high) and a smaller number of 10-storey residential buildings (30–40 m high) (133,855 inhabitants, 1.45 km from the heat and power plant). The selected area allowed for the assessment of the impact of "low" and "high" emissions on residential areas. Moreover, the nearby air quality measurement station made it possible to verify the measurement results. The second critical point is an important north-south and east-west communication junction for residential and industrial areas, and an access point to the A1 motorway. It is a single-level intersection of multi-lane roads with two-lane circular traffic, located between the high emitter (EC4) and the air quality measurement point (VIEP station). The total measurement area is about 2 ha, surrounded from the north by collective housing estates and from the south by a cemetery, as well as industrial and commercial areas (in the north-south direction, 0.8 km from the main emitter of the heat and power plant). The second selected area, which is a communication node, made it possible to refer to linear emissions from road transport, and as the area of CHP plants located closer to it, it allows for the assessment of the direct impact of EC4 on pollutant emissions. The gross development intensity index for the area ranges from 0.05 to 1.0, which defines "the intensity of development of concentrated single-family housing, areas of large housing estates, areas of large industrial plants and areas of old suburban buildings". Thus, both the VIEP station and the intersection are in the zone of influence of the pollutant emitter, i.e., the EC4 heat and power plant.

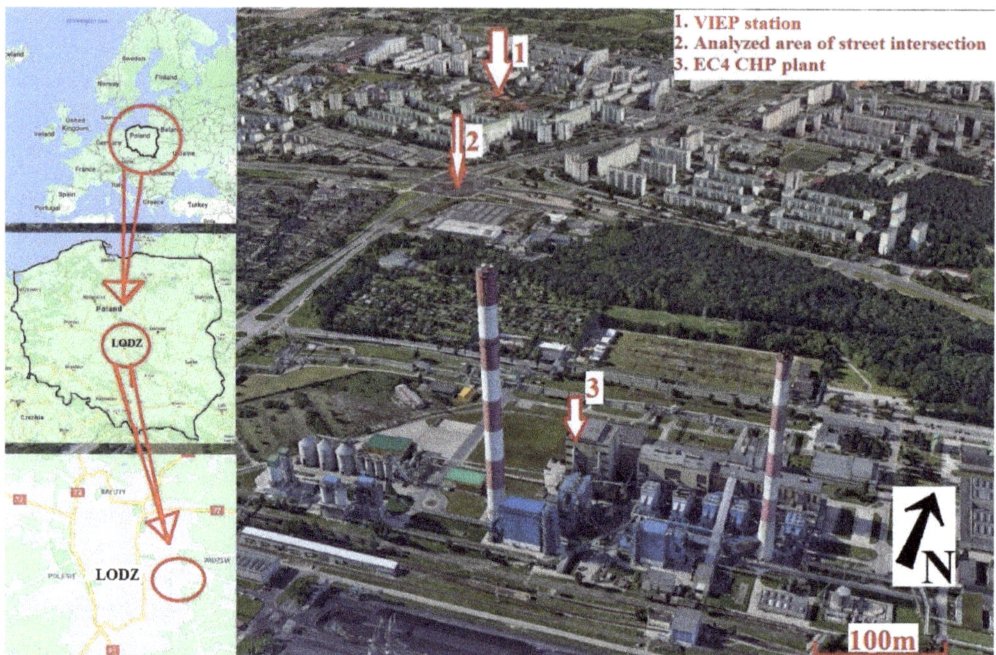

Figure 1. Map of the analyzed area and its surroundings: 1—Voivodship Inspectors of Environmental Protection (VIEP); 2—analyzed intersection of streets; 3—heat and power plant EC4 (photo background source: Google Earth Pro).

The EC4 power plant is one of two coal-fired CHP plants operating in the city of Lodz, producing electricity in a cogeneration system (thermal power 820 MW, electric power 198 MW). The main fuel used in EC4 is hard coal. The plant uses approximately 530,000 Mg/year of hard coal, in addition to 395,000 Mg/year of biomass and 400 Mg/year of light fuel. This translates into average emissions of 9.86 kg/h of PM_{10}, 4.23 kg/h of $PM_{2.5}$, and 250 kg/h of SO_2 (data from January to March 2019). The maximum average annual emissions from the plant of PM_{10}, $PM_{2.5}$ and SO_2 at 1.5 m above ground level are $0.013\ \mu g/m^3$, $0.005\ \mu g/m^3$, and $0.827\ \mu g/m^3$, respectively.

Measurements were made at three levels (20, 35, and 50 m) above the intersection, using an DJI Matrice 600 Pro unmanned aerial vehicle (UAV) (MOT < 25 kg, diameter approximately 2 m) equipped with instruments for measuring particulate matter (PM_{10}, $PM_{2.5}$, PM_1) and gas pollutants (H_2S, SO_2, and VOC). The minimum height of 20 m was chosen to ensure safe flight conditions over the traffic route, street lighting, and tram lines. The maximum height of 50 m was prescribed by the controlled traffic region (CTR) around Wladyslaw Reymont airport in Lodz, located 9 km away. A grid of points was established in the area of the intersection, where the UAV made its measurements (Figure 2B).

The measurements at a height of 20 m reveal the state of air quality in the area of so-called "Low emission" (up to about 40 m relative height). The measurements taken at a height of 50 m represent the area of "high emissions". In the vicinity of the measuring station (Figure 1, point 1), measurements were made at a height of 2 m, and then from 5 m to 50 m with a gradient of 5 m (Figure 2A).

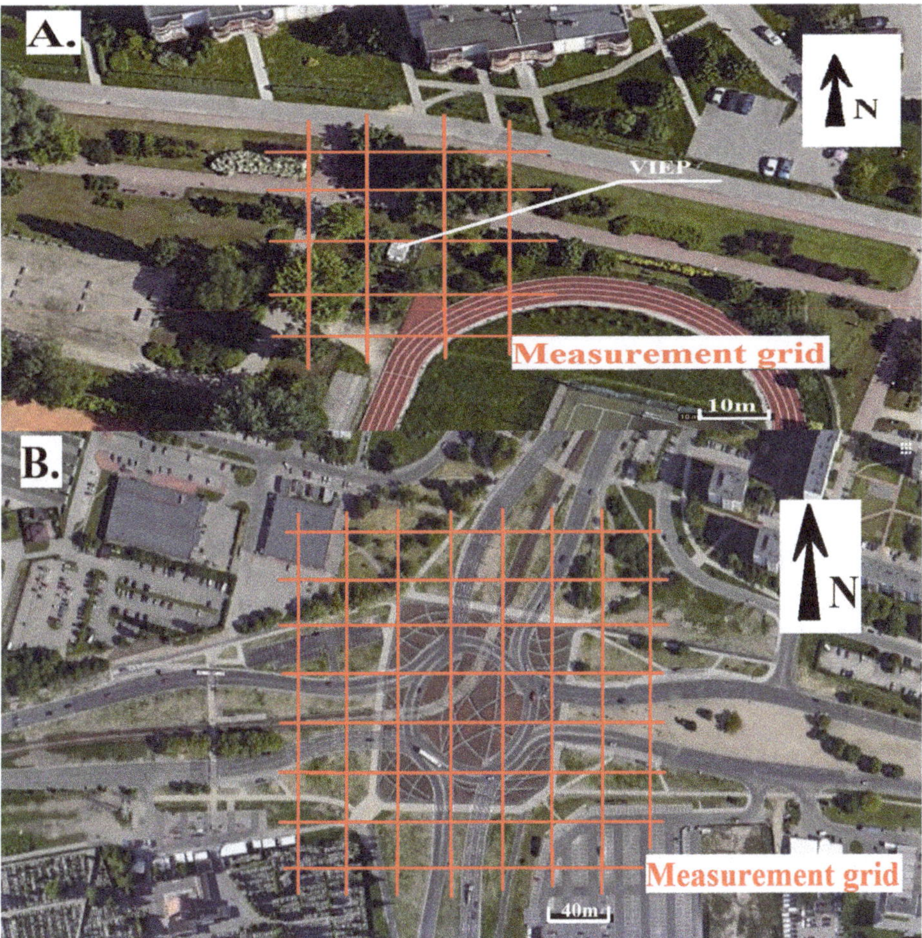

Figure 2. Location of measurement points around the: (**A**). VIEP station; (**B**). crossroads (photo background source: Google Earth Pro).

The measurement apparatus was an eight-sensor module, which enabled the quantification of particulate matter and gaseous pollutants in the atmospheric air. Using a laser scattered (LS) sensor, the module measured PM_{10}, $PM_{2.5}$, and $PM_{1.0}$ (10,000 particles per second). Metal oxide semiconductor (MOS) type sensors measured the concentration of organic solvents (Ethanol, Iso-Butane, H_2, 0–500 ppm) and odors (0.5–1000 ppm isobutanol). ElectroChemical (EC) type sensors measured H_2S (3 ppb–1 ppm), O_2 (0.20–100%) and SO_2 (0.5–2000 ppm). The measuring apparatus was equipped with a probe, 1.5 m long, which eliminated the influence of the UAV on air turbulence and, therefore, any possible impact on the results (Figure 3).

Validation of the measurement data of particulate matter was performed on the basis of data from an accredited measuring station VIEP (the method equivalent to the reference method), while the gaseous pollutants were validated in relation to the VEGA-GC microchromatograph (equipped with a thermal conductivity detector TCD, minimum concentration of 500 ppb (0.005 ppm)). The analysis was based on the results of the parameters of particulate matter PM_{10}, $PM_{2.5}$, $PM_{1.0}$, as well as H_2S, SO_2 and VOC.

Measurement data from station (VIEP) was obtained from a publicly accessible internet database [37].

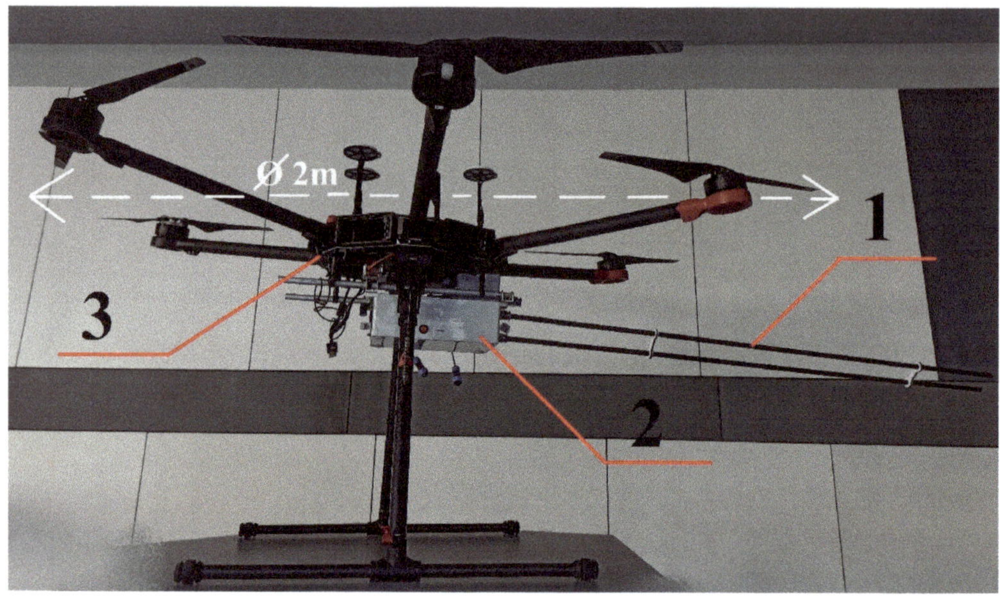

Figure 3. Unmanned aerial vehicle (UAV) with measuring equipment: 1—measuring probe; 2—measuring equipment, 3—UAV.

3. Analysis of Results

Data analysis was carried out using the ArcGIS Pro 2.6 program (Advanced license). Data containing the GPS coordinates of the measurement points from the measuring equipment were entered in the form of a vector layer in the EPSG: 2180 system. The frame of reference was changed to WGS 1984 Web Mercator for spatial analysis. The attributes of the points were the concentration values of the measured pollutant parameters. Using ArcGIS Pro software, the spatial distribution of pollutants was determined based on data interpolation. Interpolation was carried out using the Empirical Bayesian Kriging 3D method. The Kriging method has the advantage over other interpolation methods (Inverse Distance Weighting 'IDW', Triangulated Irregular Network 'TIN') that it treats the observed variable as a random variable [38,39]. The weighting factors are estimated by minimizing the sum of squared deviations for regression and using spatial autocorrelation (semi-variogram) of the examined feature. This improves the quality of spatial prediction.

4. Meteorological Conditions

Instantaneous measurements were made during the "winter period" (from 30 September 2020 to 1 April 2021). This period was chosen because it is commonly associated with increased air pollution, especially due to the combustion of solid fuels, especially hard coal. Poor air quality is one of the most significant problems in Polish cities (of the 50 most polluted cities in the EU, 36 are in Poland). The level of pollution in the "winter period" is strongly related to meteorological conditions, in particular air temperature, wind direction and wind strength. The analysis presented here focuses on three representative measurement series, taken on 11 December 2020 (series A), 18 December 2020 (series B), and 8 January 2021 (series C). The three series of measurements reflect typical conditions for the period. The measurements were all taken between 9:30 am and 12:00 am. During measurement series A and B, the weather conditions were similarly characterized by fog, low cloud cover, high relative humidity (above 80%), and average temperatures above 0. They differed in the direction of the wind. During series A, the wind was blowing from the south-east—i.e., from the EC4 pollutant emitter. During series B, a south-westerly wind predominated—i.e., from the side of the cemetery, industrial areas and single-family houses. Series C was selected for analysis because it showed a similar wind direction to series B, and a similar wind speed to series A. However, the weather conditions differed from those in both series A and series B. On the day of the measurements and on preceding days, there were freezing temperatures, frost, light snowfall, and a low and high base of cloud cover. Figure 4 shows photos of the weather conditions taken by the on-board UAV camera. The meteorological data are summarized in Table 1.

Table 1. Statement of meteorological parameters (data source: www.meteoblue.com).

Series:	Parameters		Temp.	Relative Humidity	Total Precipitation	Total Cloud Cover	Wind Speed	Wind Direction
			[2 m above gnd]		(High resol.) [sfc]		[10 m above gnd]	
	Unit		°C	%	mm	%	km/h	°
A	Min	9–12	−1.2	75	0	2.7	13.5	149
		24 h	−2.4	75	0	0	7.59	119
	Max	9–12	3.0	93	0	4.5	16.3	155
		24 h	3.4	97	0	100	16.3	191
	Average	9–12	1.2	81.8	0	3.8	15.5	151
		24 h	0.6	89.6	0	45.95	13.34	152
B	Min	9–12	3	93	0	62	3.6	220
		24 h	0	87	0	30	3.6	83
	Max	9–12	5	100	0	95	10.8	240
		24 h	5.2	98	15	95	10.8	237
	Average	9–12	4	96	0	78.5	5.9	230
		24 h	3.1	93.6	7.5	62.5	5.85	190
C	Min	9–12	−1.6	68	0	32	14.8	225
		24 h	−1.6	65	0	32	4.6	221.3
	Max	9–12	1	83	0	66	19.3	244
		24 h	1.3	94	0	100	25.5	270
	Average	9–12	−0.2	74.5	0	50.8	16.9	236
		24 h	−0.1	83.6	0	81.6	14.9	233.3

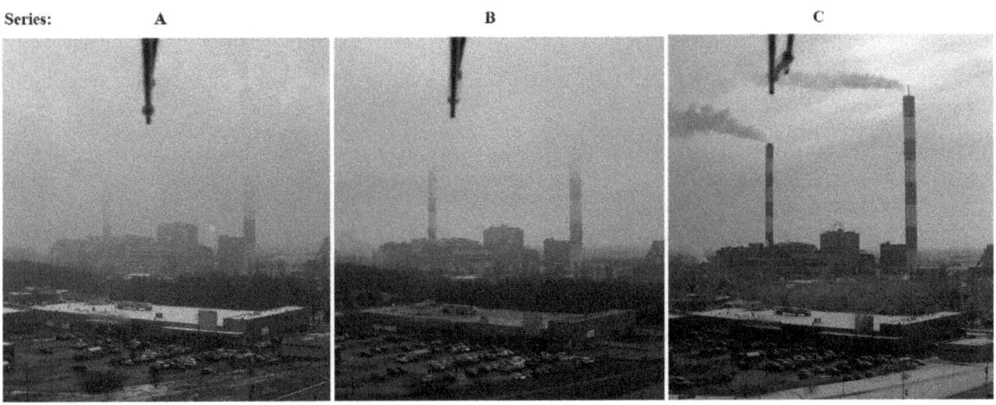

Figure 4. Pictures towards the EC4 showing the weather conditions during the series of measurements.

5. Results

Each series of measurements began at the voivodeship measuring station "1" (VIEP station). This was so that the values detected by its measuring equipment for PM_{10} and $PM_{2.5}$ could be used as a reference (Table 2). The percentage difference between the data obtained from our measurements and those of the VIEP station did not exceed the permissible level of $\pm 10\%$. The average for all series was approximately $\pm 2\%$.

Table 2. The difference between the measurement data and the data from the VIEP station.

Series		A		B		C	
	Unit	PM_{10}	$PM_{2.5}$	PM_{10}	$PM_{2.5}$	PM_{10}	$PM_{2.5}$
VIEP station data	$\mu g/m^3$	40	34	37.5	31.75	30	25.6
Measurement	$\mu g/m^3$	43.7	31.4	40.9	31.6	31.5	25
Difference	%	−9.3	7.6	−9.1	0.5	−5.0	2.3

The measurements performed by the station also allowed us to observe changes in the air quality around the area of dense multi-family housing located about 1.45 km south of the EC4 emitter and industrial facilities. According to the Chief Inspectorate for Environmental Protection [37], a PM_{10} level below 20 $\mu g/m^3$ is very good; up to 50 $\mu g/m^3$ is good; up to 80 $\mu g/m^3$ is moderate; up to 110 $\mu g/m^3$ is poor; up to 150 $\mu g/m^3$ is bad; above 150 $\mu g/m^3$ is very bad. The distribution of PM_{10} pollution observed in series C was different from those observed in series A and B. The highest concentrations, reaching 38.4 $\mu g/m^3$, were recorded at heights from 25 to 35 m above ground level. Above 35 m and below 25 m, the PM_{10} concentrations were about 30% lower and ranged from 21 to 33 $\mu g/m^3$.

The concentration of PM_{10} pollutants in series A and B were found to exceed the permissible level of 50 $\mu g/m^3$ at heights above 25 m (Figure 5). The concentrations of PM_{10} were on average 8% higher in series A than in series B. The highest concentration was 66.5 $\mu g/m^3$, which is 133% of the PM_{10} standard. Closer to the ground (<25 m), in both series A and B the dust content in the air was also above the permissible standard, with 5-perc values as high as 40.7 $\mu g/m^3$ (series A) and 37.7 $\mu g/m^3$ (series B).

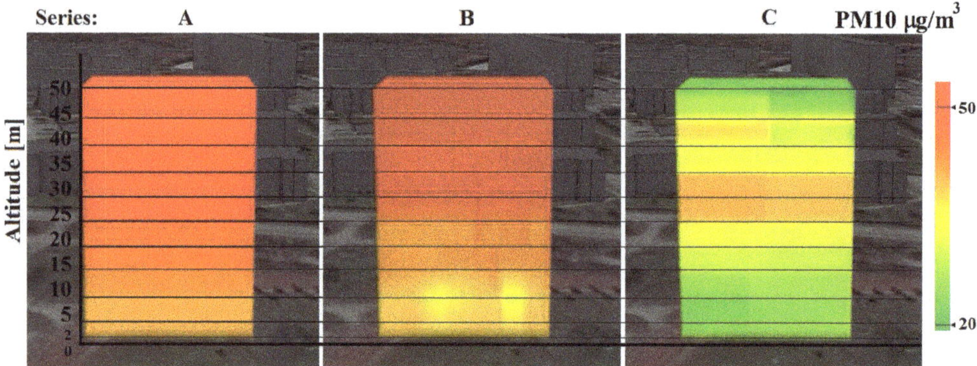

Figure 5. Spatial distribution of PM_{10} pollution with altitude for three measurement series in the VEIP station area.

Figures 6 and 7 show the measured concentrations of $PM_{2.5}$ and $PM_{1.0}$, which have much greater impact on human health. The concentration of $PM_{1.0}$ was only about 2–6% lower than the concentration of $PM_{2.5}$. During series A and B, high concentrations of these pollutants were measured at heights above 25 m. This is consistent with the measurement data for PM_{10}. At heights from 25 to 40 m, the average concentration of $PM_{2.5}$ was in the range of 30 to 43 $\mu g/m^3$. Above 40 m, the $PM_{2.5}$ concentration was more than 45 $\mu g/m^3$.

The concentration of pollutants above 25 m was between 120% and 205% of the permissible $PM_{2.5}$ level of 25 $\mu g/m^3$. Better air quality in terms of these pollutants was observed in series C. The highest concentrations of $PM_{2.5}$ and $PM_{1.0}$, at 28 and 27 $\mu g/m^3$, respectively, were measured 20–30 m above ground level. Interestingly, the lowest concentrations of $PM_{2.5}$ were recorded above 45 m (21 $\mu g/m^3$ on average). Below 10 m, the average was 25 $\mu g/m^3$.

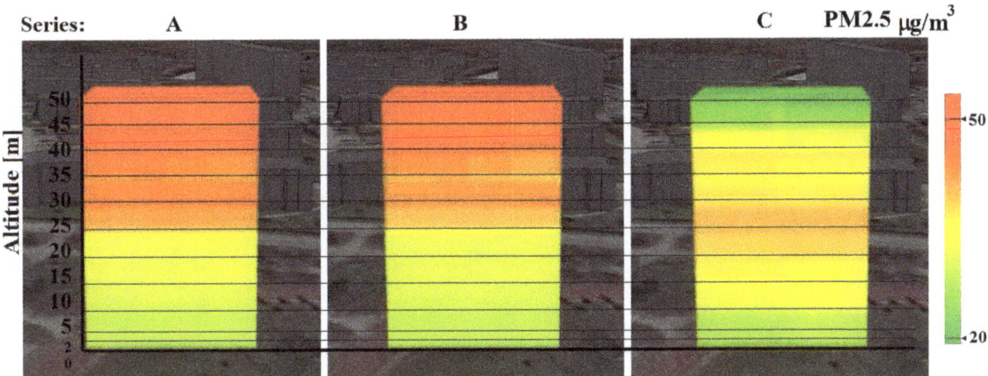

Figure 6. Spatial distribution of $PM_{2.5}$ pollution with altitude for three measurement series in the VEIP station area.

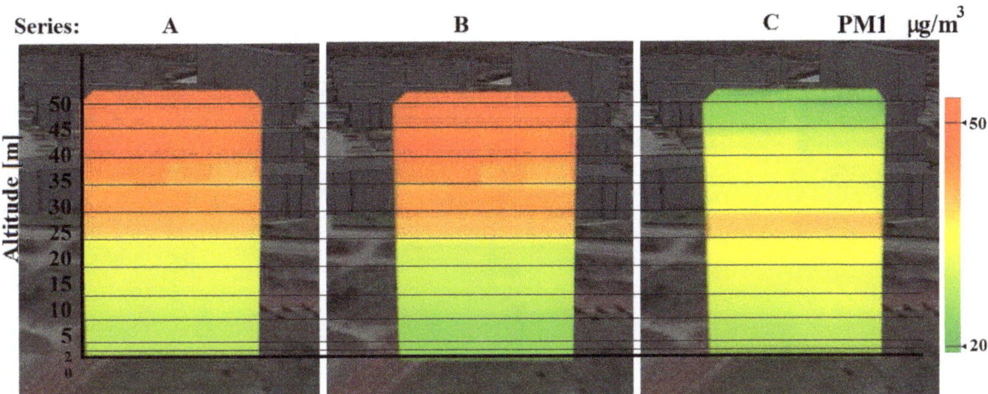

Figure 7. Spatial distribution of $PM_{1.0}$ pollution with altitude for three measurement series in the VEIP station area.

The results for particulate matter pollution in the area of the VIEP station can be explained by the temperature inversion occurring during series A and B. This phenomenon contributed to the formation of a low-suspended layer that prevented the pollutants from floating to the upper parts of the atmosphere. Based on the data in Figure 4, it is possible to estimate the height of the base of the inversion layer 100–130 m above the ground level. This causes a high concentration of PM pollutants at altitudes above 25 m. Lower concentrations of PM in the air layers below 25 m probably resulted from the presence in the measurement area of multi-family buildings with average heights of about 15–20 m and abundant greenery. In series A and B, it can be assumed that the sources of the dust pollution were low emissions (below 40 m), probably from single-family houses, industrial plants, and the main communication routes approximately 1.4 km away (Figure 8) (in the direction of the wind). Low emissions are mainly related to the heating season, as noted by Sówka [40]. In series A and B, the influence of the EC4 emitter on air quality can be excluded, because the chimney outlet is located at a height of 260 m, which is significantly above the temperature inversion layer. However, in the case of series C, there was no clear inversion layer. This translated into the presence of a noticeable air layer in the range of 20 to 30 m, characterized by lower concentrations of particulate matter pollutants. This was probably due to the formation of turbulence and the local transport of pollutants at the height of the roofs of the surrounding buildings. At higher altitudes, the pollutants dispersed, as evidenced by the fact the lowest concentrations in series C were recorded at heights above 40 m.

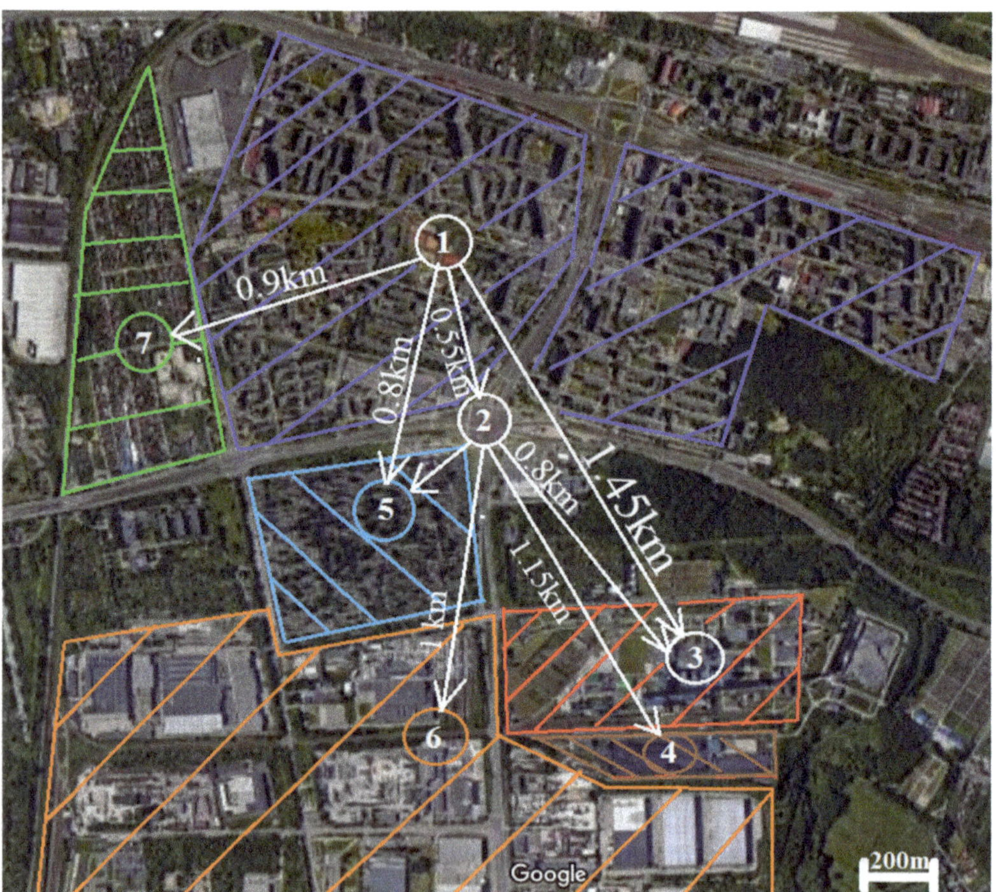

Figure 8. Map of the main areas affecting air quality: 1—the area of a multi-family housing estate with a VIEP station; 2—analyzed street intersection; 3—EC4 heat and power plant; 4—hard coal storage; 5—the cemetery area; 6—industrial areas; 7—the area of the estate of single-family houses (photo background source: Google Earth Pro).

The concentrations of gaseous pollutants were also measured in the area of VIEP stations. No clear differences in concentration versus height were observed in any of the measurement series (Table 3). This was probably due to the distance of the measurement site from the main sources of gaseous pollutants, allowing them to mix with air. The only reproducible pattern was the increased concentrations of both hydrogen sulphide and sulphur dioxide at heights below 10 m, by up to 30% compared to the average value. This probably resulted from air stagnation caused by the presence of buildings in the measurement area. In all measurement series, the concentration of pollutants at many of the measurement points was below the minimum measuring threshold of the equipment.

Only the average concentration of hydrogen sulphide significantly exceeded acceptable levels. In series A and B, the average concentration of hydrogen sulphide was about 0.036 mg/m^3, which is 77% higher than the permissible value of 0.02 mg/m^3. In series C, the concentration was as much as 750% of the permissible level, at 0.170 mg/m^3, although it should be emphasized that this is significantly below lethal levels (1400 mg/m^3). Overall, in series C the mean concentration of H_2S was five times higher than in series A and B and the concentration of SO_2 was twice as high. It can therefore be assumed that the results were influenced by the emission of pollutants from the EC4 CHP plant, especially since in series C there was no temperature inversion. The high concentrations of H_2S were likely emitted from the numerous vents from sewage systems in the densely packed residential buildings that surround the vicinity of the measuring station. This supports the results of previous research [11].

Table 3. Summary of measurement data for three series from the area of VIEP stations.

Series:	Parameters	PM_{10}	$PM_{2.5}$	PM_1	H_2S	SO_2	VOC
	Unit	µg/m^3	µg/m^3	µg/m^3	[ppm]	[ppm]	[ppm]
A	max	66.5	54.2	54.2	0.1580	0.7070	0.0020
	average	52.2	41.0	40.3	0.0235	0.0206	0.0003
	min	30.9	25.1	25.0	<min	<min	<min
	95-perc	63.6	52.1	51.4	0.1110	0.1530	0.0010
	5-perc	40.7	29.5	28.6	<min	<min	<min
B	max	61.6	54.6	54.6	0.1580	0.7070	0.0020
	average	48.4	41.2	40.6	0.0235	0.0194	0.0003
	min	28.4	25.3	25.2	<min	<min	<min
	95-perc	58.9	52.4	51.8	0.1110	0.1530	0.0010
	5-perc	37.7	29.7	28.8	<min	<min	<min
C	max	38.4	32.3	31.9	0.3280	0.8510	0.0030
	average	32.5	27.2	26.5	0.1172	0.0425	0.0010
	min	20.6	10.3	8.6	0.0030	<min	0.0010
	95-perc	36.3	31.5	31.0	0.2030	0.1460	0.0010
	5-perc	28.7	21.6	20.7	0.0320	<min	0.0010

<min—below the minimum excitation threshold of the measuring sensor.

After the tests within the VIEP station, measurements were taken at the intersection of Puszkina street and Przybyszewskiego street. Clear changes were observed in the spatial distribution of the pollutants on the plane and with changes in height. For example, in series A the highest concentration of PM_{10} (83 µg/m^3) was measured at 50 m (average 63 µg/m^3), whereas the lowest average (52 µg/m^3) was measured at an altitude 35 m (Figure 9). Concentrations higher than 50 µg/m^3 occurred in series A from the windward direction—i.e., from the open space side. On the opposite side of the intersection, where there are multi-family buildings and a larger number of trees, the concentration was up to 50% lower. The lowest concentration of PM_{10} was recorded in the upper right corner of the area of analysis, where there are large numbers of trees and a green area. The influence of this area was visible even at a height of 50 m. An identical relationship was observed in the other series of measurements (B and C). In both series B and C the wind was blowing from the direction of the cemetery and industrial facilities. In both series (B and C), the average concentration of PM pollutants was 35% lower than in series A (Table 4). It can be assumed that this was due to the roughness of the terrain in the cemetery, which is covered with tall trees that trapped or accumulated the pollutants. In series A, the wind was blowing from the EC4 side, and probably transported particulate matter pollutants directly from the hard coal storage area. In series B, no significant differences (±2%) were found between the mean concentration of PM_{10} at different heights. This was probably due to the wind speed, which was below 6 km/h and did not cause either violent air turbulence or PM_{10} transport. The highest concentration of PM_{10}, about 40 µg/m^3, was recorded in series B towards the boundary of the cemetery and the surrounding buildings. As noted in [41], buildings contribute to the accumulation of dust pollution on the leeward side. On the other hand, the lowest concentration of PM_{10} (less than 30 µg/m^3) was measured only in the green area around the intersection. Similar observations have been made with regard to the effect of traffic on the concentration of PM_{10} [42].

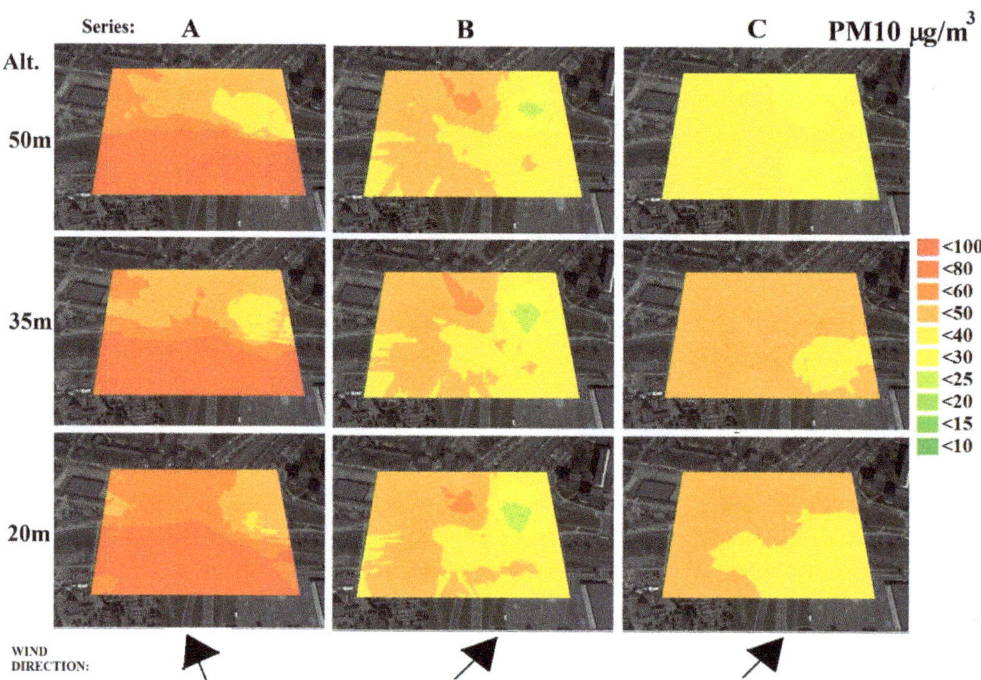

Figure 9. Spatial distribution of PM_{10} pollution at the altitude of 20, 35 and 50 m for three measurement series in the area of the street intersection.

Table 4. Measurement data from the intersection area for three selected measurement series.

Series:	Parameters	PM_{10}	$PM_{2.5}$	PM_1	H_2S	SO_2	VOC
	Unit	µg/m^3	µg/m^3	µg/m^3	[ppm]	[ppm]	[ppm]
A	max	88.6	75.3	75.1	0.2110	0.9940	0.0110
	average	57.2	48.6	48.0	0.0387	0.1993	0.0022
	min	15.1	12.9	12.9	0.0010	0.0020	0.0005
	95-perc	82.8	70.4	70.2	0.1361	0.4951	0.0070
	5-perc	23.0	19.5	18.8	0.0043	0.0319	0.0005
B	max	55.5	49.9	49.2	0.0910	0.4610	0.0140
	average	38.5	30.8	30.1	0.0288	0.0993	0.0019
	min	12.2	5.5	4.8	0.0010	0.0010	0.0010
	95-perc	54.2	46.3	45.8	0.0570	0.3800	0.0050
	5-perc	19.2	8.7	7.5	0.0039	0.0010	0.0010
C	max	43.7	37.1	36.8	0.1050	0.3990	0.0160
	average	36.6	28.7	27.9	0.0395	0.1218	0.0035
	min	25.2	12.8	12.8	0.0010	0.0040	0.0010
	95-perc	41.0	35.1	34.6	0.0750	0.1894	0.0110
	5-perc	31.2	21.3	19.6	0.0040	0.0340	0.0010

In series C, the wind was almost three times stronger than in series B, and the influence of the buildings on the PM_{10} concentration was even more visible (Figure 9). At a height of 20 m, the mean concentration of 40–50 $\mu g/m^3$ was predominantly in the area sheltered by buildings, while average concentrations below 40 $\mu g/m^3$ were measured only in the unprotected area. At a height of 35 m, practically the entire area of analysis was characterized by an average concentration of 40–50 $\mu g/m^3$, probably due to the formation of vortices. At an altitude of 50 m, the PM_{10} concentration dropped to below 40–30 $\mu g/m^3$. This can be explained by the lack of thermal inversion in series C. In series A and B, when thermal inversion occurred, there was an accumulation of pollutants at higher altitudes.

The results for $PM_{2.5}$ and PM_1 were similar to those for PM_{10} (Figures 10 and 11). A strong correlation (>0.50) was found between the concentrations of the particulate matter pollutants. In series A, the highest concentration of dust pollutants at 75 $\mu g/m^3$ was found windward from the heat and power plant and industrial areas, especially at a height of 50 m. Closer to the buildings, the PM content in the air fell by up to 40%, which may have been caused by the formation of vortices on the leeward side. As the measurement altitude decreased, the concentration of $PM_{2.5}$ and PM_1 fell by about 20% in comparison to the concentration at 50 m. This may indicate the accumulation of dust pollutants due to temperature inversion in the upper air layers. Peng [32] came to similar conclusions. In series B and C, the concentrations of $PM_{2.5}$ and PM_1 were about 30% lower than in series A. The content of pollutants in the air was below the permissible level of 25 $\mu g/m^3$ only in the wooded area. In other places, the concentration of $PM_{2.5}$ exceeded the permissible level, reaching 175% of the permissible maximum. Gao [43] reported concentrations of $PM_{2.5}$ during the winter period reaching up to 500% of the limit defined in WHO guidelines.

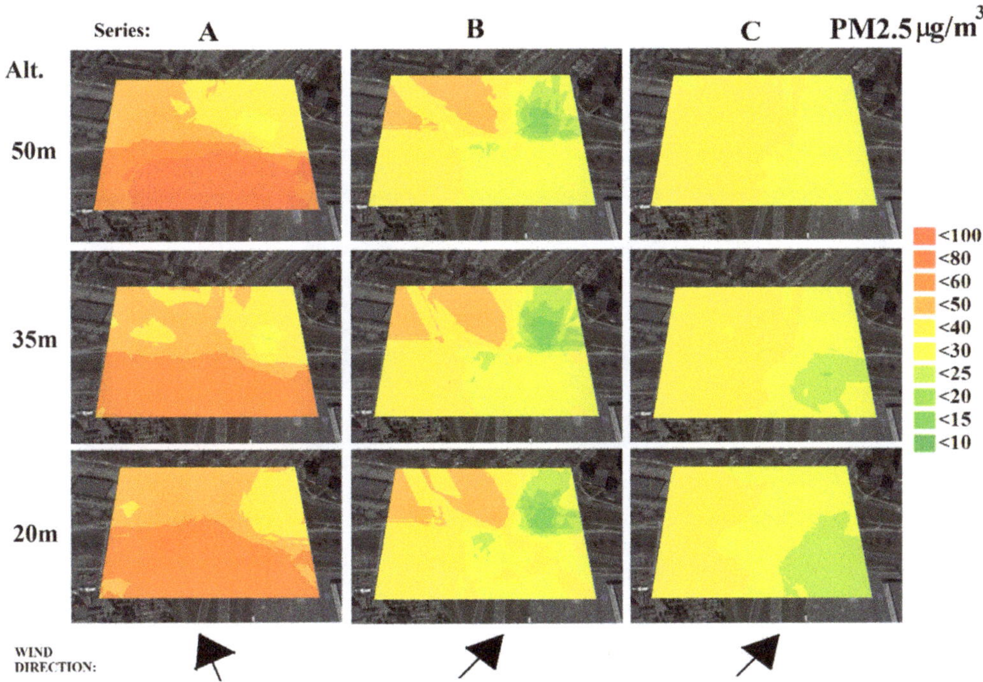

Figure 10. Spatial distribution of $PM_{2.5}$ pollution at the altitude of 20, 35 and 50 m for three measurement series in the area of the street intersection.

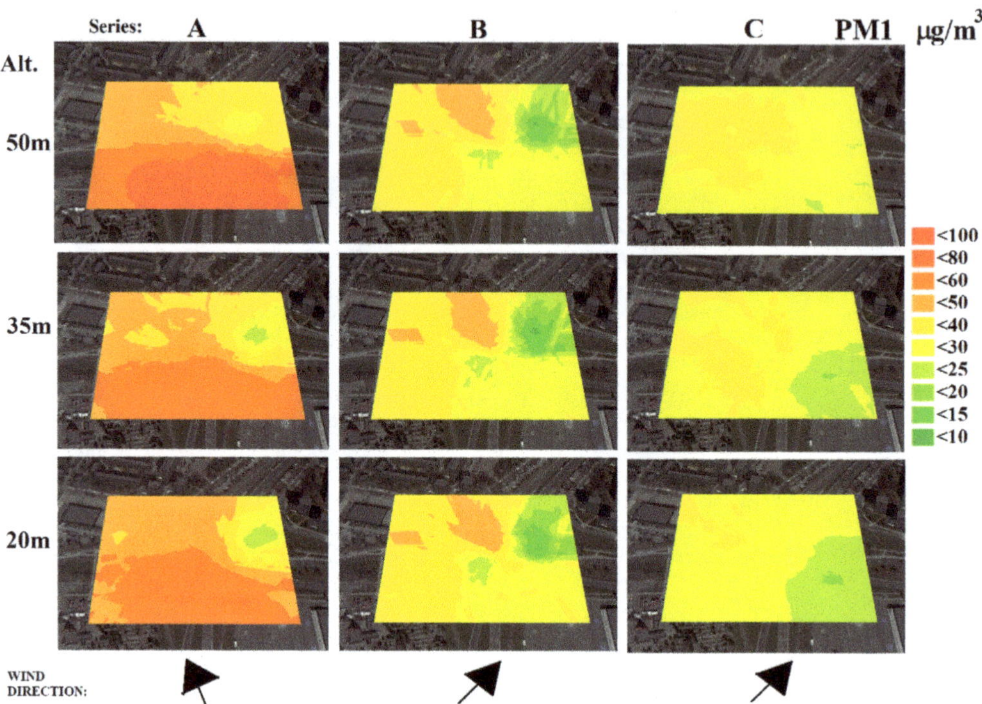

Figure 11. Spatial distribution of $PM_{1.0}$ pollution at the altitude of 20, 35 and 50 m for three measurement series in the area of the street intersection.

Sources of gaseous pollutants include technological processes, fuel combustion and heavy transport. One of the contaminants analyzed was H_2S (Figure 12). High concentrations of H_2S were observed in series A, reaching 0.20 ppm (average 0.042 ppm) at a height of 50 m. At lower altitudes, the concentration decreased by 35% to an average value of 0.027 ppm at a height of 20 m. This can be explained by the wind direction and pollution accumulation due to thermal inversion.

In contrast to series A, in series B and C the highest concentrations of H_2S (over 0.06 ppm) were recorded at 20 m (Figure 12). This is consistent with the fact that H_2S is heavier than air and, therefore, accumulates near the ground. High concentrations of H_2S have also been observed near the ground by Cichowicz and Dobrzański [44]. An interesting phenomenon was the increased concentrations of H_2S at the exits of the intersection. This can be seen by comparing the maps of the H_2S distribution (Figure 12) in series A and B with the layout of the roads at the intersection. Given the temperature inversion in series A and B, it can be assumed that the H_2S in the air came from low emissions from the traffic rand industrial plants. Kourtidis et al. [45] suggest that a source of H_2S may be emissions from car catalysts. It is also possible to exclude the CHP as a source of H_2S, as in series C (no inversion) there was no increase in the concentration compared to series A and B. It should be emphasized that in all series the average concentration of H_2S was alarmingly high, exceeding the permissible level of 0.02 mg/m^3. In series A, the average concentration was 0.058 mg/m^3 (Table 4), which is almost 300% of the permissible level [14]. In series B, the average concentration was double the permissible level. In series C, it was about 40% higher than 0.02 mg/m^3.

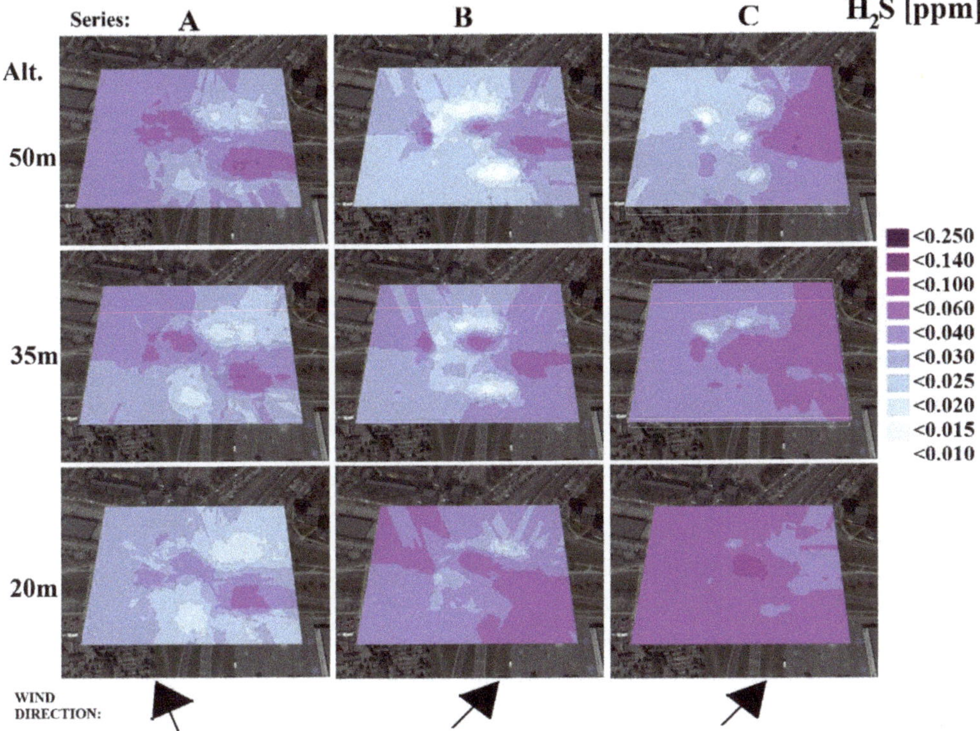

Figure 12. Spatial distribution of H_2S pollution at the altitude of 20, 35, and 50 m for three measurement series in the area of the street intersection.

Another of the analyzed gas pollutants was sulphur dioxide, which is also heavier than air (Figure 13). In series B and C, high concentrations of SO_2 (more than 0.20 ppm, which is approximately 0.57 mg/m^3) were measured at a height of 20 m. The average concentration fell at higher altitudes, by 17% and 25% in series B and C, respectively. However, in series A, where there was temperature inversion and a strong wind, the highest concentration (above 0.40 ppm) occurred mainly at a height of 50 m. Analyzing the area of the intersection, it can be seen that in series A and B there was a space with an increased SO_2 concentration compared to the surroundings. In series A, this is the exit from the intersection towards the city center, whereas in series B it is the exit from the intersection towards the residential part of the city. Note that these areas are in line with the direction air travel from the intersection. Interestingly, in the absence of thermal inversion (series C) no such relationship was observed. In most of the crossing area, the permissible SO_2 concentration of 0.35 mg/m^3 was not exceeded. Only at the exits from the intersection was the permissible concentration of SO_2 exceeded, reaching about 130%. However, increased concentrations of SO_2 are usually observed in the winter [46], probably due to the increased production of thermal energy.

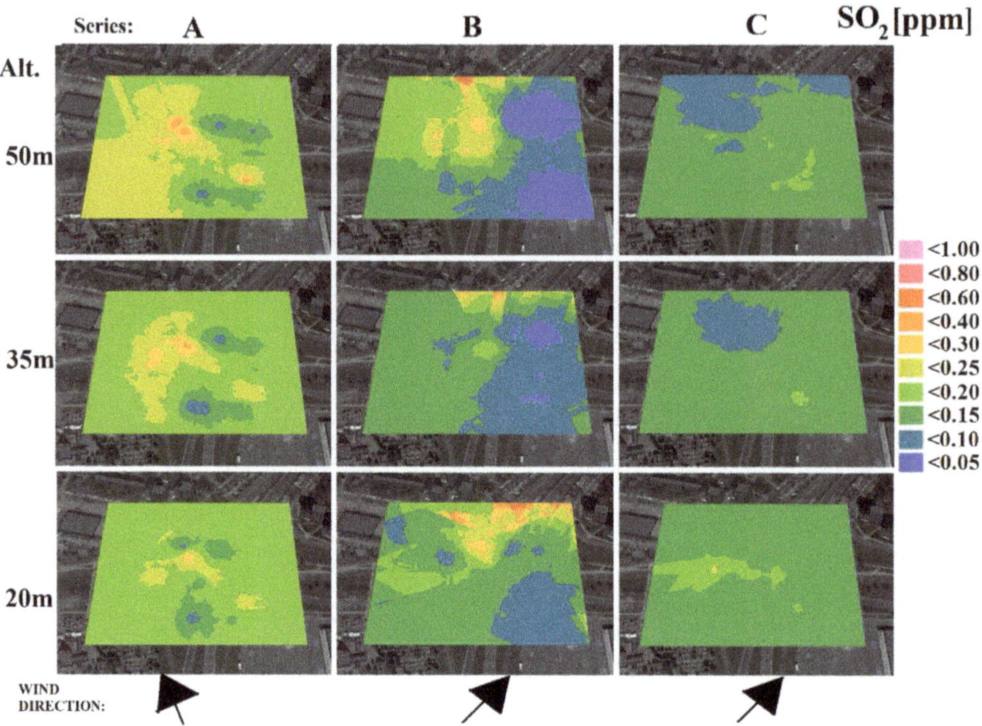

Figure 13. Spatial distribution of SO$_2$ pollution at the altitude of 20, 35, and 50 m for three measurement series in the area of the street intersection.

The final analyzed gas compound was VOC (Figure 14). There are no specific permissible levels for VOC. The distributions of VOC observed in series A and B were completely different to those in series C. Under the meteorological conditions in series A and B, with the visible influence of thermal inversion, a very low VOC concentration of less than 0.004 ppm (0.014 mg/m^3) was recorded, but in most of the studied area the value was even lower, at less than 0.002 ppm. In series A and B, up to 50% higher concentrations were recorded at a height of 20 m, averaging 0.003 ppm at 35 and 50 m. The increased concentrations of VOC in series A and B were mainly associated with the communication routes, so the source of pollution was "local" low emission. The situation was completely different in series C (Figure 14), where the highest concentrations of up to 0.016 ppm (0.056 mg/m^3) were found at an altitude of 50 m. However, at 35 m and 20 m the concentration of VOC decreased up to 7-fold. This may be evidence of the migration of VOC pollutants as a result of high emissions, although it would only be possible to test such a hypothesis under stable temperature gradient conditions.

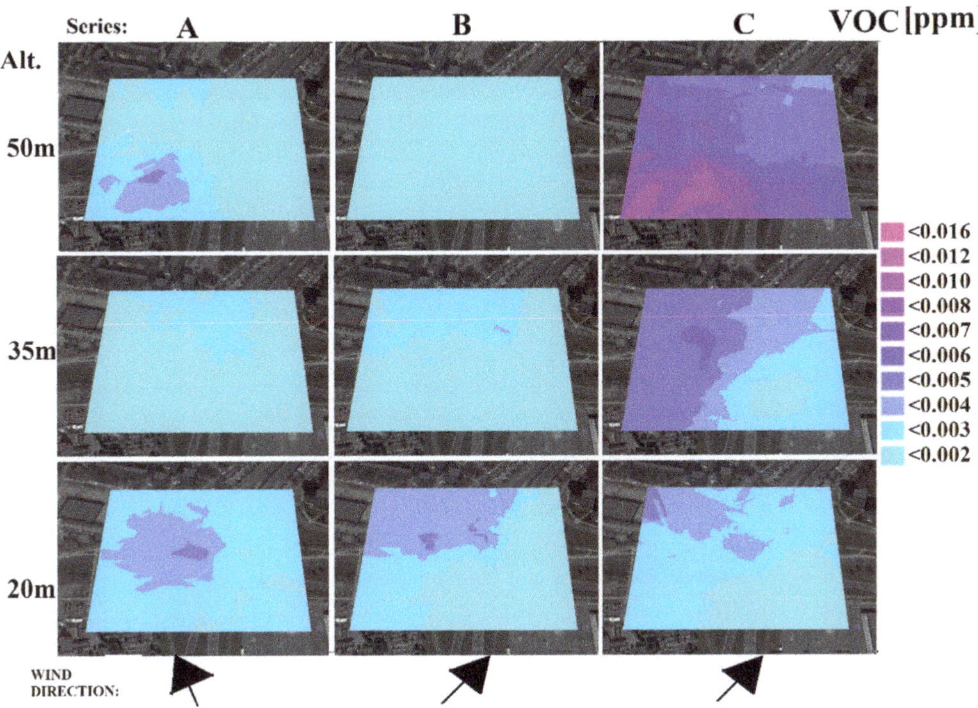

Figure 14. Spatial distribution of VOC pollution at the altitude of 20, 35, and 50 m for three measurement series in the area of the street intersection.

6. Conclusions

This study set out to perform a spatial analysis of the distribution pollutants caused by "low" and "high" emissions. In series A and B, "low" emissions were associated with high concentrations of PM, SO_2 and H_2S at heights of 30–50 m. This was probably due mainly to the occurrence of temperature inversion. In series C there was no temperature inversion, which translated into lower concentrations of pollutants and a different spatial distribution to those in series A and B. The concentrations of pollutants were higher when temperature inversion occurred (Series A and B) than when there was no temperature inversion (Series C). This indicates that the emitters/chimneys of the coal-fired heat and power plant (EC4) had little effect on air quality in the analyzed area. It can be assumed that the EC4 CHP plant only contributed to high concentrations of VOC at altitudes above 40 m.

In series C, increased levels of suspended dust in the air were recorded at heights of 25 and 35 m above ground level. This may indicate an air corridor transporting PM pollution at this altitude. In all the measurement series, a strong correlation (above 0.5) was found between the PM_{10}, $PM_{2.5}$, and PM_1 dust fractions. A particularly high correlation coefficient (close to 1) was calculated for $PM_{2.5}$ and PM_1 (Table 5). Similar observations were made by Li et al. (2015), obtaining a correlation of 0.71–0.77 between PM. In all three series, concentrations of PM were measured that exceeded permissible values by up to 175% ($PM_{2.5}$). Presumably, this was related to season, with increased production of individual and industrial thermal energy during the winter period. In the housing estate, the concentration of gaseous pollutants remained the same in the height range from 2 to 50 m. This may be explained by turbulence from buildings, which mixed the gaseous pollutants with air, and by the distance from the source.

Table 5. Correlation coefficient between measurement parameters from the intersection area for three measurement series.

	PM_{10}	$PM_{2.5}$	PM_1	H_2S	SO_2	VOC
Series A						
PM_{10}	1.00	0.98	0.98	0.07	0.11	0.27
$PM_{2.5}$	0.98	1.00	1.00	0.02	0.07	0.25
PM_1	0.98	1.00	1.00	0.02	0.08	0.24
H_2S	0.07	0.02	0.02	1.00	0.73	−0.17
SO_2	0.11	0.07	0.08	0.73	1.00	−0.14
VOC	0.27	0.25	0.24	−0.17	−0.14	1.00
Series B						
PM_{10}	1.00	0.93	0.92	0.05	0.32	0.10
$PM_{2.5}$	0.93	1.00	1.00	0.08	0.39	0.21
PM_1	0.92	1.00	1.00	0.08	0.40	0.20
H_2S	0.05	0.08	0.08	1.00	0.16	0.23
SO_2	0.32	0.39	0.40	0.16	1.00	0.61
VOC	0.10	0.21	0.20	0.23	0.61	1.00
Series C						
PM_{10}	1.00	0.70	0.68	−0.06	−0.13	0.41
$PM_{2.5}$	0.70	1.00	0.99	−0.02	0.01	0.22
PM_1	0.68	0.99	1.00	−0.03	0.02	0.22
H_2S	−0.06	−0.02	−0.03	1.00	0.43	−0.37
SO_2	−0.13	0.01	0.02	0.43	1.00	−0.13
VOC	0.41	0.22	0.22	−0.37	−0.13	1.00

The measurements taken at the intersection of communication routes, located closer to the industrial areas, showed clear differences in terms of the distribution of gaseous pollutants at different heights. In the case of H_2S and SO_2, the distribution of concentrations was consistent with the road system and the orientation towards industrial areas. This indicates that these pollutants were caused by local low emissions. High VOC concentrations were only recorded at altitudes of more than 40 m within the intersection and housing estate areas in Series C (no temperature inversion), indicating that they originated from high emissions. The permissible level for H_2S was exceeded by up to 300% both in the area of the housing estate and at the intersection. The concentrations of H_2S were higher in the housing estate than in the area around the intersection. This was probably due to the numerous drainage vents located on the roofs of buildings on the housing estate. There was no clear correlation between the levels of gaseous pollutants in the three series. A weak correlation of 0.44 was found between H_2S and SO_2 levels in the three series. There was also no correlation between dust and gaseous pollutants (correlation coefficients below 0.5).

Summing up, the spatial analysis of pollutants with the use of UAVs gives the opportunity to get acquainted with the distribution of pollutants concentration in a selected area in detail. It should be emphasized that it is not possible to perform such measurements using traditional, stationary measuring stations located at a constant height from the ground, with a low spacing density. The method of measurement and analysis proposed by the authors can be implemented all over the world, therefore, it seems to be an important supplement to the existing methods of air quality measurement. The research methods presented in the article are now successfully used by us in air quality analyzes in various areas of Poland. The most important advantage of the obtained results is their high resolution in a small area, directly answering the question of air quality.

Author Contributions: Conceptualization, R.C. and M.D.; methodology, R.C. and M.D.; software, M.D., R.C.; writing—original draft, R.C. and M.D., review and editing, R.C. All authors have read and agreed to the published version of the manuscript.

Funding: This study was conducted as part of the research project entitled "Spatial analysis of air pollution changes in the Lodz agglomeration (in Polish: Analiza przestrzenna zmian stanu zanieczyszczenia powietrza w aglomeracji łódzkiej)", which was co-financed by approximately 80% by the Provincial Fund for Environmental Protection and Water Management in Lodz (in Polish: Wojewódzki Fundusz Ochrony Środowiska i Gospodarki Wodnej w Łodzi).

Institutional Review Board Statement: Not applicable.

Informed Consent Statement: Not applicable.

Data Availability Statement: Data available on request.

Conflicts of Interest: The authors declare no conflict of interest.

References

1. Chlebowska-Styś, A.; Sówka, I.; Kobus, D.; Pachurka, Ł. Analysis of concentrations trends and origins of PM_{10} in selected European cities. In Proceedings of the 9th Conference on Interdisciplinary Problems in Environmental Protection and Engineering EKO-DOK, Boguszów-Gorce, Poland, 23–25 April 2017. [CrossRef]
2. Jeong, C.H.J.; Jon, M.; Wang, J.M.; Greg, J.; Evans, G.J. Source apportionment of particulate matter in Europe: A review of methods and results. *J. Aerosol. Sci.* **2008**, *39*, 827–849.
3. Zhao, Y.; Song, X.; Wang, Y.; Zhao, J.; Zhu, K. Seasonal patterns of PM_{10}, $PM_{2.5}$, and $PM_{1.0}$ concentrations in a naturally ventilated residential underground garage. *Build. Environ.* **2017**, *124*, 294–314. [CrossRef]
4. World Health Organization. Review of Evidence on Health Aspects of Air Pollution—REVIHAAP Project. World Health Organization 309. 2013. Available online: http://www.euro.who.int/en/health-topics/environment-and-health/air-quality/publications/2013/review-of-evidence-on-health-aspects-of-air-pollution-revihaap-project-finaltechnical-report (accessed on 1 February 2021).
5. Moreno-Silva, C.; Calvo, D.; Torres, N.; Ayala, L.; Gaitán, M.; González, L.; Rincón, P.; Susa, M.R. Hydrogen sulphide emissions and dispersion modelling from a wastewater reservoir using flux chamber measurements and AERMOD® simulations. *Atmos. Environ.* **2020**, *224*, 117263. [CrossRef]
6. Liu, N.; Liu, X.; Jayaratne, R.; Morawska, L. A study on extending the use of air quality monitor data via deep learning techniques. *J. Clean. Prod.* **2020**, *274*, 122956. [CrossRef]
7. Edginton, S.; O'Sullivan, D.E.; King, W.D.; Lougheed, M.D. The effect of acute outdoor air pollution on peak expiratory flow in individuals with asthma: A systematic review and meta-analysis. *Environ. Res.* **2021**, *192*, 110296. [CrossRef] [PubMed]
8. Pope Iii, C.A.; Burnett, R.T.; Thun, M.J.; Calle, E.E.; Krewski, D.; Ito, K.; Thurston, G.D. Lung cancer, cardiopulmonary mortality, and long term exposure to fine particulate air pollution. *J. Am. Med. Assoc.* **2002**, *287*, 1132–1141. [CrossRef]
9. Dey, S.; Di Girolamo, L.; Van Donkelaar, A.; Tripathi, S.; Gupta, T.; Mohan, M. Variability of outdoor fine particulate ($PM_{2.5}$) concentration in the Indian Subcontinent: A remote sensing approach. *Remote. Sens. Environ.* **2012**, *127*, 153–161. [CrossRef]
10. Aventaggiato, L.; Colucci, A.P.; Strisciullo, G.; Favalli, F.; Gagliano-Candela, R. Lethal Hydrogen Sulfide poisoning in open space: An atypical case of asphyxiation of two workers. *Forensic Sci. Int.* **2020**, *308*, 110122. [CrossRef]
11. Matos, V.R.; Ferreira, F.; Matos, S.J. Influence of ventilation in H2S exposure and emissions from a gravity sewer. *Water Sci. Technol.* **2020**, *81*, 2043–2056. [CrossRef]
12. Fujita, Y.; Fujino, Y.; Onodera, M.; Kikuchi, S.; Kikkawa, T.; Inoue, Y.; Niitsu, H.; Takahashi, K.; Endo, S. A fatal case of acute hydrogen sulfide poisoning caused by hydrogen sulfide: Hydroxocobalamin therapy for acute hydrogen sulfide poisoning. *J. Anal. Toxicol.* **2011**, *35*, 119–123. [CrossRef]
13. Stetkiewicz, J. Siarkowodór. Dokumentacja dopuszczalnych wielkości narażenia zawodowego. *Podstawy I Metod. Oceny Środowiska Pracy.* **2011**, *4*, 97–117.
14. Polish Journal of Laws 1966: Regulation of the Council of Ministers of 13 September 1966 on Admissible Concentrations of Substances in the Atmospheric Air. Available online: http://isap.sejm.gov.pl/isap.nsf/DocDetails.xsp?id=WDU19660420253 (accessed on 10 September 2020).
15. World Health Organization. *Air Quality Guidelines*, 2nd ed.; Hydrogen Sulfide: 2000; Chapter 6.6. Available online: https://www.euro.who.int/en/health-topics/environment-and-health/air-quality/publications/pre2009/who-air-quality-guidelines-for-europe,-2nd-edition,-2000-cd-rom-version (accessed on 1 February 2021).
16. Wu, M.X.; Basu, R.; Malig, B.; Broadwin, R.; Ebisu, K.; Gold, B.E.; Qi, L.; Derby, C.; Green, S.R. Association between gaseous air pollutants and inflammatory, hemostatic and lipid markers in a cohort of midlife women. *Environ. Int.* **2017**, *107*, 131–139. [CrossRef]
17. United States Environmental Protection Agency EPA. Available online: https://www.epa.gov/indoor-air-quality-iaq/volatile-organic-compounds-impact-indoor-air-quality (accessed on 12 September 2020).
18. Hormigos-Jimenez, S.; Padilla-Marcos, M.Á.; Meiss, A.; Gonzalez-Lezcano, R.A.; Feijó-Muñoz, J. Ventilation rate determination method for residential buildings according to TVOC emissions from building materials. *Build. Environ.* **2017**, *123*, 555–563. [CrossRef]

19. IARC. IARC (International Agency for Research on Cancer) Monographs on the Evaluation of Carcinogenic Risks to Human. 2016. Available online: http://monographs.iarc.fr/ENG/Classification/latest_classif.php (accessed on 10 December 2020).

20. Meng, M.; Zhou, J. Has air pollution emission level in the Beijing–Tianjin–Hebei region peaked? A panel data analysis. *Ecol. Indic.* **2020**, *119*, 106875. [CrossRef] [PubMed]

21. Directive 2008/50/EC of the European Parliament and of the Council of 21 May 2008 on Ambient Air Quality and Cleaner Air for Europe. OJ L 152, 11.6.2008, p. 1–44 (BG, ES, CS, DA, DE, ET, EL, EN, FR, IT, LV, LT, HU, MT, NL, PL, PT, RO, SK, SL, FI, SV) Special Edition in Croatian: Chapter 15. 2008, Volume 029, pp. 169–212. Available online: http://data.europa.eu/eli/dir/2008/50/oj (accessed on 6 July 2021).

22. World Health Organization. *Who Air Quality Guidelines for Particulate Matter, Ozone, Nitrogen Dioxide and Sulfur Dioxide. Global Update 2005. Summary of Risk Assessment*; 2005. Available online: https://www.euro.who.int/en/health-topics/environment-and-health/air-quality/publications/pre2009/air-quality-guidelines.-global-update-2005.-particulate-matter,-ozone,-nitrogen-dioxide-and-sulfur-dioxide (accessed on 1 February 2021).

23. Hosker, R.P. Flow around isolated structures and building clusters, a review. *ASHRAE Trans.* **1985**, *91*, 1672–1692.

24. Pesic, D.; Zigar, D.; Anghel, I.; Glisovic, S. Large Eddy Simulation of wind flow impact on fire-induced indoor and outdoor air pollution in an idealized street canyon. *J. Wind. Eng. Ind. Aerodyn.* **2016**, *155*, 89–99. [CrossRef]

25. Lateb, M.; Meroney, R.N.; Yataghene, M.; Fellouah, H.; Saleh, F.; Boufadel, M.C. On the use of numerical modelling for near-field pollutant dispersion in urban environments—A review. *Environ. Pollut.* **2016**, *208*, 271–283. [CrossRef] [PubMed]

26. Murena, F.; Mele, B. Effect of short–time variations of wind velocity on mass transfer rate between street canyons and the atmospheric boundary layer. *Atmos. Pollut. Res.* **2014**, *5*, 484–490. [CrossRef]

27. Kikumoto, H.; Ooka, R.A. numerical study of air pollutant dispersion with bimolecular chemical reactions in an urban street canyon using large-eddy simulation. *Atmos. Environ.* **2012**, *54*, 456–464. [CrossRef]

28. Zhang, Y.; Kwok, K.; Liu, X.-P.; Niu, J.-L. Characteristics of air pollutant dispersion around a high-rise building. *Environ. Pollut.* **2015**, *204*, 280–288. [CrossRef]

29. Zwack, L.M.; Paciorek, C.J.; Spengler, J.D.; Levy, J.I. Characterizing local traffic contributions to particulate air pollution in street canyons using mobile monitoring techniques. *Atmos. Environ.* **2011**, *45*, 2507–2514. [CrossRef]

30. Galatioto, F.; Bell, M. Exploring the processes governing roadside pollutant concentrations in urban street canyon. Environ. *Sci. Pollut. Res.* **2013**, *20*, 4750–4765. [CrossRef] [PubMed]

31. Rakowska, A.; Wong, K.C.; Townsend, T.; Chan, K.L.; Westerdahl, D.; Ng, S.; Močnik, G.; Drinovec, L.; Ning, Z. Impact of traffic volume and composition on the air quality and pedestrian exposure in urban street canyon. *Atmos. Environ.* **2014**, *98*, 260–270. [CrossRef]

32. Peng, Z.-R.; Wang, D.; Wang, Z.; Gao, Y.; Lu, S. A study of vertical distribution patterns of PM2. 5 concentrations based on 410 ambient monitoring with unmanned aerial vehicles: A case in Hangzhou, China. *Atmos. Environ.* **2015**, *123*, 357–369. [CrossRef]

33. Kalantar, B.; Mansor, S.B.; Abdul Halin, A.; Shafri, H.Z.M.; Zand, M. Multiple moving object detection from UAV videos 394 using trajectories of matched regional adjacency graphs. IEEE Trans. *Geosci. Remote Sens.* **2017**, *55*, 395. [CrossRef]

34. Lambey, V.; Prasad, A.D. A review on air quality measurement using an unmanned aerial vehicle. *Water Air Soil Pollut.* **2021**, *232*, 1–32. [CrossRef]

35. Xiong, X.; Shah, S.; Pallis, J.M. Balloon/Drone-Based Aerial Platforms for Remote Particulate Matter Pollutant Monitoring. 380 2018. Available online: https://scholarworks.bridgeport.edu/xmlui/handle/123456789/2236 (accessed on 1 February 2021).

36. Jumaah, H.J.; Mansor, S.; Pradhan, B.; Adam, S.N. UAV-based $PM_{2.5}$ monitoring for small-scale urban areas. *Int. J. Geoinform.* **2018**, *14*, 408. [CrossRef]

37. Air Quality Parameters Data. Available online: https://powietrze.gios.gov.pl/pjp/current/station_details/table/295/3/0 (accessed on 8 May 2021).

38. Singh, P.; Verma, P. A Comparative Study of Spatial Interpolation Technique (IDW and Kriging) for Determining Groundwater Quality. *GIS Geostat. Tech. Groundw. Sci.* **2019**, 43–56. [CrossRef]

39. Lima, A.; De Vivo, B.; Cicchella, D.; Cortini, M.; Albanese, S. Multifractal IDW interpolation and fractal filtering method in environmental studies: An application on regional stream sediments of (Italy), Campania region. *Appl. Geochem.* **2003**, *18*, 1853–1865. [CrossRef]

40. Sówka, I.; Chlebowska-Styś, A.; Pachurka, Ł.; Rogula-Kozłowska, W.; Mathews, B. Analysis of Particulate Matter Concentration Variability and Origin in Selected Urban Areas in Poland. *Sustainability* **2019**, *11*, 5735. [CrossRef]

41. Cichowicz, R.A.; Dobrzański, M. Indoor and Outdoor Concentrations of Particulate Matter and Gaseous Pollutants on Different Floors of a University Building: A Case Study. *J. Ecol. Eng.* **2021**, *22*, 162–173. [CrossRef]

42. Rogula-Kozłowska, W.; Rogula-Kopiec, P.; Mathews, B.; Klejnowski, K. Effects of road trac on the ambient concentrations of three PM fractions and their main components in a large Upper Silesian city. Annals of Warsaw University of Life Sciences-SGGW. *Land Reclam.* **2013**, *45*, 243–253.

43. Gao, J.; Wang, K.; Wang, Y.; Liu, S.; Zhu, C.; Hao, J.; Liu, H.; Hua, S.; Tian, H. Temporal-spatial characteristics and source apportionment of $PM_{2.5}$ as well as its associated chemical species in the Beijing-Tianjin-Hebei region of China. *Environ. Pollut.* **2018**, *233*, 714–724. [CrossRef] [PubMed]

44. Cichowicz, R.; Dobrzański, M. Spatial Analysis (Measurements at Heights of 10 m and 20 m above Ground Level) of the Concentrations of Particulate Matter (PM_{10}, $PM_{2.5}$, and $PM_{1.0}$) and Gaseous Pollutants (H_2S) on the University Campus: A Case Study. *Atmosphere* **2021**, *12*, 62. [CrossRef]
45. Kourtidis, K.; Kelesis, A.; Petrakakis, M. Hydrogen sulfide (H_2S) in urban ambient air. *Atmos. Environ.* **2008**, *42*, 7476–7482. [CrossRef]
46. Cichowicz, R.; Wielgosiński, G. Analysis of Variations in Air Pollution Fields in Selected Cities in Poland and Germany. *Ecol. Chem. Eng.* **2018**, *25*, 217–227. [CrossRef]

Article

Removal of Zn(II) and Mn(II) by Ion Flotation from Aqueous Solutions Derived from Zn-C and Zn-Mn(II) Batteries Leaching

Agnieszka Sobianowska-Turek [1,*], Katarzyna Grudniewska [2], Paweł Maciejewski [3] and Małgorzata Gawlik-Kobylińska [4]

[1] Faculty of Environmental Engineering, Wrocław University of Science and Technology, Wybrzeze Wyspianskiego 27, 50-370 Wroclaw, Poland
[2] Faculty of Chemistry, Wroclaw University of Science and Technology, Wybrzeze Wyspianskigo 27, 50-370 Wroclaw, Poland; katarzyna.grudniewska@pwr.edu.pl
[3] Faculty of Civil Engineering, Czech Technical University in Prague, Jugoslavskych partyzanu 1580/3, 160 00 Prague 6-Dejvice, Czech Republic; ppmaciej@yahoo.co.uk
[4] Command and Management Faculty, War Studies University, gen. Chrusciela 103 Avenue, 00-910 Warszawa, Poland; ggawlik2000@yahoo.com
* Correspondence: agnieszka.sobianowska-turek@pwr.edu.pl

Abstract: The Zn(II) and Mn(II) removal by an ion flotation process from model and real dilute aqueous solutions derived from waste batteries was studied in this work. The research aimed to determine optimal conditions for the removal of Zn(II) and Mn(II) from aqueous solutions after acidic leaching of Zn-C and Zn-Mn waste batteries. The ion flotation process was carried out at ambient temperature and atmospheric pressure. Two organic compounds used as collectors were applied, i.e., m-dodecylphosphoric acid 32 and m-tetradecylphosphoric 33 acid in the presence of a non-ionic foaming agent (Triton X-100, 29). It was found that both compounds can be used as collectors in the ion flotation for Zn(II) and Mn(II) removal process. Process parameters for Zn(II) and Mn(II) flotation have been established for collective or selective removal metals, e.g., good selectivity coefficients equal to 29.2 for Zn(II) over Mn(II) was achieved for a 10 min process using collector 32 in the presence of foaming agent 29 at pH = 9.0.

Keywords: ion flotation; used batteries; ecological safety; recovery; Zn(II); Mn(II)

Citation: Sobianowska-Turek, A.; Grudniewska, K.; Maciejewski, P.; Gawlik-Kobylińska, M. Removal of Zn(II) and Mn(II) by Ion Flotation from Aqueous Solutions Derived from Zn-C and Zn-Mn(II) Batteries Leaching. *Energies* **2021**, *14*, 1335. http://doi.org/10.3390/en14051335

Academic Editor: Robert Król

Received: 3 February 2021
Accepted: 22 February 2021
Published: 1 March 2021

Publisher's Note: MDPI stays neutral with regard to jurisdictional claims in published maps and institutional affiliations.

1. Introduction

Ecological aspects and an increasing depletion of polymetallic ores generate the need for research on new nonferrous materials in unused and hard-to-reach places as well as for the improvement in techniques used to recover metals from secondary sources. There is a growing interest in the recovery of metals from electrical and electronic solid wastes [1–6], batteries and accumulators [7–10], as well as car catalysts [11]. The main advantages of waste treatment are the recovery of valuable metallic materials and the prevention of the release of toxic metals into the natural environment. There are effective industrial methods for the treatment of waste batteries and accumulators; nevertheless, due to the changes in quantity and quality of waste stream composition, new innovative and ecological technologies are involved [12]. They are characterized by high and selective recovery of metals at low cost [13–15].

Hydrometallurgical methods can be effectively used for the recovery of metals from secondary sources [16]. Due to economic and environmental aspects, flotation methods stand as an alternative for recovering metals from secondary sources [17,18]. The ion flotation process is effective for diluted aqueous solutions and can be selective using specific collectors [19,20].

Sobianowska-Turek et al. [21] used flotation methods, i.e., ion flotation and solvent sublation for Zn(II) and Mn(II) removal from a dilute model and real aqueous solutions from a Zn-C and Zn-Mn black battery mass after mechanical treatment and acidic leaching. The ionizable lariat ethers were used as collectors (Table 1, compounds 1–7). Macrocyclic organic compounds possessing three different crown sizes: DB22C7 1, 2, 3, DB19C6 4, 6, 7, and DB16C5 5 were applied. As the foaming agent, Triton X-100 was used (Table 2, 29). Ion flotation from aqueous solution at pH equal to 5.0 with collectors 2 and 3 allowed the removal of Zn(II) and Mn(II) at 98%, i.e., higher than in the solvent sublation process (Zn(II) —81%, Mn(II)—85%). Ulewicz et al. [22–24] studied several proton-ionizable lariat ethers (Table 1, 5–21) for Zn(II) and Cd(II) removal from aqueous solutions in ion flotation. In [24], a correlation between lariat ethers' crown size and the selectivity of the Zn(II) over Cd(II) flotation process was found. It was also shown that there is a significant influence of specific parameters on the efficiency and selectivity of the ion flotation process. These parameters were the type of lipophilic and acidic groups, collector concentration, and pH of the aqueous solution. Ion flotation with lariat ethers 8–12 in the presence of a foaming agent 29 allowed for removal of Cd(II) and Zn(II) from dilute aqueous solutions with high efficiency. Collector 12 gave high flotation selectivity for Cd^{2+} over Zn^{2+} ($SCd/Zn \cong 7$) at pH 4.0, (Cd(II)) = (Zn(II)) = 1.0×10^{-5} M, (12) = 1.0×10^{-4} M, (29) = 2.0×10^{-4} M. Ulewicz et al. [22] applied lariat ethers 13–21 for Zn(II) and Cd(II) removal from aqueous solution. The results proved the influence of the crown ether cavity size, collector concentration, and pH of aqueous solution on the efficiency of Zn(II) and Cd(II) removal. Maciejewski et al. [25] found the correlation between the maximal percent removal of metals and the following parameters: initial concentrations of lariat ether and the non-ionic foaming agent, the pH of the aqueous solution as well as hydrophilic–lipophilic balance (HLB). In addition, Ulewicz et al. [23] studied other collectors, i.e., 22–28, for Zn(II) and Cd(II) flotation from aqueous solutions. A correlation between structural variability of ethers concentration of alkali metal cations, collector concentration, pH of aqueous phase, and the efficiency of metals removal was confirmed. They also studied the HLB of ethers impact on selectivity of Zn(II) and Cd(II) in ion flotation. Kozlowski et al. [26] compared the effectiveness of ion flotation and transport across polymeric inclusion membranes for Zn(II) and Cd(II) recovery from aqueous solutions. The experiments were carried out for aqueous chloride solutions containing an equimolar mixture of both cations, with the use of two classical collectors/carriers: i.e., anionic—30 and cationic—31 (Table 2). Finally, they found that the use of a cationic surfactant 31, causes the selectivity coefficient of Cd/Zn to increase with the increasing concentration of chloride ions (from dilute aqueous solutions both metals at concentrations of 1×10^{-5} M).

Table 1. Collectors used in Zn(II) and Cd(II) ion flotation process in the quoted works [22–26].

Structural Formula	Number of the Chemical Compound	-R	-X	-Y-
	1	$-C_4H_9$		$-(CH_2)_2O(CH_2)_2O(CH_2)_2O(CH_2)_2-$
	2	$-C_7H_{15}$		
	3	$-C_{10}H_{21}$	$-O(CH_2)_3SO_3Na$	
	4			$-(CH_2)_2O(CH_2)_2O(CH_2)_2-$
	5	$-C_7H_{15}$		$-(CH_2)_2O(CH_2)_2-$
	6		$-CH_2CONHSO_2CF_3$	
	7		$-OCH_2COOH$	$-(CH_2)_2O(CH_2)_2O(CH_2)_2-$
	8	$-H$	$-O(CH_2)_3SO_3Na$	-
	9	$-C_3H_7$	$-O(CH_2)_3SO_3Na$	-
	10	$-C_{10}H_{21}$	$-O(CH_2)_3SO_3Na$	-
	11	$-C_{10}H_{21}$	$-OCH_2CO_2H$	-
	12	$-C_4H_9$	$-OCH_2PO(OH)(OC_2H_5)$	-

Table 1. *Cont.*

Structural Formula	Number of the Chemical Compound	-R	-X	-Y-
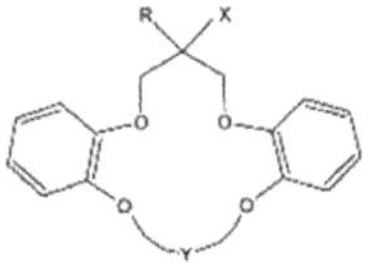	13	$-C_{10}H_{21}$	$-OCH_2COOH$	$-(CH_2)_2-$
	14	$-C_{10}H_{21}$	$-O(CH_2)_3SO_3Na$	$-(CH_2)_2O(CH_2)_2-$
	15	$-C_3H_7$	$-O(CH_2)_3SO_3Na$	$-(CH_2)_2O(CH_2)_2-$
	16	-H	$-O(CH_2)_3SO_3Na$	$-(CH_2)_2O(CH_2)_2-$
	17	$-C_4H_9$	$-OCH_2COOH$	$-(CH_2)_2O(CH_2)_2-$
	18	-H	$-OCH_2PO(OH)(OC_2H_5)$	$-(CH_2)_2O(CH_2)_2-$
	19	$-CH_3$	$-OCH_2COOH$	$-(CH_2)_2O(CH_2)_2O(CH_2)_2-$
	20	$-C_2H_5$	$-OCH_2COOH$	$-(CH_2)_2O(CH_2)_2O(CH_2)_2-$
	21	$-C_3H_7$	$-OCH_2COOH$	$-(CH_2)_2O(CH_2)_2O(CH_2)_2-$
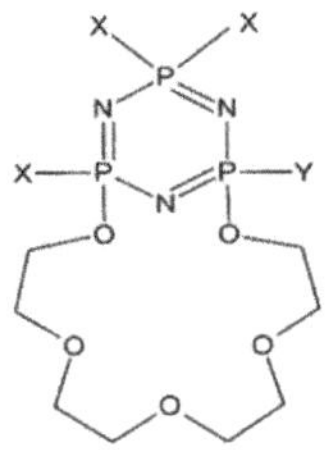	22	-	X: 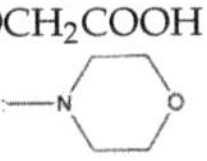 —N(morpholine)	$-NH(CH_2)_3OH$
	23	-	X: 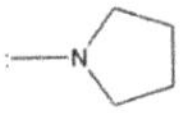 —N(pyrrolidine)	$-NH(CH_2)_3OH$
	24	-	X: 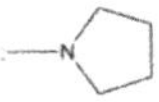 —N(pyrrolidine)	Y: —NH—CH$_2$—(pyridine) 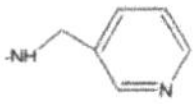
	25	-	$-O(CH_2CH_2O)_3CH_3$	
	26	-	X = Y: 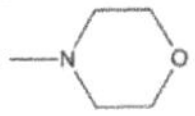 —N(morpholine)	
	27	-	X = Y: 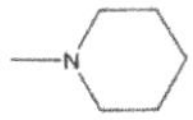 —N(piperidine)	
	28	-	X = Y: 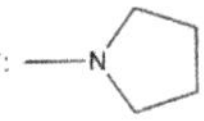 —N(pyrrolidine)	

Table 2. Additional compounds used in Zn(II) and Cd(II) ions flotation in the quoted works [26–28].

Structural Formula	Chemical Compound Name	Number/Function of Chemical Compound
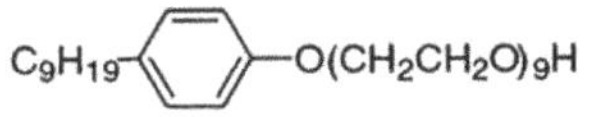 C_9H_{19}—⟨benzene⟩—$O(CH_2CH_2O)_9H$	1,1,3,3-tetramethylbutyl phenyl polyethylene glycol ether	29 foaming agent
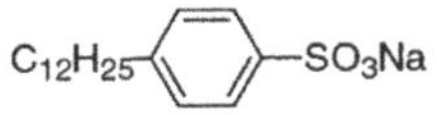 $C_{12}H_{25}$—⟨benzene⟩—SO_3Na	sodium dodecylbenzenesulfonate	30 anionic collector
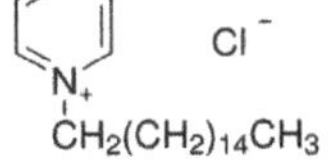 ⟨pyridinium⟩ Cl^-, N^+—$CH_2(CH_2)_{14}CH_3$	cetylpyridinium chloride	31 cationic collector

Selective removal of Zn(II) and Ag(I) in ion flotation with a classical collector from diluted aqueous solutions was done by Charewicz et al. [27]. Anionic collectors used in the process were 30 and ammonium tetradecyl sulfate, while as a cationic collector, they used 31. The Zn(II) can be selectively removed from acidic dilute aqueous solutions of both metals because Zn^{2+} has a much greater affinity for anionic collectors than Ag^+. They indicated that collective removal of both metal ions is possible in the presence of thiosulphate or thiocyanate ligands. For the first ligand, silver(I) ions are a mixture of $[Ag(S_2O_3)]^-$ and $[Ag(S_2O_3)_2]^{3-}$ form, while the Zn(II) remain in the aqueous phase as Zn^{2+}. However, a partial removal of Zn(II) and Ag(I) ions was obtained in the presence of cyanide ligands, where Zn(II) ions were in the cyanide anionic form, and the removal was 92–98%, which is higher than for Ag(I) (73–82%). Jurkiewicz [29] studied the influence of collectors, such as chloride, bromide, iodide, and cetyltrimethylamine thiocyanate, for the removal of anionic complexes of Zn(II) forms from aqueous solutions in the ion flotation process. It was shown that the removal of Zn(II) ions depends on the nature of the ligand. He also indicated that the amount of zinc–thiocyanate complexes decreased with the type of acid in the following order: $HClO_4 < H_3PO_4 < HCl < H_2SO_4 <$ no acid. Walkowiak [28] conducted research on the Zn(II) and Mn(II) removal from an equimolar mixture of cations: $Zn^{2+}, Mn^{2+}, Co^{2+}, Fe^{3+}, Cr^{3+}$. The selective removal of Zn(II), Mn(II) in a flotation process with anionic collectors (sodium dodecylsulfonate or 30) in the presence of other metals in aqueous solutions increased in the order: $Mn^{2+} < Zn^{2+} < Co^{2+} < Fe^{3+} < Cr^{3+}$. Trivalent cations have a better affinity for the anionic collector since they have higher ionic potential values.

The aim of the experimental work in this article was the Zn(II) and Mn(II) flotation process from model and real solutions after acid leaching of black battery mass originating from spent Zn-C and Zn-Mn batteries. It allowed the determination of the factors influencing the efficiency and selectivity of the process.

2. Materials and Methods

2.1. Apparatus and Measuring Equipment

The main component of the apparatus for ion flotation was a glass flotation column, size 45.7 cm (height), 2.4 cm (diameter), equipped with a sintered glass sparger possessing a hole diameter of 20–30 μm. Above the porous sinter, there was an injection-capable valve in the column for introducing a collector and/or a foaming agent. The argon gas surface was introduced into the flotation column by a gas flow meter and aqueous scrubber. The foam generated in the process was collected in a tank. The process was carried out at ambient temperature (20–21 °C), at atmospheric pressure, and constant argon gas flow (12 cm^3/min). The Zn(II) and Mn(II) concentrations in feed and residual solutions (model and real) were measured by atomic absorption spectrometry on a VARIAN®, Palo Alto, California, USA SpectrAA 20 Plus spectrometer.

2.2. Characteristics of Reagents and Tested Solutions

The initial concentrations of metal cations in base model aqueous solution (both metals mixture) were Zn(II)—0.00168 M and Mn(II)—0.00091 M. The known weight ratio of Zn(II) to Mn(II) content, equal to 2.14:1.00, was a reference to the number of ions in the solution derived from the acidic non-reducing leaching of the black battery mass [21] which was carried as follows. This solution was prepared by dissolving chemically pure compounds $ZnSO_4 \cdot 7H_2O$ and $MnSO_4 \cdot 5H_2O$ of respective sample weights in demineralized water (conductivity 5.0 μS at 20 °C). The material used to prepare the real aqueous solution was a mixture of paramagnetic and diamagnetic fractions mixed in a 2:1 weight ratio of Zn-C and Zn-Mn batteries generated after mechanical treatment and leaching with sulfuric acid [30]. Leaching was carried out using 500 g portions of batteries mixture. The tested material was flooded with distilled water (1.0 dm^3) and leached with 95% sulfuric acid, added in the amount estimated to be necessary for leaching zinc and manganese, and the amount necessary to balance the H^+ ions of the appropriate reduction reaction with a small excess. The leaching process was carried without heating with an average temperature of 55 °C. The measured concentration of Zn(II) and Mn(II) in the aqueous solution after leaching was 55.0 and 25.7 g/dm^3, respectively [21]. The initial concentration of the cation mixture in the real aqueous solution was measured by ASA.

Two organic reagents, m-dodecylphosphoric acid 32 and m-tetradecylphosphoric acid 33 were used in the ion flotation process in ethanol solutions (Table 3) at initial concentrations equal to 5.0×10^{-3} M in the feed solution. A non-ionic foaming agent was used, Triton X-100 29 (1,1,3,3-tetramethylbutyl phenyl polyethylene glycol ether), in ethanol solution at an initial concentration equal to 2.0×10^{-4} M in an aqueous solution.

The sulfuric acid and ammonia solutions (at concentrations: 0.1, 0.5, and 1.0 M) were used to adjust the pH of the feed solutions. The collectors 32 and 33 were obtained at the Wroclaw University of Technology, Faculty of Chemistry University. The other reagents used were purchased from Avantor Performance Materials S.A., Gliwice, Poland (former POCH), all were of analytical purity.

Table 3. Characteristics of organic acids used for research as collectors for the process of ionic flotation.

Structural Formula	Total Formula	Chemical Compound Name	Molar Mass, [u]	Number/Function of Chemical Compound
	$C_{12}H_{27}O_4P$	m-dodecylphosphoric acid	266.31	**32** collector
	$C_{14}H_{31}O_4P$	m-tetradecylphosphoric acid	294.36	**33** collector

2.3. Research Methodology

Flotation of Zn(II) and Mn(II) from a dilute aqueous model and real solutions was carried out in the apparatus presented in Figure 1. The initial volume of each aqueous solution was 100 cm^3, and initial concentrations of Zn(II) and Mn(II) (in the mixture) were 0.00168 and 0.00091 M, respectively. In the real solutions experiments, the feed solutions were obtained after acidic leaching of Zn-C and Zn-Mn batteries, and the concentrations of Zn(II) and Mn(II) in the solution were 55.0 and 25.7 g/dm^3, respectively. The pH feed solutions were adjusted with ammonia or sulfuric acid to pH values from 2.0 to 11.0. One cubic centimeter volume of collector and foaming agent was injected into the floated aqueous solution. The initial concentrations of the collector and foaming agent in the aqueous solutions were 5.0×10^{-3} M and 2.0×10^{-4} M, respectively. The ion flotation experiments were carried out for 5, 10, and 20 min, after which the samples of the residual solutions were collected to be analyzed. The argon gas was saturated with water, and the flow rate was maintained at 12 cm^3/min through a sintered glass sparger. During the process, a foaming formation of the solutions was observed, and the pH of the residual solution was measured. The concentrations of Zn(II) and Mn(II) in the residual flotation solution were measured by atomic adsorption spectrometry to determine the maximal percent removal, which is described by Equation (1):

$$W = \left(1 - \frac{cr}{ci}\right) * 100\% \tag{1}$$

where W is the maximal percent removal, c_i is the initial ion concentration in mol/dm^3, c_r is the ion concentration in the residual solution in mol/dm^3.

Selective removal of Zn(II) and Mn(II) is describe by the selectivity Coefficient (2):

$$S_{M1/M2} = \frac{W_{M1}}{W_{M2}} \tag{2}$$

where $S_{M1/M2}$ is the selectivity coefficient for metal ions of the first and the second metal, W_{M1} and W_{M2} are the maximal percent removal for the first and the second metal.

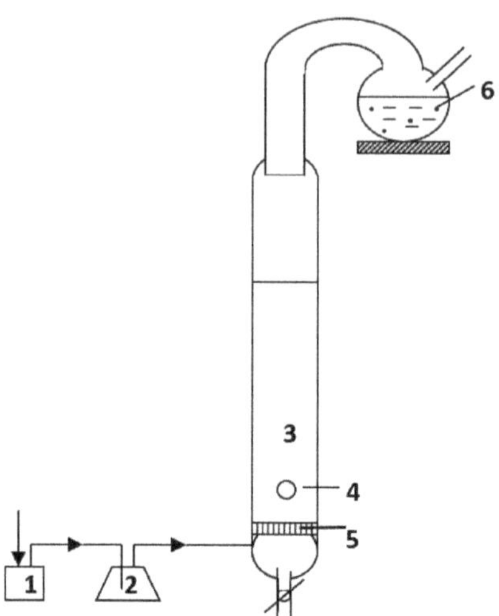

Figure 1. Scheme of ion flotation apparatus. 1—gas flow regulator, 2—aqueous scrubber, 3—flotation column, 4—injection capable valve (for collector), 5—sintered glass sparger, 6—foam receiver.

3. Results and Discussion

3.1. Ion Flotations from Model Aqueous Solutions

- Parameters influencing Zn(II) and Mn(II) flotation removal with collector 32

In the preliminary experiments, the influence of the time of the Zn(II) and Mn(II) flotation process was tested for collector 32 at pH = 5.0. The ion flotation tests were carried out for 5, 10, and 20 min, based on which it was found that the process runs within 10 min. In subsequent studies, flotations were carried out for 10 min.

The preliminary test with collector 32 in the ion flotation process indicated that this collector had a sufficient foaming capacity and can be used in the process without the addition of a foaming agent. Studied collector 32 was an ionogenic surfactant and consisted of two groups: hydrophilic—polar phosphate acid residue -OPO(OH)$_2$ and lipophilic—a non-polar aliphatic one. The phosphate group of collectors can form ion-collector connections due to the presence of labile hydrogens.

- Effect of pH solutions

In the first step of experiments, the influence of aqueous solution pH on Zn(II) and Mn(II) removal was tested for a pH equal to 2.0, 4.0, 5.0, 7.0, 9.0, 10.0, and 11.0. Ion flotation was carried out with collector 32 (5.0×10^{-3} M in aqueous solutions) in a 10 min process at initial concentrations (in the mixture): Zn(II)—0.00168 M and Mn(II)—0.00091 M. There was no foaming agent 29 used. The results of the Zn(II) and Mn(II) flotations are summarized in Table 4.

Table 4. Results of Zn(II) and Mn(II) flotation using collector 32 in a 10 min process with and without foaming agent 29 at a pH from 2.0 to 11.0; (Zn) = 0.00168 M, (Mn) 0.00091 M, (32) = 5.0×10^{-3} M, (29) = 2.0×10^{-4} M.

pH	Without Foaming Agent 29			With Foaming Agent 29		
	$W_{Zn(II)}$,%	$W_{Mn(II)}$,%	$S_{Zn/Mn}$	$W_{Zn(II)}$,%	$W_{Mn(II)}$,%	$S_{Zn/Mn}$
2.0	80.4	75.4	1.07	40.1	53.1	0.76
4.0	82.1	55.9	1.47	99.6	19.4	5.13
5.0	92.9	43.6	2.13	77.4	16.1	4.81
7.0	99.7	73.8	1.35	91.7	8.4	10.9
9.0	50.7	68.8	0.74	96.3	3.3	29.2
10.0	37.1	54.7	0.68	77.2	60.4	1.28
11.0	60.1	53.9	1.12	86.8	54.4	1.60

The Zn(II) and Mn(II) removal depends on the pH value, which can give a metals' separation effect. The removal of Zn(II) increased from 80.4% to 99.7%, with an increase in pH from 2.0 to 7.0. A further increase in pH to 11 resulted in a rapid decrease in Zn(II) removal, which may be related to a change in the ionic form of Zn(II). At pH 8.0 to 10.0, these ions pass in aqueous solutions from a simple Zn^{2+} to $Zn(OH)^+$ and $Zn(OH)_2$, while in the pH ranges from 10.0 to 13.0, it is in the form of $Zn(OH)_3^-$ and $Zn(OH)_4^{2-}$ anions which possess a significant influence on the process [31]. In strongly alkaline aqueous solutions the Zn(II) removal increased from 37.1% (pH = 10.0) to 60.1% (pH = 11.0).

The flotation removal of Mn(II) in the whole range of tested pH was above 50%. The highest removal (75.4%) for Mn(II) was achieved at pH = 2.0 after the 10 min process. The results of Zn(II) and Mn(II) ion flotation in the pH range studied show differences, which can be used for selective removal of both metal ions, but the selectivity of the process was low. At pH = 5.0, the value of the selectivity coefficient was 2.13. The best collective removal of Zn(II) (99.7%) and Mn(II) (73.8%) after 10 min was obtained at pH 7.0.

- Effect of foaming agent presence

In addition, Zn(II) and Mn(II) flotation was also conducted with collector 32 in the presence of Triton X-100 29, at a concentration equal to 2.0×10^{-4} M. The function of the foaming agent was to develop a liquid–gas surface and to form a stable foam over the solution, which is crucial in the ion flotation process. The removal of Zn(II) and Mn(II) was performed with 32 (initial concentration of 5.0×10^{-5} M in feed solution) in the presence of foaming agent 29 (2.0×10^{-4} M) in the pH range of 2.0 to 11.0 for 10 min. The results are presented in Table 4. Despite a significant improvement in the interfacial surface in the solution and a stable foam over the solution, the addition of foaming agent caused a decrease in Zn(II) removal at pH = 2.0, 5.0, and 7.0. Comparing the results for Zn(II) removal with and without the addition of foaming agent, it can be indicated that the addition of 29 lowers Zn(II) removal by half (from 80.4% to 40.1%) at pH = 2.0. At pH = 5.0, the removal of Zn(II) decreased by 17.5%, while for a neutral solution, the reduction in Zn(II) removal rate was 8.0%. In an acidic environment, the addition of 29 increased Zn(II) removal to 99.6% only at pH 4.0. On the other hand, in alkaline solutions, the foaming agent addition increased the Zn(II) removal, at pH = 9.0, from 50.7% to 96.3%, at pH = 10.0, from 37.1% to 77.2%, and at pH = 11.0, from 60.1% to 86.8%. It was probably related to the observed higher foaming intensity during the flotation process.

For Mn(II) flotation, the addition of 29 significantly decreased metal removal with pH = 9.0. The results were very low in comparison to process without foaming agent: by 22.3% for pH = 2.0, by 36.5% for pH = 4.0, by 27.5% for pH = 7.0, and by 65.5% for pH = 9.0. This situation may be caused by the collector's flotation and thus lowering of the concentration of its ionized forms in the solution, which also reduces the number of ion-collector connections.

Therefore, it is possible to obtain relatively good selectivity for Zn(II) over Mn(II) after 10 min of the flotation process. At pH = 7.0, the selectivity coefficients were equal to 10.9 and at pH = 9.0–29.2 (see Table 4). Finally, it can be concluded that the addition of a foaming agent to the ion flotation processes decreases collective removal of metals but increases selective removal of Zn(II) over Mn(II).

- Parameters influencing Zn(II) and Mn(II) flotation removal with collector 33

In the preliminary experiments, the influence of the time of the Zn(II) and Mn(II) flotation process was tested for collector 33 at pH = 5.0. The ion flotation tests were carried out for 5, 10, and 20 min, based on which it was found that the process runs within 10 min. In subsequent studies, flotations were carried out for 10 min.

- Effect of pH solution

The preliminary test with collector 33 in the ion flotation process indicated that the collector had a sufficient foaming capacity and could be used without the addition of a foaming agent. Studied collector 33 (similar to 32) was an ionogenic surfactant and consisted of two groups: hydrophilic—polar phosphate acid residue -OPO(OH)$_2$ and lipophilic—a non-polar aliphatic one. In comparison to 32, collector 33 has a longer lipophilic group with two carbons in the alkyl chain, which caused an increase in lipophilic properties of the compound. The phosphate group of the collector is capable of forming ion-collector connections due to the presence of labile hydrogens.

The schema of experiments was similar to in the previous chapter (for collector 32). In the first step, the influence of the aqueous pH solution on Zn(II), Mn(II) removal was tested at pH 2.0, 4.0, 5.0, 7.0, 9.0, 10.0, and 11.0. Ion flotation was carried out with collector 33 (5.0×10^{-3} M in feed solution) in 10 min from the mixture of both metals at initial concentrations of Zn(II)—0.00168 M, Mn(II)—0.00091 M. There was no addition of 29. The results of the Zn(II) and Mn(II) flotation process are summarized in Table 5.

Table 5. Results of Zn(II) and Mn(II) flotation using collector 33 in a 10 min process with and without foaming agent 29 at pH from 2.0 to 11.0; (Zn) = 0.00168 M, (Mn) 0.00091 M, (33) = 5.0×10^{-3} M, (29) = 2.0×10^{-4} M.

pH	Without Foaming Agent 29			With Foaming Agent 29		
	$W_{Zn(II)}$,%	$W_{Mn(II)}$,%	$S_{Zn/Mn}$	$W_{Zn(II)}$,%	$W_{Mn(II)}$,%	$S_{Zn/Mn}$
2.0	83.3	59.9	1.39	38.3	57.8	0.66
4.0	80.4	87.5	0.92	81.7	85.2	0.96
5.0	90.4	20.2	4.48	82.4	94.5	0.87
7.0	67.1	87.5	0.77	98.6	67.6	1.46
9.0	58.5	76.6	0.76	61.1	74.2	0.82
10.0	99.1	73.2	1.35	59.5	72.7	0.82
11.0	91.6	59.4	1.54	89.5	31.0	2.89

An influence of pH solutions on Zn(II) and Mn(II) removal with 33 had a variable character for each metal. The Zn(II) removal in the 10 min process slightly increased with pH value from 83.3% to 90.4% at pH = 5, then it systematically fell to 58.5% at pH = 9.0. A further increase in the pH value boosted Zn(II) removal to 99.1% at pH = 10.0 then slightly decreased to 91.6% at pH = 11.0. What is important, at pH = 7.0 and 9.0, the removal decreased slightly, which according to Eh-pH diagrams, may be related to the change in simple Zn^{2+} to $Zn(OH)_2$, while, in strongly alkaline solutions, pH = 10.0 and 11.0, the removal increased, which may be related to the ionic form of the metal $Zn(OH)_3^{-}$.

The Mn(II) removal with collector 33 in the acidic solutions for the 10 min process were discontinuous, because the noticed results were 59.9% at pH = 2.0, 87.5% at pH = 4.0, and 20.2% at pH = 5.0. In the neutral solutions, Mn(II) removal rebound to 87.5%, similar to pH = 4.0, while the result was the same. Then the Mn(II) removal decreased linearly with the pH value to 59.5 at pH = 11. A quick analysis of the results indicates significant differences in the Zn(II) and Mn(II) flotation vs. pH solution value and process time. This situation allows the description of the optimal conditions for collective or selective removal studied cations. For example, collective removal is effective at pH = 4.0 in a 10 min process, which gave an 80.4% and 87.5% removal for Zn(II) and Mn(II), respectively. On the other hand, selective Zn(II) removal from a mixture of both metals is possible at a pH = 5.0 in a 10 min process, with a selectivity coefficient $S_{Zn/Mn}$ = 4.48.

- The influence of a foaming agent

The results of the Zn(II) and Mn(II) flotation process with collector 33 confirm sufficient foaming capacity, but in some experiments, the foam was very low. Therefore, in the experiments, the addition of foaming agent 29 at a concentration of 2.0×10^{-4} M was used in the feed solution in the 10 min process. According to results presented in Table 5, it can be concluded that Zn(II) and Mn(II) removals in the presence of 29 were often worse than without 29, despite strong solutions foaming in all range of tested pH solutions. The noticed relations were different for each metal. For example, the results for Zn(II) removal in the presence 29 were better (by 31.5%) only at pH = 7.0, at other pH values, the results were similar (pH = 4.0; 5.0; 9.0; 11.0) or worse (pH = 2.0; 10.0, by −45.0% and by −39.6%, respectively).

The results for Mn(II) removal in the presence of 29 were a few percent worse or similar at pH = 2.0, 4.0, 9.0, 10.0. In two cases, completely different results were obtained. The removal of Mn(II) at pH 5.0 in the presence of 29 was better by 74.3% and at pH 11.0 was worse by 28.4%. During Mn(II) flotations in the presence of 29 in an alkaline reaction, a lack of permanent foam formation was observed, which probably contributed to low results. These differences in the Zn(II) and Mn(II) removal (from a mixture of both metals) allowed their separation. For example, Zn(II) selective removal was possible at pH = 5.0 with 33 in a 10 min process (without 29), with a selectivity coefficient $S_{Zn/Mn}$ = 4.78. For Mn(II), selective removal with 33 in the presence of 29 was possible at pH = 2.0 in a 10 min process, with $S_{Mn/Zn}$ = 1.51.

3.2. Real Solution

The real solutions obtained after acid leaching of Zn-C and Zn-Mn batteries had a content of Zn(II) and Mn(II) in the solution of 55.0 and 25.7 g/dm^3, respectively. The series of experiments with model solutions allowed the determination of the conditions for the selective Zn(II) removal from real solutions (mixture of both metals). The ion flotation with the real solution had been limited to the use of the collector 33 only and the following process conditions: pH 4.0, 7.0, and 9.0 in a 10 min process duration. The results are shown in Table 6. The selected conditions of the process enabled the selective separation of Zn(II), while Mn(II) remained in solution. Selective Zn(II) removal in a 10 min process from real solution was the best at pH = 9.0, and the selectivity coefficient for Zn(II)/Mn(II) was equal to 22.56. At pH = 4.0 and 7.0, the selectivity coefficients $S_{Zn/Mn}$ were 6.28 and 8.90, respectively. Analysis of the collectors' 32 and 33 properties at this stage is difficult and requires additional research in the field of physical surface physics, in particular the measurement of surface tension, hydrophilic–lipophilic balance. The preliminary results are very promising and are the starting point for further research on hydrometallurgical selective metals recovery from used cells of first Zn-C and Zn-Mn types, which are still the largest stream of used batteries on the European market [32].

Table 6. Influence of pH on the degree of Zn(II) and Mn(II) cations separation with the use of collector 33 after 10 min process from the real solution without foaming agent 29 at pH from 4.0 to 9.0; (Zn) = 0.00168 M, (Mn) 0.00091 M, (33) = 5.0×10^{-3} M.

pH	$W_{Zn(II)}$, %	$W_{Mn(II)}$, %	$S_{Zn/Mn}$
4.0	94.2	15.0	6.28
7.0	89.0	10.0	8.90
9.0	97.0	4.3	22.56

4. Conclusions

The preliminary studies on the Zn(II) and Mn(II) flotation with 32 and 33 compounds proved that the substances can be considered as new collectors for selective and/or collective Zn(II) and Mn(II) removal. The research involved model and real solutions (after acid leaching of black battery mass originating from used Zn-C and Zn-Mn batteries). Through a series of experiments, it was possible to distinguish factors determining the efficiency and selectivity of the floatation process. Based on the findings, the following conclusions can be made.

- The main parameters which affect Zn(II) and Mn(II) flotation are the type of collector, feed solution pH, concentration of collector and/or foaming agent, and process duration.
- The collective Zn(II) and Mn(II) removal from dilute aqueous solutions in the ion flotation process with using collector 32 or 33 depended on the pH solution. High efficiency Zn(II) removal with collector 32 is possible in the pH range from 2.0 to 7.0 in a 10 min ion flotation process. It has been shown that the maximal percent removal of Zn(II) increased from 80.4% (at pH = 2.0) to 99.7% (at pH = 7.0). In the same experiments, results for Mn(II) removal were as follow: 75.4% and 73.8% for pH 2.0 and 7.0, respectively. On the other hand, effective removal of Zn(II) using second collector, 33, required alkaline solutions, and the observed removal was in the range of 99.1% (pH = 10.0) and 91.6% (pH = 11.0). In the same conditions, results for Mn(II) were 73.2% (pH = 10.0) and 59.4% (pH = 11.0).
- The addition of foaming agent 29 affected Zn(II) removal in particular. Despite a significant improvement in the interfacial surface in the solution and a stable foam over the solution, the addition of 29 caused a decreased removal of both metals. Comparing the results for Zn(II) removal (using collector 32), it can be noticed that addition of 29 lowers Zn(II) removal by half (from 80.4% to 40.1%) at pH = 2.0. On the other hand, in an alkaline solution (using collector 32), the presence of a foaming agent increased the Zn(II) removal, at pH = 9.0, from 50.7% to 96.3%, at pH = 10.0, from 37.1% to 77.2%, and at pH = 11.0, from 60.1% to 86.8%. It was probably related to the observed higher foaming intensity during the flotation process. However, in the case of the second metal, it did not significantly affect the Mn(II) removal. Generally, in some cases, the addition of an agent is beneficial, especially in an alkaline feed solution.
- The time of conducting the ion flotation process also had a significant influence on both metals removal. Comparing the results for Zn(II) and Mn(II) removal in 5 and 20 min processes, it can be concluded that the prolongation of the flotation time affects the results negatively, especially in the case of Zn(II), e.g., removal of Zn(II) with collector 32 in a 5 min process at pH = 5.0 was always over 80%, while in the 20 min process, it did not exceed 20%. Too long a process duration (above 10 min) negatively affects the removal of both metals.
- It was found that the ion flotation process with collectors 32 or 33 can be an alternative to currently known processes in the recovery of metals from waste chemical energy sources. It is also possible to use ion flotation in a hybrid process, as one of the stages of the currently used hydrometallurgical recycling processes. There is also potential to regenerate a waste stream of a solution that would be purified and recycled back into the process.

Author Contributions: Conceptualization, K.G., A.S.-T. and P.M.; methodology, K.G.; formal analysis, K.G.; investigation, K.G.; resources, A.S.-T., P.M., M.G.-K.; writing; writing—review and editing, M.G.-K. and P.M.; supervision, M.G.-K. All authors have read and agreed to the published version of the manuscript.

Funding: This research received no external funding.

Informed Consent Statement: Informed consent was obtained from all subjects involved in the study.

Data Availability Statement: The data presented in this study are available in the paper.

Conflicts of Interest: The authors declare no conflict of interest.

References

1. Bigum, M.; Damgaard, A.; Scheutz, C.; Christensen, T.H. Environmental impacts and resource losses of incinerating misplaced household special wastes (WEEE, batteries, ink cartridges and cables). *Resour. Conserv. Recycl.* **2017**, *122*, 251–260. [CrossRef]
2. Kohl, C.A.; Gomes, L.P. Physical and chemical characterization and recycling potential of desktop computer waste, without screen. *J. Clean Prod.* **2018**, *184*, 1041–1051. [CrossRef]
3. Polat, O.; Capraz, O.; Gungor, A. Modelling of WEEE recycling operation planning under uncertainty. *J. Clean Prod.* **2018**, *180*, 769–779. [CrossRef]
4. Zabłocka-Malicka, M.; Szczepaniak, W.; Rutkowski, P.; Ochromowicz, K.; Leśniewicz, A.; Chęcmanowski, J. Decomposition of the ISA-card under steam for valorized polymetallic raw material. *J. Anal. Appl. Pyrolysis* **2018**, *130*, 256–268. [CrossRef]
5. Awasthi, A.K.; Li, J. An overview of the potential of eco-friendly hybrid strategy for metal recycling from WEEE. *Resour. Conserv. Recycl.* **2017**, *126*, 228–239. [CrossRef]
6. Li, J.; Lopez, B.N.; Liu, L.; Zhao, N.; Yu, K.; Zheng, L. Regional or global WEEE recycling. Where to go? *Waste Manag.* **2013**, *33*, 923–934. [CrossRef]
7. Dutta, T.; Kim, K.-H.; Deep, A.; Szulejko, J.E.; Vellingiri, K.; Kumar, S.; Kwon, E.E.; Yun, S.-T. Recovery of nanomaterials from battery and electronic wastes: A new paradigm of environmental waste management. *Renew. Sustain. Energy Rev.* **2018**, *82*, 3694–3704. [CrossRef]
8. Urbańska, W. Recovery of Co, Li and Ni from spent Li-ion batteries by inorganic and/or orgnic reducer assistend leaching metod. *Minerals* **2020**, *10*, 555.
9. Petranikova, M.; Ebin, B.; Mikhailova, S.; Steenari, B.-M.; Ekberg, C. Investigation of the effects of thermal treatment on the leachability of Zn and Mn from discarded alkaline and Zn C batteries. *J. Clean Prod.* **2018**, *170*, 1195–1205. [CrossRef]
10. Innocenzi, V.; Ippolito, N.M.; De Michelis, I.; Prisciandaro, M.; Medici, F.; Vegliò, F. A review of the processes and lab-scale techniques for the treatment of spent rechargeable NiMH batteries. *J. Power Sources* **2017**, *362*, 202–218. [CrossRef]
11. Gil, S.; Bialik, W.; Saternus, M.; Fornalczyk, A. Thermal Balance of the magneto-hydro-dynamic pump for recovery of platinum group metals from spent auto catalysts. *Arch. Metall. Mater.* **2016**, *61*, 253–256.
12. Sobianowska-Turek, A. Hydrometallurgical recovery of metals: Ce, La, Co, Fe, Mn, Ni and Zn from the stream of used Ni-MH cells. *Waste Manag.* **2018**, *77*, 213–219. [CrossRef]
13. Dominguez-Benetton, X.; Varia, J.C.; Pozo, G.; Modin, O.; Ter Heijne, A.; Fransaer, J.; Rabaey, K. Metal recovery by microbial electro-metallurgy. *Prog. Mater. Sci.* **2018**, *94*, 435–461. [CrossRef]
14. Mylarappa, M.; Venkata Lakshmi, V.; Vishnu Mahesh, K.R.; Nagaswarupa, H.P.; Prashantha, S.C.; Shravana Kumara, K.N.; Siddeswara, D.M.K.; Raghavendra, N. Resource Recovery and Material Characterization of Metals from Waste Li-ion Batteries by an Eco-Friendly Leaching Agent. *Mater. Today Proc.* **2017**, *4*, 12215–12222. [CrossRef]
15. Torkaman, R.; Asadollahzadeh, M.; Torab-Mostaedi, M.; Ghanadi Maragheh, M. Recovery of cobalt from spent lithium ion batteries by using acidic and basic extractants in solvent extraction process. *Sep. Purif. Technol.* **2017**, *186*, 318–325. [CrossRef]
16. Jha, M.K.; Kumari, A.; Panda, R.; Rajesh Kumar, J.; Yoo, K.; Lee, J.Y. Review on hydrometallurgical recovery of rare earth metals. *Hydrometallurgy* **2016**, *165*, 2–26. [CrossRef]
17. Bi, P.Y.; Dong, H.R.; Dong, J. The recent progress of solvent sublation. *J. Chromatogr. A* **2010**, *1217*, 2716–2725. [CrossRef] [PubMed]
18. Sobianowska, K.; Walkowiak, W.; Kozłowski, C. Principles and applications of solvent sublation—A review. *Ars. Separatoria. Acta* **2009**, *7*, 23–38.
19. Grudniewska, K.L. *Ion Flotation Cs$^+$, Sr^{2+}, Ba^{2+} and Co^{2+} Ion Flotation from Dilute Aqueous Solutions with Macrocyclic Compounds*; Wrocław University of Technology: Wrocław, Poland, 2012.
20. Lemlich, R. *Adsorptive Bubble Separation Techniques*; Elsevier Science & Technology: Saint Louis, MO, USA, 2012.
21. Sobianowska-Turek, A.; Ulewicz, M.; Sobianowska, K. Ion flotation and solvent sublation of zinc (II) and manganese (II) in the presence of proton-ionizable lariat ethers. *Physicochem. Probl. Miner. Process.* **2016**, *52*, 1048–1060. [CrossRef]
22. Ulewicz, M.; Walkowiak, W.; Bartsch, R.A. Ion flotation of zinc (II) and cadmium (II) with proton-ionizable lariat ethers—Effect of cavity size. *Sep. Purif. Technol.* **2006**, *48*, 264–269. [CrossRef]
23. Ulewicz, M.; Walkowiak, W.; Brandt, K.; Porwolik-Czomperlik, I. Ion Flotation of Zinc (II) and Cadmium (II) in the Presence of Side-Armed Diphosphaza-16-Crown-6 Ethers. *Sep. Sci. Technol.* **2003**, *38*, 633–645. [CrossRef]
24. Ulewicz, M.; Walkowiak, W.; Jang, Y.; Kim, J.S.; Bartsch, R.A. Ion flotation of cadmium (II) and zinc (II) in the presence of proton-ionizable lariat ethers. *Anal. Chem.* **2003**, *75*, 2276–2279. [CrossRef] [PubMed]
25. Maciejewski, P.; Ulewicz, M.; Robak, W.; Walkowiak, W. Lariat ethers with a novel proton-ionisable groups: New generation of collectors in ion flotation process. *Int. J. Environ. Waste Manag.* **2011**, *8*. [CrossRef]
26. Kozłowski, C.A.; Ulewicz, M.; Walkowiak, W. Removal of zinc and cadmium ions from aqueous chloride solutions by ion flotation and liquid membranes. *Physicochem. Probl. Miner. Process.* **2000**, *34*, 141–151.
27. Charewicz, W.; Hoławiecka, B.; Walkowiak, W. *Selective Flotation of Zinc(II) and Silver(I) Ions from Dilute Aqueous Solutions*; Wrocław University of Technology: Wrocław, Poland, 1998.
28. Walkowiak, W. Mechanism of Selective Ion Flotation. 1. Selective Flotation of Transition Metal Cations. *Sep. Sci. Technol.* **1991**, *26*, 559–568. [CrossRef]
29. Jurkiewicz, K. The removal of zinc from solutions by foam separation, I. Foam separation of complex zinc anions. *Int. J. Miner. Process.* **1990**, *28*, 173–187. [CrossRef]

30. Sobianowska-Turek, A.; Szczepaniak, W.; Żurek, A.; Satora, A.; Starowicz, A. The Method of Utilizing the Waste Mass of Zn-C and Zn-Mn Batteries Left after the Removal of the Ferromagnetic Fraction. PL Patent 220853 B1, 25 August 2011.
31. Polat, H.; Erdogan, D. Heavy metal removal from waste waters by ion flotation. *J Hazard. Mater.* **2007**, *148*, 267–273. [CrossRef] [PubMed]
32. Commision, E. Batteries & Accumulators. Available online: https://ec.europa.eu/environment/waste/batteries/index.htm (accessed on 3 February 2021).

Article

Investigation of the Physico-Chemical Properties of the Products Obtained after Mixed Organic-Inorganic Leaching of Spent Li-Ion Batteries

Weronika Urbańska [1],*[iD] and Magdalena Osial [2][iD]

1 Department of Environmental Engineering, Wrocław University of Science and Technology, Wybrzeże Wyspiańskiego 27, 50-370 Wrocław, Poland
2 Faculty of Chemistry, University of Warsaw, Pasteura 1 Street, 02-093 Warsaw, Poland; mosial@chem.uw.edu.pl
* Correspondence: weronika.urbanska@pwr.edu.pl

Received: 9 November 2020; Accepted: 17 December 2020; Published: 20 December 2020

Abstract: Lithium-ion batteries are currently one of the most important mobile energy storage units for portable electronics such as laptops, tablets, smartphones, etc. Their widespread application leads to the generation of large amounts of waste, so their recycling plays an important role in environmental policy. In this work, the process of leaching with sulfuric acid for the recovery of metals from spent Li-ion batteries in the presence of glutaric acid and hydrogen peroxide as reducing agents is presented. Experimental results indicate that glutaric-acid application improves the leaching performance compared to the use of just hydrogen peroxide under the same conditions. Obtained samples of leaching residues after mixed inorganic-organic leaching were characterized with Scanning Electron Microscopy, Fourier Transform Infrared Spectroscopy, and X-ray diffraction.

Keywords: acid leaching; battery recycling; Li-ion batteries; metal recovery; raw material sustainable use

---

1. Introduction

Lithium-ion batteries and accumulators are used in a wide variety of devices—mobile phones, laptops, tablets, but also children's toys and medical equipment. In general, batteries made our life easier and revolutionized the world—thanks to the possibility of using cells as an electricity source, many devices have become mobile, making them more attractive. Complex battery packs are also used in the automotive industry—as a power source for modern hybrid or electric cars. Despite their many advantages, the widespread use of batteries has some drawbacks, mainly due to the quite complex composition of the cells. Lithium-ion batteries are made up of a graphite anode and a cathode, made out of a $Li_xMe_yO_n$ type compound—the most commonly used is lithium cobaltate, $LiCoO_2$. Lithium-ion cells also contain an organic electrolyte, which is usually lithium salts dissolved in a mixture of organic solvents (e.g., salts—$LiClO_4$, $LiPF_6$, $LiBF_4$; solvents—dimethylsulfoxide, propylene carbonate, diethyl carbonate). In addition, the electrodes are separated from each other by a synthetic separator—usually a polyolefin membrane, which enables the transfer of lithium ions while at the same time protecting against a short circuit that may occur as a result of direct contact of the anode with the cathode [1–5]. Each component of the Li-ion battery generates large amounts of waste, while the most hazardous ones are compounds containing heavy metals such as cobalt.

Several studies confirm that one of the main components, the lithium cobaltate, has a negative influence on the environment as well as on living organisms. It induces oxidative DNA damage causing mutagenic effect and inflammatory response, mainly caused by the cobalt release [6]. Many data show the role of Co(II) ions released from batteries, which causes the formation of OH-radical species by a

Fenton-like reaction, leading to chromosomal breaks and the formation of cancer cells [7,8]. Due to that effect, battery waste containing cobalt becomes dangerous for aquatic systems and humans, generating a huge need for recycling of the spent batteries. Compared to the production of new batteries, the processing of the waste batteries is a much more complicated and costly process. Nevertheless, due to the constant increase in the amount of this waste, technologies for recycling spent batteries to recover metals contained in them have been developed, and they are used in industrial practice around the world. The following processes are distinguished: mechanical separation, thermal treatment (pyrometallurgy) and acid leaching (hydrometallurgy). Due to the technological, environmental and economic benefits, the acid leaching of spent Li-ion batteries for cobalt and lithium recovery is widely studied in the literature. The commonly used leaching agents are inorganic acids, especially sulfuric, hydrochloric, phosphoric and nitric acids [9–12], and organic acids, e.g., citric, succinic, acetic and malic acids [13–16]. Cathode powder leaching processes are usually performed with the use of reducing agents, the presence of which may cause the metals contained in the battery mass to take a bivalent form (mainly Co(III) to Co(II)), soluble in acid solutions [17]. Most often, hydrogen peroxide is used for this purpose [18–22]. However, in recent years, there has been an increasing interest among scientists in the search for organic reducing agents, e.g., ascorbic acid [23–25]. This trend is triggered by environmental concerns, and it offers more effective metal recovery from spent Li-ion batteries.

During the acid leaching of the powder obtained from spent Li-ion batteries, many chemical reactions take place between the tested material and the added reagents (leaching agents and reducing agents), resulting in the formation of various products. In the literature, there are examples of reactions occurring during the leaching of battery powder ($LiCoO_2$ cathode), where both the leaching agent and the reducing agent are inorganic compounds (e.g., H_2SO_4 and H_2O_2) [26]. The reactions taking place between the powder material and various organic acids are more complicated. Nayaka G. P. et al. investigated the leaching process with iminodiacetic and maleic acids with the addition of ascorbic acid [24]. The authors paid attention to the formation of products in the form of lithium and cobalt complexes (the initially colorless solution turns pink over time, which indicates the transition of Co^{3+} to complex forms; lithium complexes are not colored), as well as the reducing effect of ascorbic acid on cobalt ions—a reduction from Co^{3+} to Co^{2+} (in the case of using only iminodiacetic acid in the leaching solution, two forms of cobalt were present: Co^{3+} and Co^{2+}, while for maleic acid alone, only unreduced cobalt was found—Co^{3+}). Nayaka G. P. et al. similarly presented the changes taking place during the processes, the results of which have been presented in their publications on similar topics (leaching experiments: citric acid with ascorbic acid, tartaric acid with ascorbic acid, adipic acid/nitryldiacetic acid) [25,27,28]. However, no specific record of the course of these reactions has been proposed due to the need to study the structures and chemical composition of powders after leaching each time in order to determine the compounds present in the material and their forms.

This article is focused on the determination of the chemical composition and form of compounds present in battery powders after leaching in the presence of 1.5 M H_2SO_4 as a leaching agent with or without reducing agents: 30% solution of H_2O_2 and/or $C_5H_8O_4$ (glutaric acid). The rates of metal recovery obtained as a result of acidic reductive leaching and the effects of quantitative and qualitative analysis performed with the use of the following methods: ICP-OES, SEM-EDS, FT-IR and XRD are presented and discussed.

2. Materials and Methods

In this study, various types of Li-ion batteries were used. They were obtained from numerous sources of commercial origins. They were dedicated for laptops from different manufacturers, such as Lenovo, Toshiba or Hewlett-Packard. Initially, upstream material was dismantled with the manual procedure described widely in literature, including removal of both steel and plastic cases that cover the batteries [29–31]. Anode and cathode were crushed carefully and sieved to separate particular elements such as plastic film, scrap paper, outer metallic body and membrane. Obtained battery powder was collected and used for further research. The obtained mass was subjected to XRD and SEM analyses as well as a mineralization method. To mineralize, 0.5 g of the tested waste battery powder was spread wet in an open system using the DigiPREP Jr. mineralization system (parameters: 10.00 mL 65% HNO_3, digestion time: 5 h, temperature: 120 °C). The obtained solution was evaporated to a volume of approximately 0.5 mL, transferred quantitatively to a plastic container and supplemented with deionized water to 50 g. At the same time, a blank sample was prepared and included in the final results. The analyses were performed in parallel in three repetitions. Then the metal concentrations in the solution were determined using ICP-OES (Agilent 720).

The separated powder from spent Li-ion batteries was subjected to the leaching. Acid leaching was performed in 1.5 M sulfuric acid (96% analytical grade, STANLAB), as the reducing agents glutaric acid $C_5H_8O_4$ (5% *w/v*) and hydrogen peroxide H_2O_2 (30% analytical grade, STANLAB; 0.9% *v/v*) were used. An experiment without the addition of reducing agents was also performed in order to compare its results with samples 1, 2 and 3, and to determine the effect of reducers on the obtained recovery degrees of the tested metals. The ratio of the solid phase to the liquid phase was 1/10 in each case. Other assumed leaching parameters were: temperature—90°C, time—2 h and mixing speed—500 rpm (Table 1). The methodology of these experiments has also been demonstrated in previous studies where the results of leaching research were compared with different process parameters [32].

Table 1. Summary of the best operational conditions for leaching the battery powder.

Sample	Volume of 1.5 M H_2SO_4	Concentration of Reducing Agent	Time (h)	Slurry Density (*w/v*)	Temp. (°C)	Rotation Rate (rpm)
Sample 1	100 mL	0.9% H_2O_2 (*v/v*)	2	10/100	90	500
Sample 2	100 mL	5% $C_5H_8O_4$ (*w/v*)	2	10/100	90	500
Sample 3	100 mL	0.9% H_2O_2 (*v/v*) + 5% $C_5H_8O_4$ (*w/v*)	2	10/100	90	500
Sample 4	100 mL	-	2	10/100	90	500

After leaching was complete, the samples were vacuum-filtered. The content of metals Al, Ca, Co, Cr, Cu, Fe, Li, Mn, Na, Ni, Si and Zn in the leachates was determined (ICP-OES, Agilent 720) and their percent recovery degrees were determined in relation to the initial concentrations in the tested material. The samples of battery powder after leaching with reducing agents (no. 1, 2 and 3) were dried at 105 °C for 24 h in order to characterize the metal content. Then, the dried material was analyzed with microscopic and spectroscopic techniques. To determine the morphology of powder the FE-SEM Merlin (Zeiss) equipped with a Gemini II column was used. The device worked in low kV value range (0.5–1.5 kV) and low probe current 10–20 µA. EDS was used for elemental analysis of the measured samples. Complementary to the EDS analysis, for determination of the chemical composition of samples the FT-IR analysis was performed with Nicolet 8700 in range 4000-400 cm^{-1}. The powder X-ray diffraction patterns were recorded with a powder diffraction X-ray diffractometer (PXRD) X'PERT Phillips with PW 1830 generator with CuKα radiation with line λ = 1.5405980 Å and a scan rate of 0.05° per minute in 0.016° steps covering the 2θ angle range from 20° to 130°. Measurements were performed at room temperature using DHN software.

3. Results and Discussion

3.1. Mechanical Treatment

As a result of the mechanical processing of individual cells, the following fractions were obtained: ferromagnetic (metals, e.g., steel housing), diamagnetic (plastics, paper) and paramagnetic (anode and cathode—battery powder) constituting the material used for further research. The components of the entire spent cell and a single lithium-ion battery separated as a result of mechanical processing are presented in Figure 1.

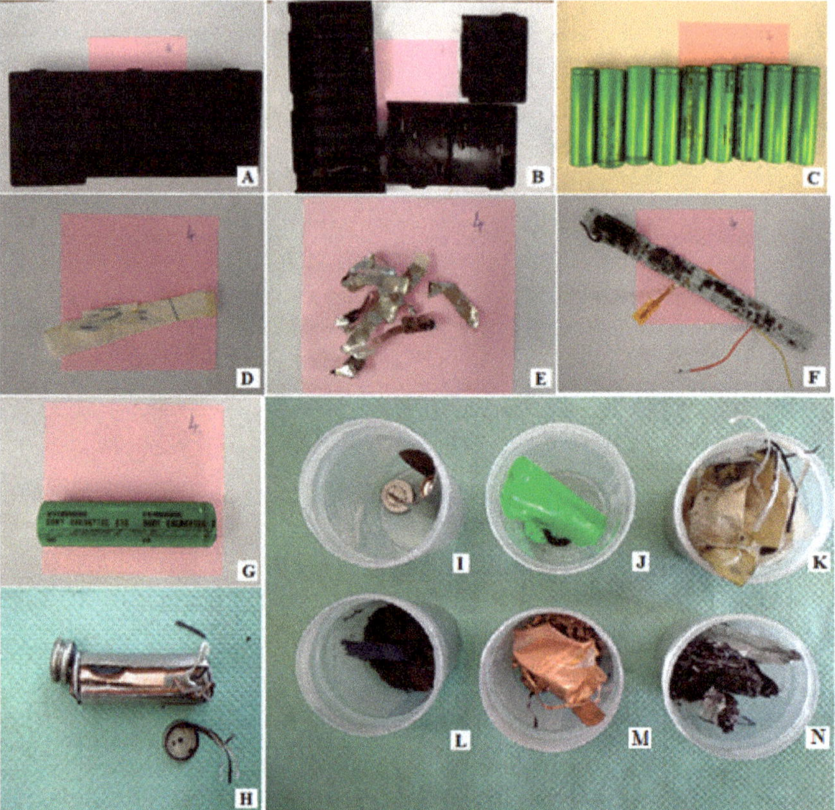

Figure 1. Mechanical treatment of a single spent lithium-ion cell. Stage I: (**A**) spent lithium-ion battery; (**B**) plastic case; (**C**) lithium-ion cells; (**D**) paper; (**E**) metals; (**F**) PCB and wires. Stage II: (**G**) single spent lithium-ion cell; (**H**) cell after cutting; (**I**) metal case; (**J**) cover and other plastics; (**K**) separator; (**L**) battery powder; (**M**) copper foil; (**N**) aluminum foil.

From the presented results, it can be concluded that the entire lithium-ion battery forms a package of individual (in this case nine) cells. Both the complete battery and a single cell are composed of many different components, such as plastics, paper or metals. The most valuable raw material is the battery mass (anode and cathode powders), rich in metals that can be successfully recovered. Battery powder constitutes more than 36% of the total mass of a single spent Li-ion cell, and it was used for further laboratory experiments. The obtained results are very similar to those in the literature [33].

3.2. Quantitative and Qualitative Analysis of Battery Powder before Leaching

In order to investigate the qualitative composition of the tested material, the battery powder was subjected to XRD analysis, which showed that the main components in the material are carbon and the cathode compound—$LiCoO_2$. Then the powder material was mineralized, as was the content of metal ions: Al, Ca, Co, Cr, Cu, Fe, Li, Mn, Na, Ni, Si and Zn in the obtained solution (ICP-OES, Agilent 720). It was specified that the tested battery powder contains all the marked metals, mostly cobalt, lithium and nickel (Table 2).

Table 2. Metal content in the sample of powder from the spent Li-ion batteries [32].

Metal	Al	Ca	Co	Cr	Cu	Fe	Li	Mn	Ni	Si	Zn
C (g/kg)	0.93	0.40	256.0	0.005	3.59	0.27	33.20	0.560	14.40	3.76	0.07

3.3. Acidic Reductive Leaching

As a result of the acidic reductive leaching of the tested battery powder, multiple solutions with a similar raspberry color, but different saturation, were obtained. This could already indicate the degrees of leaching of individual metals, especially cobalt responsible for the pink color of the samples. The contents of the metals in the obtained solutions were determined and presented in the form of percent recovery degrees (Table 3).

Table 3. Degrees of metal recovery from solutions after acidic reductive leaching.

Metal		Al	Ca	Co	Cr	Cu	Fe	Li	Mn	Ni	Si	Zn
	Sample 1	71.61	48.73	1.95	67.49	83.53	87.35	84.19	56.42	80.25	71.68	87.58
Recovery, %	Sample 2	10.64	24.58	32.30	66.25	82.12	99.45	75.86	31.22	38.77	43.30	47.66
	Sample 3	53.66	35.98	59.37	84.60	97.36	100.00	79.81	62.39	81.75	70.25	93.93
	Sample 4	49.76	34.09	30.94	47.67	79.96	93.44	72.32	31.10	37.61	58.52	50.55

Based on the results, it can be concluded that all tested metals were leached to some extent in each of the samples (see Table 3). In the case of aluminum, the highest degree of recovery was obtained for sample 1 (reducer 30% solution of H_2O_2), i.e., 71.61%; slightly lower—53.66% for sample 3, where two reducing agents were dosed; and 49.76% for sample 4 (an experiment without reducing agents). The lowest Al recovery was noted for sample 2—10.64%. Therefore, it can be concluded that the addition of glutaric acid does not significantly improve the Al leaching effects and when used with hydrogen peroxide, it may even inhibit the action of the inorganic reducing agent. Similar relationships were also obtained for calcium (1—48.73%, 2—24.58%, 3—35.98%, 4—34.09%); and silicon (1—71.68%, 2—43.30%, 3—70.25%, 4—58.52%). Relatively high levels of lithium recovery were achieved in each of the samples. The greatest amount of lithium was leached in sample 1, where the hydrogen peroxide as a reducing agent (84.19%) was used. A slightly lower degree of recovery (79.81%) was obtained in sample 3 (reducing agents—hydrogen peroxide and glutaric acid). The lowest results were recorded in samples 2 and 4 (respectively: 75.86% and 71.21%). For the other metals, i.e., Co, Cr, Cu, Fe, Mn, Ni and Zn, the highest levels of their recovery were determined for sample 3 ($H_2O_2 + C_5H_8O_4$), which confirms the assumed synergistic action of two chemical compounds used as reducers, in the tested reaction environment. Taking into account all the received results of the analyses, it can be seen that the addition of a reducing agent to the leaching process allows for obtaining higher recovery rates for most of the metals contained in the leached waste battery powder. However, the type of reagent used as a reducing agent is important. It is also worth noting that in sample 1 a very low leaching degree of cobalt was recorded (1.95%), while the use of glutaric acid as a reducing agent in sample 2 increased the recovery degree of this metal to 32.30%. Nevertheless, literature data confirm the positive influence of the presence of hydrogen peroxide used as a reducing agent on the obtained degrees of recovery of cobalt ions (reduction of Co (III) to the more easily leached form—Co (II)) [26], which could not be reproduced in the presented laboratory studies. This may be due to differences in the initial preparation of the research material and the assumed parameters of the leaching process, especially the adopted too-low dose of hydrogen peroxide. Such an assumption was made in previously conducted research [32]. The effect of an H_2O_2 dose on the sulfuric acid leaching process was investigated, where glutaric acid was also used as a reducing agent. By increasing the dose of H_2O_2 to 15 mL, it was possible to achieve almost 90% cobalt recovery, with about 96% lithium and 91% nickel recovery.

Thus, the hypothesis, that increasing the volume of H_2O_2 contributes to higher Co recovery degrees, was confirmed [32].

As mentioned before, the battery powder contained the most cobalt, lithium and nickel in its composition. The recovery of these metals from spent Li-ion batteries is most commonly studied in the research. Comparing the obtained results for Co and Ni, it can be clearly stated that the highest levels of their recovery (59.37% and 81.75%, respectively) were obtained in sample 3 leached in the presence of two reducing agents. For lithium, the highest result was obtained for sample 1, however, it is not much higher than in sample 3. Therefore, it can be concluded that taking into account the leaching of all three metals, the process parameters for sample 3 are optimal and give the best results—high efficiency of leaching of metals such as Co, Li and Ni. Detailed considerations on this subject are included in the publication [32]. Moreover, the simultaneous use of hydrogen peroxide and glutaric acid as reducing agents also allowed obtaining the highest recovery degrees of most of the tested materials: Cr (84.60%), Cu (97.36%), Fe (100%), Mn (62.39%) and Zn (93.93%). Next, the leaching residues were investigated to determine the morphology of samples, their chemical composition and crystallinity.

3.4. Morphological Studies

The morphology of leaching residues was investigated with scanning electron microscopy (SEM). Initially, SEM analysis was performed for waste battery powder before leaching. As can be seen in Figure 2, the sample has uniform morphology that is mainly based on a granulelike structure. The sample has homogeneous morphology within the whole surface.

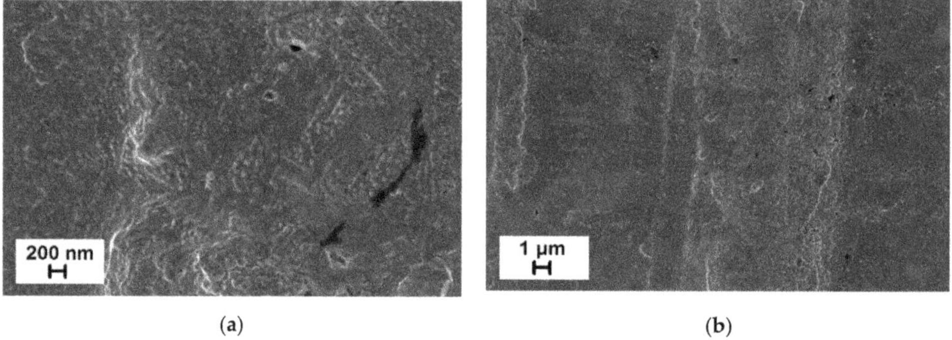

(a)　　　　　　　　　　　　　　　　(b)

Figure 2. SEM images of waste battery powder before leaching (sample 0), where (**a**) and (**b**) are presented in different scale.

Figure 3 shows images obtained for sample 1 consisting of irregular structures having a different shape and size. The morphology of the sample is similar to the whole bulk structure. As can be seen in Figure 3a, small particles having a grainlike structure about 150–200 nm are visible, however some of them tend to agglomerate, forming wirelike objects. Following images show irregular structures that seem to be crystals. They are present in the whole sample. Figure 3c reveals large objects with a round shape that tends to peel onto their surface. Based on the EDS analysis presented at Figure 3d as the map of elements the smallest granulelike objects consist mainly of cobalt or its compounds. The Following crystal-like structures consist of fluorine, while the oxygen accompanies almost all objects in the sample except the largest ones that are carbon coming from the graphite in the spent battery powder.

Figure 3. SEM images of sample 1, where (**a**–**c**) show different scales, while (**d**) is an EDS map of elements.

Following images recorded for sample 2 are presented in Figure 4. As can be seen in Figure 4a,b, this sample has a completely different shape than sample 1, and the main objects seem to be crystal-like. That sample looks more compact, and grainlike structures are not visible. Figure 4c reveals large objects that are present in the whole morphology. Their size ranges from about 1 μm up to larger than 10 μm. EDS analysis indicates that the sample is heterogeneous. Additionally, in comparison to the previous sample, it also contains manganese or its compounds that were not leaching within the procedure used for sample 2.

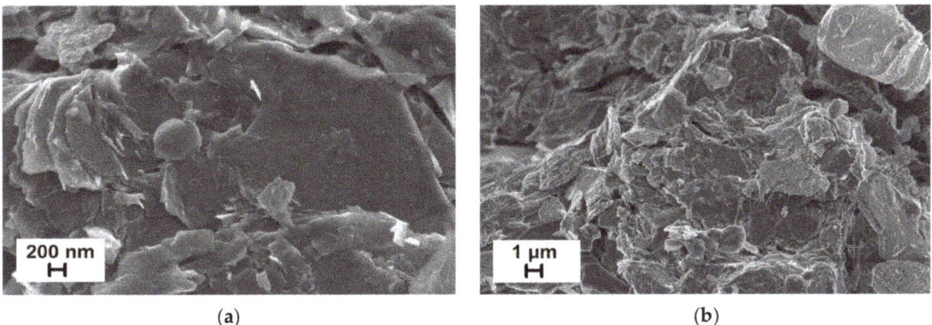

Figure 4. *Cont.*

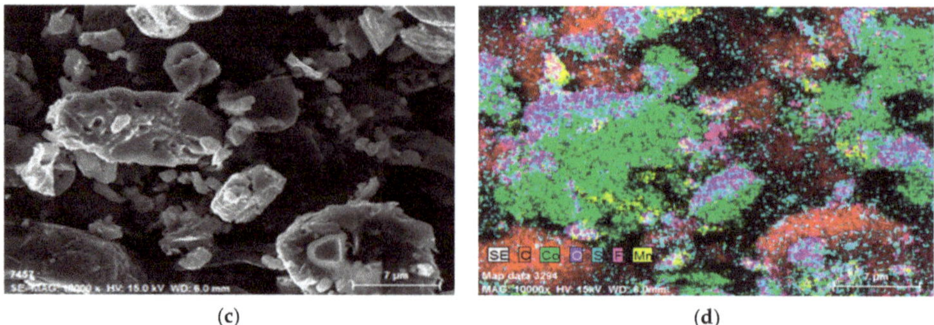

Figure 4. SEM images of sample 2, where (**a**–**c**) show different scales, while (**d**) is an EDS map of elements.

The SEM images for sample 3 are presented in Figure 5. The morphology is different from samples 1 and 2. As can be seen in Figure 5a–c, the sample seems to be the most uniform, consisting of rough and compact structures. The objects present in the sample have a size from a few μm to more than 10 μm. The EDS map presented in Figure 5d reveals the elemental composition of the sample, showing large areas of carbon, looking like the substrate and objects that contain cobalt, oxygen, fluoride and even some nitrogen or their compounds. As visible, the surface of carbon is covered uniformly with sulfur that comes from the sulfuric acid during the leaching process.

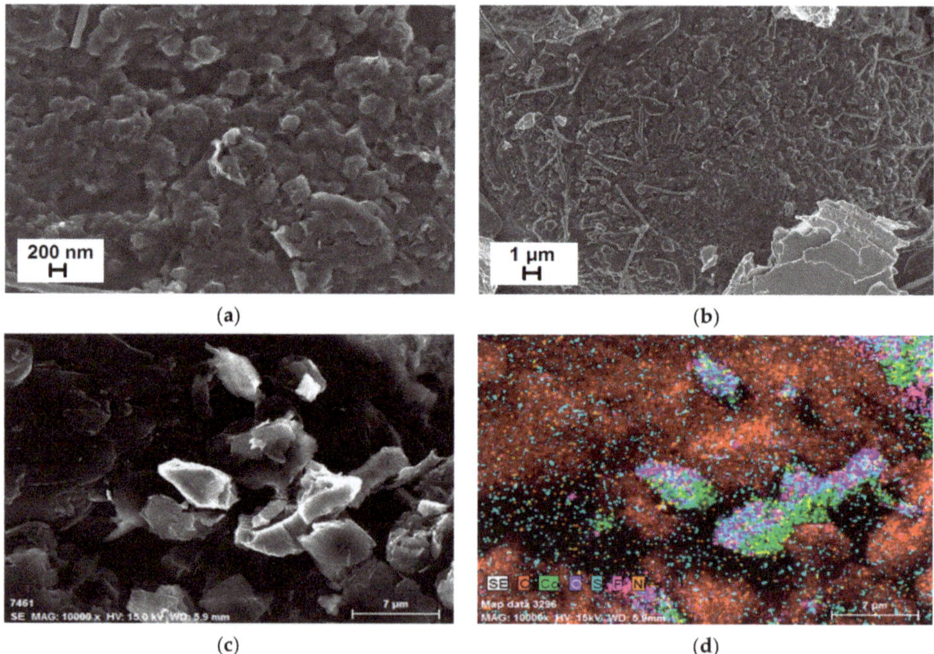

Figure 5. SEM images of sample 3, where (**a**–**c**) show different scales, while (**d**) is an EDS map of elements.

Meshram P. et al. [34] investigated leaching residues consisting of $LiCoO_2$, $Li_2CoMn_3O_8$ and $(Li_{0.85}Ni_{0.05})(NiO_2)$ that were leached with H_2SO_4, and revealed that the morphology of the final product is very complex. Similarly to our results, the particles of $LiCoO_2$ are much smaller after leaching. Additionally, the final cobalt content was smaller, confirming successive recovery of that element. Lv W. et al. [35] additionally used ammonium chloride as a reducer to treat an initial sample of $Li(Ni_{0.5}Co_{0.2}Mn_{0.3})O_2$, showing that after leaching, the morphology is amorphous, and there is a lower content of Co, Ni and Mn, and a higher content of carbon. This corresponds with our experimental results. Chen X. et al. [36] treated waste battery powder consisting mainly of the $LiCoO_2$ with sulfuric acid and an organic reducer, such as glucose, saccharose and cellulose. Their research indicated that the sample of leaching residue has a more regular structure with well-defined crystallites after leaching. The efficiency of the leaching was better than without the application of organic additives. All the above-mentioned studies showed a heterogeneous product that required deep characterization with spectroscopic techniques. Therefore, to understand the chemical composition of the products obtained within this work, the FT-IR analysis was performed.

3.5. FT-IR Analysis

As shown in Figure 6, the FT-IR was employed to characterize the possible chemical composition of the obtained leaching residues. Spectra were recorded in the range of wavenumber from 4000 cm^{-1} to 400 cm^{-1} with a resolution of 5 cm^{-1}. Pure KBr pellets were used as reference material. Spectra were recorded for all three samples.

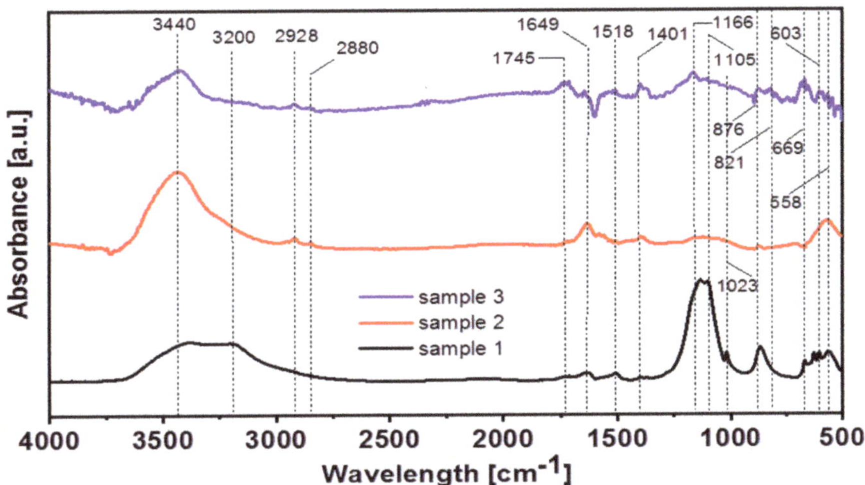

Figure 6. FT-IR spectra of battery waste powder for sample 1 that was leached with H_2SO_4 and H_2O_2, sample 2 that was leached with H_2SO_4 and $C_5H_8O_4$, and sample 3 that was leached with H_2SO_4, $C_5H_8O_4$ and H_2O_2.

The broad bands located around 3440 cm^{-1}, 3200 cm^{-1} are characteristic for vibrations of -OH group, which is absorbed onto the surface of the sample and originates from water. The band that can be ascribed to δH_2O is visible at 1649 cm^{-1} [37,38]. The two bands at 2928 cm^{-1} and 2880 cm^{-1} are characteristic for CH_2 and CH_3, respectively [39,40]. The band at 1649 cm^{-1} can be assigned to C=O stretching vibration [39], while 1745 cm^{-1} can be ascribed to stretching of -COOH [41]. Two bands at 1518 cm^{-1} and 1401 cm^{-1} may correspond to the asymmetric and symmetric vibrations of the -COO$^-$ [42,43]. The band located at 1166 cm^{-1} confirms the presence of asymmetric stretch of S=O in -SO$_3$H groups. This comes from the leaching process with sulfuric acid. The 1105 cm^{-1} band corresponds to symmetric stretch [44,45]. The band at 1023 cm^{-1} is characteristic of the C-O vibration [46,47]. Two small bands located at 876 cm^{-1} and 821 cm^{-1} can be attributed to -C-H vibration.

Following bands at 669 cm^{-1} and 558 cm^{-1} can be assigned to νCo-O vibrations that confirm the formation of Co$_2$O$_2$ nanocrystals [48–50]. The band at 603 cm^{-1} is unclear, and it can be ascribed to Co-O stretching or -OH vibration from the water molecule [50,51].

As can be seen in Figure 6, the spectrum recorded for sample 1 (black curve) that was leached with H$_2$SO$_4$ and H$_2$O$_2$ reveals strong bands at 1166 cm^{-1} and 1105 cm^{-1}. These two are the most intensive among all samples. Obtained results correspond with previous analyses that indicated the presence of sulfur or its compounds with EDS mapping and formation of Li$_2$SO$_4$ within the XRD studies. The presence of a strong band in the spectra confirms the formation of the chemical reaction of the acid with the substrate and formation of a compound with S=O groups. Following samples also have these bands, while their intensity confirms lower content of sulfur-based compounds. The bands characteristic to the carbon-based compounds have the highest intensity for sample 3, which confirms the most efficient leaching and corresponds with the above-mentioned analyses.

Based on the recorded spectra it is clear that depending on the leaching method, the chemical composition and content of species change. Procedure 2 seems to be the least efficient towards the recovery of metals, while the most efficient one is procedure 3, where both glutaric acid and hydrogen peroxide were used as reducing agents. Following measurements were performed for a deeper understanding of the chemical composition of the species formed in the samples after the leaching process.

3.6. Crystallographic Structure

The XRD patterns were recorded in the diffraction angle (2θ) range from 15° to 85°, with a scan rate of 1°/min. The diffractogram that is presented in Figure 7 shows the patterns for the sample before leaching. The observed X-ray reflecting Bragg angle positions of LiCoO$_2$ for (003) is about 19.5° [49,52]. The following pattern located at 37.3° angle can be described as (001), while the next pattern that has high intensity of about 45° is characteristic to the (104) plane.

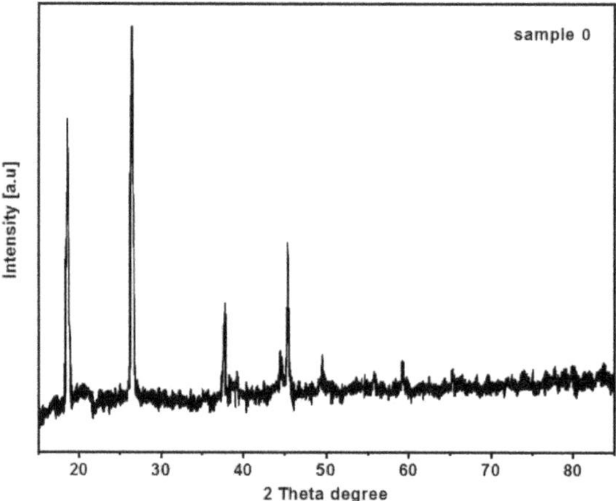

Figure 7. XRD pattern of battery waste powder before leaching (sample 0).

Figure 8 shows the reflection patterns recorded for samples after acid treatment. All samples are highly polycrystalline, and the sharp peaks confirm the formation of well-defined crystallites. The highest intensity of the reflection patterns was observed for sample 2, where the leaching agents were sulfuric and glutaric acids. The lowest intensity was observed for sample 3, where the hydrogen peroxide was used additionally. It suggests that without hydrogen peroxide the leaching process is less effective than with it.

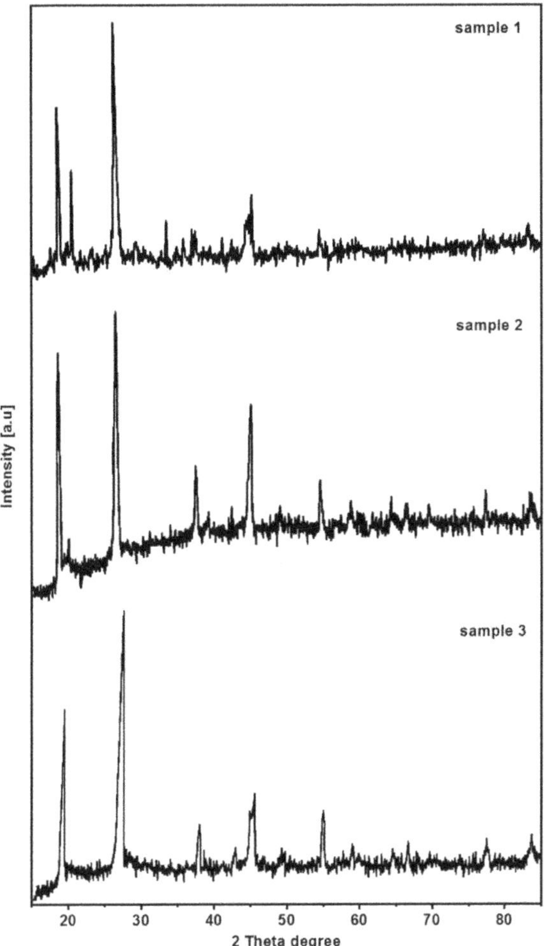

Figure 8. XRD pattern of battery waste powder for sample 1 that was leached with H_2SO_4 and H_2O_2, sample 2 that was leached with H_2SO_4 and $C_5H_8O_4$, and sample 3 that was leached with H_2SO_4, $C_5H_8O_4$ and H_2O_2.

XRD data revealed that sample 1 presented in Figure 8 has multiple patterns with predominant peaks at the angles: 18.7°, 19.5°, 27° and about 45°. The peak located at 18.7° diffraction angle can be ascribed to Co_3O_4 (111), along with lower peaks at 37.5° corresponding to the (111), 39.1° for (222), while 45° can be attributed to the (400) plane or (104) that is characteristic to $LiCoO_2$ [53,54]. Based on the literature, this peak could be also attributed to the $CoSO_4$ [55]; however, based on the EDS analysis it can be also ascribed to the LiF [56–58]. As can be seen, the pattern that spears at all samples about 19.5° is ascribed to $LiCoO_2$ for (003) [49,52], while 37.3° angle can be described as (001), 39.1° as (102) as well as the above mentioned (222) plane for Co_3O_4. The peak around the 55° angle can be ascribed to the presence of carbon in the sample. The peak at 27° can be ascribed to the carbon [59] or formation of Li_2SO_4 under the sulfuric acid treatment, which may be also ascribed to the small peak at 42.1° [60,61].

Application of glutaric acid instead of hydrogen peroxide leads to the changes in the XRD patterns. The main difference for sample 2 is presented in Figure 8. The difference is the lack of the peak at the 19.5° angle that corresponds to the (003) $LiCoO_2$. However, other peaks have a higher intensity than for sample 1. Additionally, the presence of the peak corresponding to the C (102) is more visible. The diffractogram also reveals a pattern for carbon at about a 77.5° angle. XRD patterns obtained for sample 3 reveal similar crystallinity to sample 2, while the intensity of reflection patterns is slightly lower.

Diffractograms confirm that the main components of all samples are $LiCoO_2$ and Co_3O_4. The results are in good agreement with JCPDS data no. 75-0532 that are indexed to the hexagonal structure $LiCoO_2$, JCPDS data no. 43-1003 for literature data [62–65].

All results of XRD analyses were normalized between 0 to 1 values. As can be seen, the peak visible at 26.6° corresponds to the carbon (JCPDS 98-061-7290). The presence of carbon comes from the graphite-base anode. The high content of $LiCoO_2$ as well as Co_3O_4 in all samples suggests that the leaching agents did not completely dissolve inorganic species in the waste battery powders.

4. Conclusions

In this paper, the influence of the leaching reagent on the metallic component recovery was investigated. Sulfuric acid was used as a leaching agent, while different reducing agents were used: sample 1—treatment with H_2SO_4 and H_2O_2, sample 2—treatment with H_2SO_4 and $C_5H_8O_4$, and sample 3—treatment with H_2SO_4, $C_5H_8O_4$ and H_2O_2. Application of the ICP-OES technique revealed that procedure 2 was the least effective towards metal recovery. In this procedure H_2O_2 was used as a reducing agent. The most effective procedure was observed for sample 3, where both hydrogen peroxide and glutaric acid were used.

The results of morphological studies by SEM and elemental analysis with EDS indicated that the samples were highly heterogeneous with large carbon content. A sample that was leached with both inorganic and organic reducer was the most regular, and the leached residue mainly had large objects that were carbon. The smaller crystallites came from an inorganic phase containing metals. Moreover, the results of FT-IR and XRD analysis revealed that the main components of the waste Li-ion battery powders and leaching residues were $LiCoO_2$, Co_3O_4 and carbon originating from the graphite in the battery. The content of $LiCoO_2$ and Co_3O_4 was the largest in the sample that was leached with use of glutaric acid and hydrogen peroxide in the same bath. Thus, this work presents a successful attempt to use glutaric acid in the presence of hydrogen peroxide towards metal recovery from spent Li-ion batteries.

Author Contributions: Conceptualization, W.U. and M.O.; methodology, W.U. and M.O.; software, W.U. and M.O.; validation, W.U. and M.O.; formal analysis, W.U. and M.O.; investigation, W.U. and M.O.; resources, W.U. and M.O.; data curation, W.U. and M.O.; writing—original draft preparation, W.U. and M.O.; writing—review and editing, W.U. and M.O.; visualization, W.U. and M.O.; supervision, W.U. and M.O.; project administration, W.U. and M.O.; funding acquisition, W.U. and M.O. All authors have read and agreed to the published version of the manuscript.

Funding: This research received no external funding.

Institutional Review Board Statement: Not applicable.

Informed Consent Statement: Not applicable.

Data Availability Statement: The data presented in this study are available on request.

Acknowledgments: The authors would like to thank Jakub Lewandowski and Karol Masztalerz for technical assistance.

Conflicts of Interest: The authors declare no conflict of interest.

References

1. Czerwiński, A. *Akumulatory, Baterie, Ogniwa*; WKŁ: Warszawa, Poland, 2018. (In Polish)
2. Moćko, W.; Szmidt, E. Recovery Technologies of Co and Li from spent lithium-ion cells. *Arch. Waste Manag. Environ. Prot.* **2012**, *14*, 1–10.
3. Scrosati, B.; Garche, J. Lithium batteries: Status, prospects and future. *J. Power Sources* **2010**, *195*, 2419–2430. [CrossRef]

4. Goodenough, J.B.; Park, K.-S. The Li-Ion Rechargeable Battery: A Perspective. *J. Am. Chem. Soc.* **2013**, *135*, 1167–1176. [CrossRef] [PubMed]

5. Tarascon, J.M.; Armand, M. Issues and challenges facing rechargeable lithium batteries. *Nature* **2001**, *414*, 359–367. [CrossRef] [PubMed]

6. Sirnoval, V.; Scagliarini, V.; Murugadoss, S.; Tomatis, M.; Yakoub, Y.; Turci, F.; Hoet, P.; Lison, D.; van den Brule, S. LiCoO$_2$ particles used in Li-ion batteries induce primary mutagenicity in lung cells via their capacity to generate hydroxyl radicals. *Part. Fibre. Toxicol.* **2020**, *17*, 6. [CrossRef] [PubMed]

7. De Boeck, M.; Lison, D.; Kirsch-Volders, M. Evaluation of the in vitro direct and indirect genotoxic effects of cobalt compounds using the alkaline comet assay. Influence of interdonor and interexperimental variability. *Carcinogenesis* **1998**, *19*, 2021–2029. [CrossRef]

8. Kang, D.H.P.; Chen, M.; Ogunseitan, O.A. Potential Environmental and Human Health Impacts of Rechargeable Lithium Batteries in Electronic Waste. *Environ. Sci. Technol.* **2013**, *47*, 10–5495. [CrossRef]

9. Lee, C.K.; Rhee, K.I. Preparation of LiCoO$_2$ from spent lithium-ion batteries. *J. Power Sources* **2002**, *109*, 17–21. [CrossRef]

10. Kang, J.; Senanayake, G.; Sohn, J.; Shin, S.M. Recovery of cobalt sulfate from spent lithium ion batteries by reductive leaching and solvent extraction with Cyanex 272. *Hydrometallurgy* **2010**, *100*, 168–171. [CrossRef]

11. Joulié, M.; Lacournet, R.; Billy, E. Hydrometallurgical process for the recovery of high value metals from spent lithium nickel cobalt aluminum oxide based lithium-ion batteries. *J. Power Sources* **2014**, *247*, 551–555. [CrossRef]

12. Zhuang, L.; Sun, C.; Zhou, T.; Li, H.; Dai, A. Recovery of valuable metals from LiNi$_{0.5}$Co$_{0.2}$Mn$_{0.3}$O$_2$ cathode materials of spent Li-ion batteries using mild mixed acid as leachant. *Waste Manag.* **2019**, *85*, 175–185. [CrossRef] [PubMed]

13. Golmohammadzadeh, R.; Rashchi, F.; Vahidi, E. Recovery of lithium and cobalt from spent lithium-ion batteries using organic acids: Process optimization and kinetic aspects. *Waste Manag.* **2017**, *64*, 244–254. [CrossRef]

14. Li, L.; Qu, W.; Zhang, X.; Lu, J.; Chen, R.; Wu, F.; Amine, K. Succinic acid-based leaching system: A sustainable process of recovery of valuable metals from spent Li-ion batteries. *J. Power Sources* **2015**, *282*, 443 544–551. [CrossRef]

15. Gao, W.; Song, J.; Cao, H.; Lin, X.; Zhang, X.; Zheng, X.; Zhang, Y.; Sun, Z. Selective recovery of valuable metals from spent lithium-ion batteries—Process development and kinetics evaluation. *J. Clean. Prod.* **2018**, *178*, 833–845. [CrossRef]

16. He, L.P.; Sun, S.Y.; Mu, Y.Y.; Song, X.F.; Yu, J.G. Recovery of lithium, nickel, cobalt and manganese from spent lithium-ion batteries using L-tartaric acid as leachant. *ACS Sustain. Chem. Eng.* **2017**, *5*, 714–721. [CrossRef]

17. Yao, Y.; Zhu, M.; Zhao, Z.; Tong, B.; Fan, Y.; Hua, Z. Hydrometallurgical processes for recycling spent lithium-ion batteries: A critical review. *ACS Sustain. Chem. Eng.* **2018**, *6*, 13611–13627. [CrossRef]

18. Meshram, P.; Abhilash; Pandey, B.D.; Mankhand, T.R.; Deveci, H. Comparison of Different Reductants in Leaching of Spent Lithium Ion Batteries. *J. Met.* **2016**, *68*, 2613–2623. [CrossRef]

19. Bertuol, D.A.; Machado, C.M.; Silva, M.L.; Calgaro, C.O.; Dotto, G.L.; Tanabe, E.H. Recovery of cobalt from spent lithium-ion batteries using supercritical carbon dioxide extraction. *Waste Manage.* **2016**, *51*, 245–251. [CrossRef]

20. Zheng, Y.; Long, H.L.; Zhou, L.; Wu, Z.S.; Zhou, X.; You, L.; Yang, Y.; Liu, J.W. Leaching procedure and kinetic studies of cobalt in cathode materials from spent lithium ion batteries using organic citric acid as leachant. *Int. J. Environ. Res.* **2016**, *10*, 159–168. [CrossRef]

21. Fan, B.; Chen, X.; Zhou, T.; Zhang, J.; Xu, B. A sustainable process for the recovery of valuable metals from spent lithium-ion batteries. *Waste Manag. Res.* **2016**, *34*, 474–481. [CrossRef]

22. Santana, I.L.; Moreira, T.F.M.; Lelis, M.F.F.; Freitas, M.B.J.G. Photocatalytic properties of Co$_3$O$_4$/LiCoO$_2$ recycled from spent lithium-ion batteries using citric acid as leaching agent. *Mater. Chem. Phys.* **2017**, *190*, 38–44. [CrossRef]

23. Peng, C.; Hamuyuni, J.; Wilson, B.P.; Lundstrom, M. Selective reductive leaching of cobalt and lithium from industrially crushed waste Li-ion batteries in sulfuric acid system. *Waste Manag.* **2018**, *76*, 582–590. [CrossRef] [PubMed]

24. Nayaka, G.P.; Pai, K.V.; Manjanna, J.; Keny, S.J. Use of mild organic acid reagents to recover the Co and Li from spent Li-ion batteries. *Waste Manag.* **2016**, *51*, 234–238. [CrossRef] [PubMed]

25. Nayaka, G.P.; Zhang, Y.; Dong, P.; Wang, D.; Zhiu, Z.; Duan, J.; Li, X.; Lin, Y.; Meng, Q.; Pai, K.V.; et al. An environmental friendly attempt to recycle the spent Li-ion battery cathode through organic acid leaching. *J. Environ. Chem. Eng.* **2019**, *7*, 102854. [CrossRef]

26. Aaltonen, M.; Peng, C.; Wilson, B.P.; Lundstrom, M. Leaching of Metals from Spent Lithium-Ion Batteries. *Recycling* **2017**, *2*, 20. [CrossRef]

27. Nayaka, G.P.; Manjanna, J.; Pai, K.V.; Vadavi, R.; Keny, S.J.; Tripathi, V.S. Recovery of valuable metal ions form the spent lithium-ion battery using aqueous mixture of mild organic acid as alternative to mineral acids. *Hydrometallurgy* **2015**, *151*, 73–77. [CrossRef]

28. Nayaka, G.P.; Pai, K.V.; Santhosh, G.; Manjanna, J. Dissolution of cathode active material of spent Li-ion batteries using tartaric acid and ascorbic acid mixture to recover Co. *Hydrometallurgy* **2016**, *161*, 54–57. [CrossRef]

29. Dorella, G.; Mansur, M.B. A study of the separation of cobalt from spent Li-ion battery residues. *J. Power Sources* **2007**, *170*, 210–215. [CrossRef]

30. Nayl, A.A.; Elkhashab, R.A.; Badawy, S.M.; El-Khateeb, M.A. Acid leaching of mixed spent Li-ion batteries. *Arab. J. Chem.* **2017**, *10*, 3632–3693. [CrossRef]

31. Mantuano, D.P.; Dorella, G.; Elias, R.C.V.; Mansur, M.B. Analysis of a hydrometallurgical route to recover base metals from spent rechargeable batteries by liquid–liquid extraction with Cyanex 272. *J. Power Sources* **2006**, *159*, 1510–1518. [CrossRef]

32. Urbańska, W. Recovery of Co, Li, and Ni from Spent Li-Ion Batteries by the Inorganic and/or Organic Reducer Assisted Leaching Method. *Minerals* **2020**, *10*, 555. [CrossRef]

33. Heydarian, A.; Mousavi, S.M.; Vakilchap, F.; Baniasadi, M. Application of a mixed culture of adapted acidophilic bacteria in two-step bioleaching of spent lithium-ion laptop batteries. *J. Power Sources* **2018**, *378*, 19–30. [CrossRef]

34. Meshram, P.; Pandey, B.D.; Mankhand, T.R. Recovery of valuable metals from cathodic active material of spent lithium ion batteries: Leaching and kinetic aspects. *Waste Manag.* **2015**, *45*, 306–313. [CrossRef] [PubMed]

35. Lv, W.; Wang, Z.; Cao, H.; Zheng, X.; Jin, W.; Zhang, Y.; Sun, Z. A sustainable process for metal recycling from spent lithium-ion batteries using amonnium chloride. *Waste Manag.* **2018**, *79*, 545–553. [CrossRef] [PubMed]

36. Chen, X.; Guo, C.; Ma, H.; Li, J.; Zhou, T.; Cao, L.; Kang, D. Organic reductants based leaching: A sustainable process for the recovery of valuable metals from spent lithium ion batteries. *Waste Manag.* **2018**, *75*, 459–468. [CrossRef] [PubMed]

37. Pinto, P.S.; Lanza, G.D.; Ardisson, J.D.; Lago, R.M. Controlled Dehydration of $Fe(OH)_3$ to Fe_2O_3: Developing Mesopores with Complexing Iron Species for the Adsorption of β-Lactam Antibiotics. *J. Braz. Chem. Soc.* **2019**, *30*, 310–317. [CrossRef]

38. Xia, X.; Tu, J.; Zhang, Y.; Mai, Y.; Wang, X.; Gu, C.; Zhao, X. Freestanding Co_3O_4 nanowire array for high performance supercapacitors. *RSC Adv.* **2012**, *2*, 1835–1841. [CrossRef]

39. Allaedini, G.; Muhammad, A. Study of influential factors in synthesis and characterization of cobalt oxide nanoparticles. *J. Nanostructure Chem.* **2013**, *3*, 77. [CrossRef]

40. Wang, J.; He, Y.; Yang, Y.; Xie, W.; Ling, X. Research on quantifying the hydrophilicity of leached coals by ftir spectroscopy. *Physicochem. Probls. Miner. Process.* **2017**, *53*, 227–239. [CrossRef]

41. Tannenbaum, R.; Zubris, M.; David, K.; Ciprari, D.; Jacob, K.; Jasiuk, I.; Dan, N. FTIR Characterization of the Reactive Interface of Cobalt Oxide Nanoparticles Embedded in Polymeric Matrices. *J. Phys. Chem. B* **2006**, *110*, 2227–2232. [CrossRef]

42. Noriega, S.; Subramanian, A. Consequences of Neutralization on the Proliferation and Cytoskeletal Organization of Chondrocytes on Chitosan-Based Matrices. *Int. J. Carbohydr. Chem.* **2011**, 809743. [CrossRef]

43. Uznanski, P.; Zakrzewska, J.; Favier, F.; Kazmierski, S.; Bryszewska, E. Synthesis and characterization of silver nanoparticlesfrom (bis)alkylamine silver carboxylate precursors. *J. Nanopart. Res.* **2017**, *19*, 12. [CrossRef] [PubMed]

44. Funda, S.; Ohki, T.; Liu, Q.; Hossain, J.; Ishimaru, Y.; Ueno, K.; Shirai, H. Correlation between the fine structure of spin-coated PEDOT:PSS and the photovoltaic performance of organic/crystalline-silicon heterojunction solar cells. *J. Appl. Phys.* **2016**, *120*, 033103. [CrossRef]

45. Fila, D.; Hubicki, Z.; Kołodyńska, D. Recovery of metals from waste nickel-metal hydride batteries using multifunctional Diphonix resin. *Adsorption* **2019**, *25*, 367–382. [CrossRef]

46. Li, X.; Wang, Y.; Xie, X.; Huang, C.; Yang, S. Dehydration of fructose, sucrose and inulin to 5-hydroxymethylfurfural over yeast-derived carbonaceous microspheres at low temperatures. *RSC Adv.* **2019**, *9*, 9041–9048. [CrossRef]

47. Bounaas, M.; Bouguettoucha, A.; Chebli, D.; Reffas, A.; Harizi, I.; Rouabah, F.; Amrane, A. High efficiency of methylene blue removal using a novel low-cost acid treated forest wastes, Cupressus semperirens cones: Experimental results and modeling. *Particul. Sci. Technol.* **2019**, *37*, 504–513. [CrossRef]

48. Zhang, F.; Yuan, C.; Lu, X.; Zhang, L.; Zhang, X.; Che, Q. Facile growth of mesoporous Co_3O_4 nanowire arrays on Ni foam for high performance electrochemical capacitors. *J. Power Sources* **2012**, *203*, 250–256. [CrossRef]

49. Freitas, B.G.A.; Siqueira, J.M., Jr.; da Costa, L.M.; Ferreira, G.B.; Resende, J.A.L.C. Synthesis and Characterization of $LiCoO_2$ from Different Precursors by Sol-Gel Method. *J. Braz. Chem. Soc.* **2017**, *28*, 2254–2266. [CrossRef]

50. Lakshmanan, R.; Gangulibabu; Bhuvaneswari, D.; Kalaiselvi, N. Temperature Dependent Surface Morphology and Lithium Diffusion Kinetics of $LiCoO_2$ Cathode. *Met. Mater. Int.* **2012**, *18*, 249–255. [CrossRef]

51. Xie, J.; Cao, H.; Jiang, H.; Chen, Y.; Shi, W.; Zheng, H.; Huang, Y. Co_3O_4-reduced graphene oxide nanocomposite as an effective peroxidase mimetic and its application in visual biosensing of glucose. *Anal. Chim. Acta* **2013**, *796*, 92–100. [CrossRef]

52. Qi, C.; Koenig, G.M. High-Performance $LiCoO_2$ Sub-Micrometer Materials from Scalable Microparticle Template Processing. *ChemistrySelect* **2016**, *1*, 3992–3999. [CrossRef]

53. Santiago, E.I.; Andrade, A.V.C.; Paiva-Santos, C.O.; Bulhoes, L.O.S. Structural and electrochemical properties of $LiCoO_2$ prepared by combustion synthesis. *Solid State Ion.* **2003**, *158*, 91–102. [CrossRef]

54. Zeng, X.; Li, J.; Shen, B. Novel approach to recover cobalt and lithium from spent lithium-ion battery using oxalic acid. *J. Hazard. Mater.* **2015**, *295*, 112–118. [CrossRef] [PubMed]

55. Jha, A.K.; Jha, M.K.; Kumari, A.; Sahu, S.K.; Kumar, V.; Pandey, B.D. Selective separation and recovery of cobalt from leach liquor of discarded Li-ion batteries using thiophosphinic extractant. *Sep. Purif. Technol.* **2013**, *104*, 160–166. [CrossRef]

56. Zhang, L.; Zhang, K.; Shi, Z.; Zhang, S. LiF as an Artificial SEI Layer to Enhance the High-Temperature Cycle Performance of $Li_4Ti_5O_{12}$. *Langmuir* **2017**, *33*, 11164–11169. [CrossRef]

57. Babkair, S.S.; Azam, A. Color Centers Formation in Lithium Fluoride Nanocubes Doped with Different Elements. *J. Nanomater.* **2013**, *872074*, 1–8. [CrossRef]

58. Yu, X.Q.; Sun, J.P.; Tang, K.; Li, H.; Huang, X.J.; Dupont, L.; Maier, J. Reversible lithium storage in LiF/Ti nanocomposites. *Phys. Chem. Chem. Phys.* **2009**, *11*, 9497–9503. [CrossRef]

59. Ungár, T.; Gubicza, J.; Ribárik, G.; Pantea, C.; Zerda, T.W. Microstructure of carbon blacks determined by X-ray diffraction profile analysis. *Carbon* **2002**, *40*, 929–937. [CrossRef]

60. Tao, S.; Zhan, Z.; Wang, P.; Meng, G. Chemical stability study of Li_2SO_4 on the operation condition of a H_2/O_2 fuel cell. *Solid State Ion.* **1999**, *116*, 29–33. [CrossRef]

61. Kohl, M.; Bruckner, J.; Bauer, I.; Althues, H.; Kaskel, S. Synthesis of highly electrochemically active Li_2S nanoparticles for Lithium-sulfur-Batteries. *J. Mater. Chem. A* **2015**, *3*, 16307–16312. [CrossRef]

62. Pan, A.; Wang, Y.; Xu, W.; Nie, Z.; Liang, S.; Nie, Z.; Wang, C.; Cao, G.; Zhang, J. High-performance anode based on porous Co_3O_4 nanodiscs. *J. Power Sources* **2014**, *255*, 125–129. [CrossRef]

63. Lu, Y.; Wang, J.; Zeng, S.; Zhou, L.; Xu, W.; Zheng, D.; Liu, J.; Zeng, Y.; Lu, X. An ultrathin defect-rich Co_3O_4 nanosheet cathode for high-energy and durable aqueous zinc ion batteries. *J. Mater. Chem. A* **2019**, *7*, 21678–21683. [CrossRef]

64. Rahman, M.M.; Wang, J.Z.; Deng, X.L.; Li, Y.; Liu, H.K. Hydrothermal synthesis of nanostructured Co_3O_4 materials under pulsed magnetic field and with an aging technique, and their electrochemical performance as anode for lithium-ion battery. *Electrochim. Acta* **2009**, *55*, 504–510. [CrossRef]

65. Zhou, H.; Lv, B.; Wu, D.; Xu, Y. Synthesis of the polycrystalline Co3O4 nanowires with excellent ammonium perchrolate catalytic decomposition property. *Mater. Res. Bull.* **2014**, *60*, 492–497. [CrossRef]

Publisher's Note: MDPI stays neutral with regard to jurisdictional claims in published maps and institutional affiliations.

energies

Article

Cascade Membrane System for Separation of Water and Organics from Liquid By-Products of HTC of the Agricultural Digestate—Evaluation of Performance

Agnieszka Urbanowska [1,*], Małgorzata Kabsch-Korbutowicz [1], Christian Aragon-Briceño [2], Mateusz Wnukowski [3], Artur Pożarlik [2], Lukasz Niedzwiecki [3], Marcin Baranowski [3], Michał Czerep [3], Przemysław Seruga [4,5], Halina Pawlak-Kruczek [3], Eduard Bramer [2] and Gerrit Brem [2]

[1]   Department of Water and Wastewater Treatment Technology, Faculty of Environmental Engineering, Wroclaw University of Science and Technology, Wybrzeże Wyspiańskiego 27, 50-370 Wrocław, Poland; malgorzata.kabsch-korbutowicz@pwr.edu.pl

[2]   Department of Thermal and Fluid Engineering, University of Twente, Postbus 217, 7500 AE Enschede, The Netherlands; c.i.aragonbriceno@utwente.nl (C.A.-B.); a.k.pozarlik@utwente.nl (A.P.); e.a.bramer@utwente.nl (E.B.); g.brem@utwente.nl (G.B.)

[3]   Department of Energy Conversion Engineering, Faculty of Mechanical and Power Engineering, Wroclaw University of Science and Technology, Wybrzeże Wyspiańskiego 27, 50-370 Wrocław, Poland; mateusz.wnukowski@pwr.edu.pl (M.W.); lukasz.niedzwiecki@pwr.edu.pl (L.N.); marcin.baranowski@pwr.edu.pl (M.B.); michal.czerep@pwr.edu.pl (M.C.); halina.pawlak@pwr.edu.pl (H.P.-K.)

[4]   Department of Bioprocess Engineering, Wroclaw University of Economics, Komandorska 118/120, 53-345 Wrocław, Poland; przemyslaw.seruga@ue.wroc.pl

[5]   Zakład Gospodarowania Odpadami GAĆ Sp. z o.o., Gać 90, 55-200 Oława, Poland; przemyslaw.seruga@zgo.org.pl

*   Correspondence: agnieszka.urbanowska@pwr.edu.pl

Citation: Urbanowska, A.; Kabsch-Korbutowicz, M.; Aragon-Briceño, C.; Wnukowski, M.; Pożarlik, A.; Niedzwiecki, L.; Baranowski, M.; Czerep, M.; Seruga, P.; Pawlak-Kruczek, H.; et al. Cascade Membrane System for Separation of Water and Organics from Liquid By-Products of HTC of the Agricultural Digestate—Evaluation of Performance. *Energies* **2021**, *14*, 4752. https://doi.org/10.3390/en14164752

Academic Editor: Adam Smoliński

Received: 13 July 2021
Accepted: 2 August 2021
Published: 5 August 2021

Publisher's Note: MDPI stays neutral with regard to jurisdictional claims in published maps and institutional affiliations.

Abstract: New regulations aimed at curbing the problem of eutrophication introduce limitations for traditional ways to use the by-product of anaerobic digestion—the digestate. Hydrothermal carbonisation (HTC) can be a viable way to valorise the digestate in an energy-efficient manner and at the same time maximise the synergy in terms of recovery of water, nutrients, followed by more efficient use of the remaining carbon. Additionally, hydrothermal treatment is a feasible way to recirculate recalcitrant process residues. Recirculation to anaerobic digestion enables recovery of a significant part of chemical energy lost in HTC by organics dissolved in the liquid effluent. Recirculating back to the HTC process can enhance nutrient recovery by making process water more acidic. However, such an effect of synergy can be exploited to its full extent only when viable separation techniques are applied to separate organic by-products of HTC and water. The results presented in this study show that using cascade membrane systems (microfiltration (MF) → ultrafiltration (UF) → nanofiltration (NF)), using polymeric membranes, can facilitate such separation. The best results were obtained by conducting sequential treatment of the liquid by-product of HTC in the following membrane sequence: MF 0.2 μm → UF PES 10 → NF NPO30P, which allowed reaching COD removal efficiency of almost 60%.

Keywords: digestate; biogas plant; hydrothermal carbonisation; membrane processes; water recovery

1. Introduction

Biogas production is a proven way to incorporate, in practice, energy-to-waste strategies using various sources of organic waste [1–6]. Typically, a footprint of 8 ha/MW of installed power is required to handle and store typical biogas plants, which introduces high costs [7]. Digestate management could be a feasible way to minimise the footprint. Thermal drying, followed by pelletising, could decrease the volume of the digestate significantly [8]. Nonetheless, it comes at the cost of using valuable energy, e.g., using a part of the produced biogas [8], with open-air drying being the only energy-saving option. At the state-of-the-art biogas plant, a liquid obtained after mechanical dewatering of digestate is stored in lagoons, which can be counted as a significant part of the installation's footprint. Literature reported organic and nutrient removal process to solve the problem of toxicity with biofilms for more economic nitrogen removal [9]. In wastewater treatment, it is stated that biofilm process is present and has degradation possibilities for biogas effluent treatment and is cited in newer publications regarding ORP control and electricity production combined with wastewater treatment [10]. The parameters such as pH, alkalinity, and inorganic carbon content significantly influence the process [11,12]. Such technologies could be used to tackle the problem of nitrogen removal, using extremophilic conditions, tolerating bacteria [13–15]. The liquid, rich in nutrients, is stabilised and can be later used for fertilisation purposes. However, significant limitations to this practice are being introduced by the European Nitrates Directive (91/676/EEC) [16], which aims at curbing the eutrophication problem in the EU [16]. Therefore, relatively large areas are needed for spreading, complicating the logistics of such solutions. Moreover, dewatering effluent's storage inevitably leads to water loss through evaporation, which nowadays is becoming an increasingly scarce resource in the agriculture sector during climate change [17–20]. Additionally, digestate still contains significant amounts of recalcitrant organic matter [21], which prompted many investigations regarding increasing its availability for further digestion to produce additional amounts of biogas [22–26]. Furthermore, nowadays higienisation of digestate or removal of ammonia is also considered as important issues [27,28].

Hydrothermal carbonisation (HTC) has been recently pointed out as a possibility regarding synergetic recovery of recalcitrant organic matter and water from digestate [29]. HTC is a thermal valorisation process, typically performed at temperatures typically ranging between 180 and 260 °C [30–32], in subcritical water, at elevated pressure, which comes from water vapour pressure as well as from gaseous products of HTC [33–35]. The use of HTC can enhance mechanical dewatering, as reported for various wet types of biomass [36–39]. At the temperatures of HTC, the ionic constant of water is a subject of significant increase, nearly doubling at the upper range, compared to ambient temperatures [40]. At such a temperature range, water behaves as a non-polar solvent [40]. A multitude of reactions coinciding with the output of multiple different products can be considered typical for HTC of different types of biomass [41,42]. These reactions do not represent consecutive steps but instead form parallel networks of different reaction pathways [43]. Hydrolysis usually takes place in relatively low temperatures [44]. Dehydration and decarboxylation are also named as reactions that are significant for the HTC process [44,45].

The number of hydroxyl groups decreases due to dehydration [44,46,47], subsequently causing a lower O/C ratio. Decreased amount of carboxyl and carbonyl groups, due to decarboxylation, also slightly decreases solid products' O/C ratio [44,46,47]. Intermediates become substrates of polymerisation, condensation and aromatisation [43,47,48], which are also instrumental in an aggregation of carbonaceous microspheres [43]. Microspheres precipitate, thus forming secondary hydrochar [43,49].

A decrease in hydroxyl groups is crucial in making hydrothermally carbonised biomass more hydrophobic [50]. This could effectively lower the equilibrium moisture content of biomass [51] and make physical dewatering easier [44,52], which has the potential of significant energy savings since thermal drying of biomass is an energy-intensive process [53–55]. The ability to decrease the O/C ratio is beneficial when valorisation is performed, aiming to improve subsequent pyrolysis [56–60]. Moreover, some studies reported relatively easy pelletising of hydrochars [61]. Studies suggested the application of hydrothermal processing in biorefineries [62], opening novel routes for the production of chemicals, such as HMF [63], fertilisers and soil amendments [64–68]. Additionally, improvement in terms of properties relevant from the point of view of solid fuels could be observed, when hydrochars are compared to raw biomass, prior to HTC treatment [69–72].

This makes hydrothermal carbonisation a prospective valorisation process for low-quality biomass with relatively high moisture content [73–76].

Overall, recirculation of the liquid by-products or its subsequent use for maximum energy recovery and increased systemic efficiency offers many opportunities [43]. Several studies report the composition of liquid HTC by-products [77–79], with many indicating potential synergies between HTC and anaerobic digestion (AD) [52,80,81] or application of the effluent in microbial fuel cells [82].

However, there is a significant gap, regarding the separation of the organic fraction of liquid HTC by-products from water, in order to maximise the recirculation and, at the same time, maintaining the possibility to keep the optimum solid loading of anaerobic digestion reactor, which could be problematic for AD feedstocks with low solid content. Literature sources suggest positive influence of HTC treatment on subsequent membrane separation, with the possible effect of synergy [83]. Promising results were obtained by the authors of his manuscript on ultrafiltration of the HTC effluent, with a significant reduction of COD in the permeate [84]. This study is a consequent follow-up, and its purpose is further optimisation of membrane separation of organic fraction of HTC effluent, using different configurations of a cascade membrane system.

2. Materials and Methods

The research was conducted for the liquid fraction of the digestate from an agricultural biogas plant located in the Silesia region, Poland. Samples were taken at the outlet of the reactor prior to the mechanical dewatering stage. The digestate was stored in a cooler, at the temperature of 5 °C, before the HTC experiment. Figure 1 shows a diagram of the experimental setup. The autoclave vessel, equipped with an external heating mantle that consisted of band heaters, was filled with 3.8 dm^3 of wet digestate, which had a solids content of 10.1%. Radwag MAX2A analyser (Radom, Poland), with a scale resolution of 0.001 g and a maximum sample mass of 50 g, was used to determine the moisture content of the digestate, and dry solids content was obtained by difference. Moisture content test was performed at 105 °C. The sample's mass was considered to be in equilibrium when the first derivative of the mass (dm/dt) was equal to or smaller than 1 mg/min.

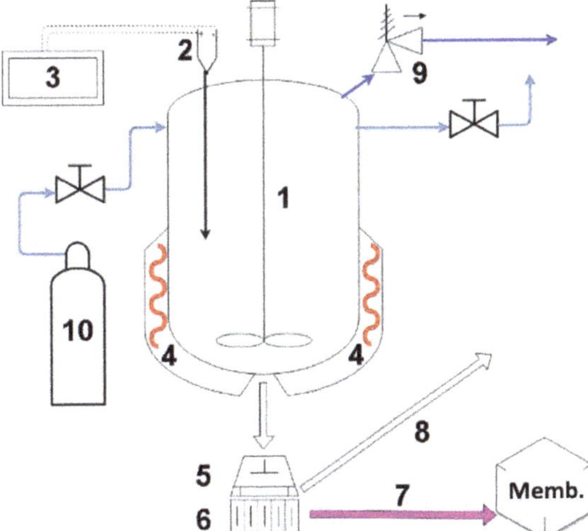

Figure 1. Experimental setup (1—autoclave; 2—type K thermocouple; 3—PLC controller; 4—heating mantle, with band heaters; 5—cotton cloth; 6—colander; 7—HTC effluent; 8—hydrochar, after separation; 9—pressure relief valve; 10—nitrogen for purging; Memb.—membrane purification followed by analyses).

The temperature of the heating mantle was controlled by a PLC. Additionally, the temperature was measured inside of the reactor by a K type thermocouple inserted close to the centre of the reactor. Setpoint temperature of 190 °C was chosen, with PID controller keeping the temperature within ±5 °C, as lower temperatures of the typical HTC temperature range (180 °C and 260 °C) are recommended for digestates of agricultural origins by various literature sources [85,86]. HTC pressure was measured by an analogue gauge, and during the experiment, it was close to the water vapour saturation pressure. The residence time in the reactor was 30 min. Reaching the setpoint temperature, the inside of the reactor triggered the time measurement. After 30 min of the process, the mantle was turned off, and the setup was left for cooling overnight. The colander, with clean cotton cloth, was used for the subsequent separation of the effluent and hydrochar. The cooled material was drained for approximately 20 min to let all of the effluents be drained.

The sample of the liquid by-products of HTC, obtained by draining, was also analysed using a gas chromatograph connected to a mass spectrometer (GC-MS), i.e., chromatograph 7820-A and 5977B MSD spectrometer produced by Agilent Technologies (Santa Clara, CA, USA). In the chromatograph, the Stabilwax-DA column from Restek was used. Helium, with a flow rate of 1.5 cm^3/min, was used as a carrier gas. The column's heating program was set in the following order: achieve 50 °C in 5 min, subsequently heat up with a ramp of 10 °C/min until 200 °C and hold for another 20 min. Compounds were automatically identified using NIST-14 MS library by comparing the mass spectra (a minimum match factor of 80% was taken into account). The MS scanning range was m/z 10–450, with a frequency of 1.7 scans/s. The gain factor and EM Volts were 0.5 and 1348.5, respectively. The temperature of MS source and quadrupole was 230 °C and 150 °C, respectively.

Measurements were performed in triplicates for each sample. The calibration curve was done for four different concentrations of each compound, with each concentration done in triplicates. The uncertainty of quantitative GC-MS analysis was calculated for each compound using the standard deviation of the sample, based on an entire population, using the residual standard deviation as the measure of the fitting quality of the calibration curve. The characteristics of the digestate liquid fraction after HTC from the rural biogas plant are presented in Table 1. The assessment of sequential purification possibility of the liquid fraction of the agricultural digestate after HTC was carried out by combining three stages of membrane separation: microfiltration, ultrafiltration and nanofiltration. Figure 2 depicts the experimental matrix. First, an experiment with single microfiltration (MF) membrane was performed and the sample of the permeate was taken (A—Figure 2). Then four more experiments were performed using different combinations of microfiltration and ultrafiltration (UF) membranes, with samples being taken after ultrafiltration (B, C, D, E—Figure 2). This was followed by eight experiments with different combinations of microfiltration, ultrafiltration and nanofiltration (NF) membranes, with samples being taken at the end of the cascade (F, G, H, I, J, K, L, M—Figure 2).

Table 1. Composition of the liquid digestate fraction from the rural biogas plant after HTC.

Index	Value
pH	7.2
Conductivity, mS/cm	14.95
Total suspended solids, mg/dm^3	3950
Chemical oxygen demand (COD), mg O_2/dm^3	38.595
5-day biochemical oxygen demand (BOD$_5$), mg O_2/dm^3	12.320
Dissolved organic carbon (DOC), mg C/dm^3	23.070
Na, mg/dm^3	521.3
K, mg/dm^3	1966.5
Ca, mg/dm^3	104.7
Mg, mg/dm^3	101.9
Fe, mg/dm^3	15.9
Mn, mg/dm^3	1.5

Table 1. *Cont.*

Index	Value
Cu, mg/dm^3	0.545
Zn, mg/dm^3	3.977
Hg, mg/dm^3	0.0029
Co, mg/dm^3	0.069
Ni, mg/dm^3	0.147

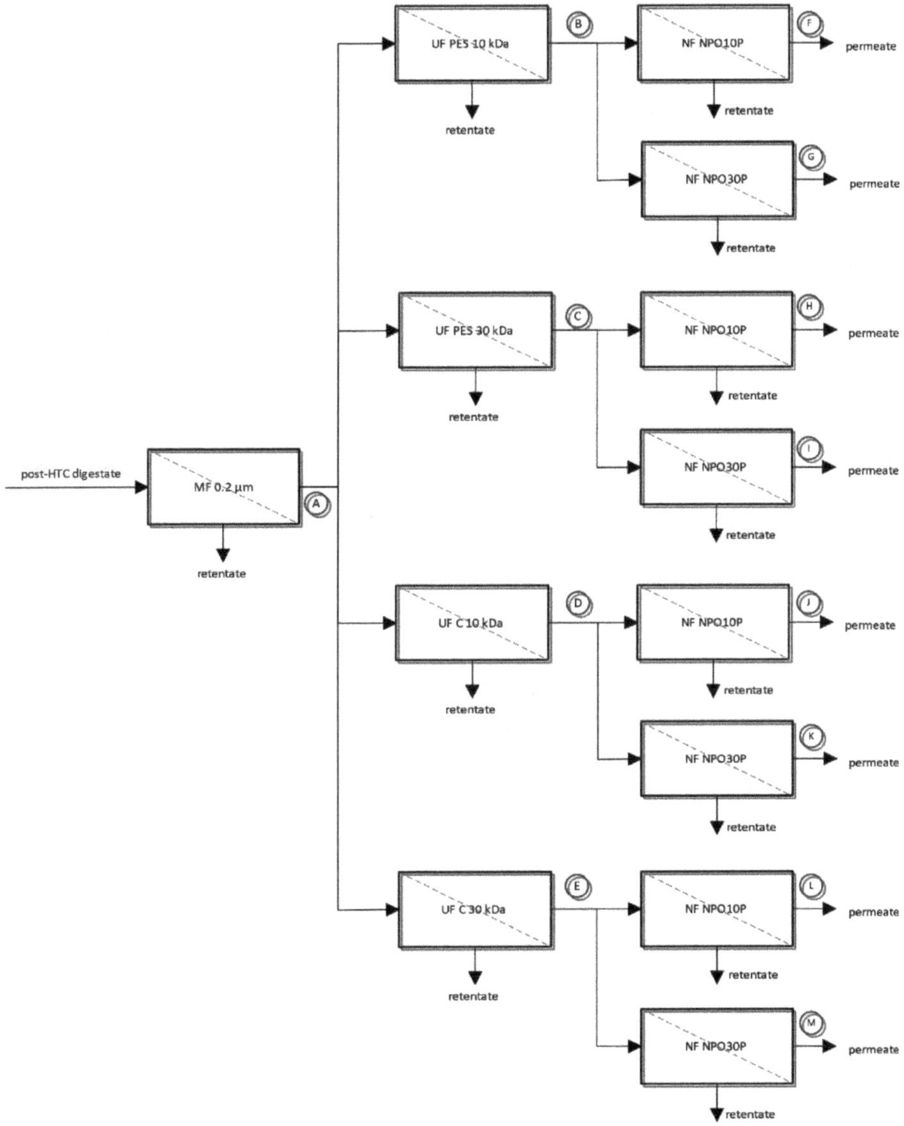

Figure 2. Diagram of sequential membrane separation of HTC liquid fraction.

The separation efficiency was measured by the value of removal rate (R), determined from the equation:

$$R = \left(1 - \frac{c_p}{c_f}\right) \cdot 100\% \tag{1}$$

where: c_p—concentration of contaminants in permeate, g/m^3, c_f—contaminant concentration in the feed, g/m^3.

The following types of flat sheet membranes were used in the research:

- MF membrane with a pore size of 0.2 μm made of polypropylene, 25.4 μm thick and 60% porosity (Hoechst Celgard Corporation),
- Four types of UF membranes (PES 10, PES 30, C 10, C 30) (Microdyn Nadir),
- Two types of NF membranes (NPO10P and NPO30P) (Microdyn Nadir).

Properties of UF and NF membranes used in the study are presented in Tables 2 and 3, respectively.

Table 2. UF membranes used in experiments.

Membrane Symbol	Membrane Material	Cut-Off, kDa [87]	Mean Pore Radius, nm [88]	Contact Angle, ° *	Polarity, % [88]	H_2O Flux, m^3/m^2d (0.4 MPa) *
PES 10	Polyethersulphone	10	2.04	52.0	44.27	6.6
PES 30		30	8.38			13.3
C 10	Regenerated cellulose	10	5.01	37.8	49.92	2.7
C 30		30	12.55			21.5

* based on our own research.

Table 3. NF membranes used in experiments.

Membrane Type	Membrane Material	Na_2SO_4 Retention [87]	Cut-Off, kDa [87]	Contact Angle, ° *	H_2O Flux, m^3/m^2d (0.4 MPa) *
NP010P	Polyethersulfone	25–40%	1040–1400	57.5	1.3
NP030P		80–95%	520–700	58.6	0.25

* based on our own research.

The purification of the liquid fraction of the HTC liquid fraction, after hydrothermal carbonisation of the digestate from the agricultural biogas plant with the use of pressure membrane processes was carried out on a test stand equipped with an Amicon 8400 cell (Millipore, Burlington, MA, USA)—effective filtration surface 45.3 cm^2. It allows for a dead-end filtration process and is designed to work with flat membranes. To ensure equal concentration in the entire volume of the solution the filtration cell was placed on a magnetic stirrer. The transmembrane pressure (TMP) in the range of 0.1–0.4 MPa was used in the tests. Nitrogen was used to generate necessary pressure, with limiting error of 0.01 MPa. The following two parameters were used: volume flux of the permeate (J) and organic substances retention coefficient (R) to estimate the separation and transport properties of the membrane under study. The quotient of the permeate flux (J) to the redistilled water flux (J_0) was also calculated to determine the relative permeability (J/J_0) of the membranes and thus, the membrane fouling susceptibility.

The purification of the liquid fraction of the agricultural digestate was carried out in various variants. The sequential use of the membranes resulted in further analysis:

- A—solution after MF membrane;
- B—solution after MF followed by UF using a PES 10 kDa membrane;
- C—solution after MF followed by UF using a PES 30 kDa membrane;
- D—solution after MF followed by UF using a C 10 kDa membrane;
- E—solution after MF followed by UF using a C 30 kDa membrane;
- F—solution after MF followed by UF using a PES 10 kDa membrane and NF using NPO10P membrane;

- G—solution after MF followed by UF using a PES 10 kDa membrane and NF using NPO30P membrane;
- H—solution after MF followed by UF using a PES 30 kDa membrane and NF using NPO10P membrane;
- I—solution after MF followed by UF using a PES 30 kDa membrane and NF using NPO30P membrane;
- J—solution after MF followed by UF using a C 10 kDa membrane and NF using NPO10P membrane;
- K—solution after MF followed by UF using a C 10 kDa membrane and NF using NPO30P membrane;
- L—solution after MF followed by UF using a C 30 kDa membrane and NF using NPO10P membrane;
- M—solution after MF followed by UF using a C 30 kDa membrane and NF using NPO30P membrane.

The organic compound concentration in the permeates expressed as COD, 5-day BOD (BOD5) and DOC was used to determine the efficiency of the process. Standard methods: dichromate and dilution [89] was used to measure COD and BOD5, respectively. The HACH IL550 TOC-TN analyser (Loveland, CO, USA) was used to measure the DOC concentration.

3. Results and Discussion

Taking into account the fact that the main objective of the research was to separate organic substances present in the digestate liquid fraction after HTC and to recover water from it, which could be used, e.g., in agriculture, special attention was paid to the quality of permeates obtained after successive membrane filtration stages. As an initial step, total organic compounds (DOC) and their biodegradable fraction (BOD5), as well as chemical oxygen demand (COD), were analysed. Subsequently, the content of selected compounds in the permeates was determined using GC-MS analyses. Characteristics of these solutions are presented in Figure 3.

Analysing the effectiveness of the sequential purification of the liquid fraction of the agricultural digestate after HTC, it can be concluded that the final quality of the analysed solution was determined by the combination of separation properties of individual membranes. The combination of microfiltration (MF), ultrafiltration (UF) and nanofiltration (NF) allows for a significant increase in the effectiveness of the purification of the raw solution (liquid fraction of the agricultural digestate after HTC), compared to the effects obtained for independent membrane processes. This effect has been observed for all analysed cases regardless of the membrane cut-off or its material (Figure 4). Overall, it could be stated that proper configuration of the cascade of membranes significantly outperformed single ultrafiltration membrane, investigated in the previous study [84].

The use of microfiltration enables the separation of both colloids and fine suspensions, as well as some macromolecular compounds and microorganisms. For example, at the transmembrane pressure of 0.4 MPa, the obtained values of R_{DOC}, R_{BOD5} and R_{COD} were respectively: 12.3%, 13.0% and 3.0%. The redirection of the microfiltration permeate to further purification in the ultrafiltration process showed an increase in the removal efficiency of organic compounds from the analysed digestate due to the combination of sieving mechanisms of both types of analysed membranes. The analysis of the obtained results suggests that the efficiency of digestate purification is generally determined by the cut-off of ultrafiltration membranes. It was observed that with the increase of membrane pore diameter, the efficiency of organic compounds removal decreased. The explanation of this phenomenon is the fact that at a higher membrane cut-off value, larger particles of organic compounds penetrated into the permeate. Moreover, it was noted that the membrane material (polyethersulfone or regenerated cellulose) did not determine the effectiveness of the purification of the tested digestate. For example, a 10 kDa polyethersulfone membrane allowed retention rate for DOC of 32.3%, for BOD_5 35.4%, and for COD 26.9% to be achieved, whereas for a regenerated cellulose membrane (10 kDa), reductions in these parameters of 29.8%, 35.2%, and 23%, respectively, were obtained. The use of a 30 kDa membrane resulted in retention values of DOC, BOD_5 and COD in purified samples of the digestate of 27.2%, 20.5% and 19.9% and 24.9%, 25.0% and 19.0%, respectively for PES and C membranes (TMP 0.4 MPa).

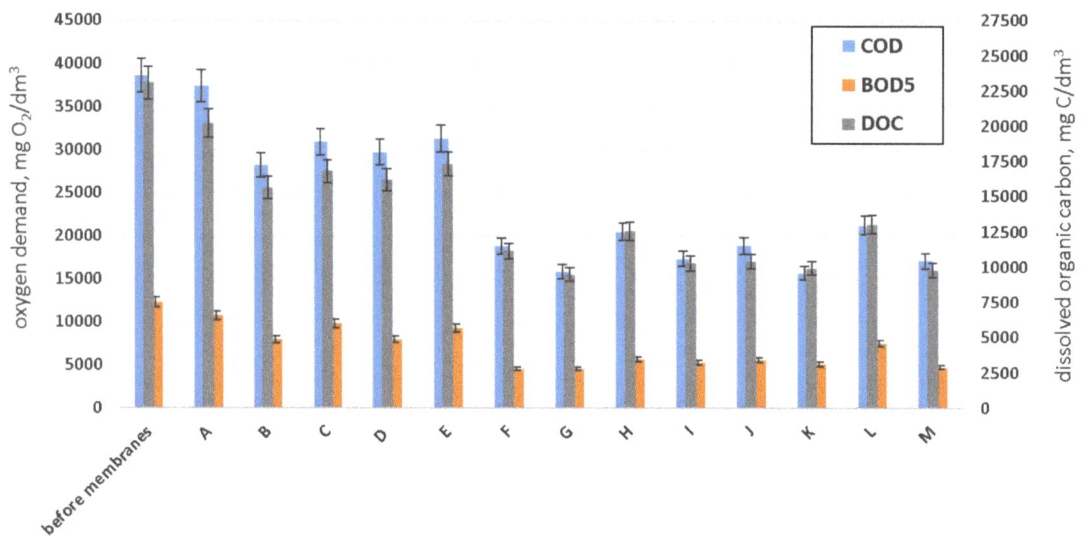

Figure 3. Properties of the solution after separation using different configurations of membranes (A: MF 0.2 μm; B: MF 0.2 μm → UF PES 10 kDa; C: MF 0.2 μm → UF PES 30 kDa; D: MF 0.2 μm → UF C 10 kDa; E: MF 0.2 μm → UF C 30 kDa; F: MF 0.2 μm → UF PES 10 kDa → NF NPO10P; G: MF 0.2 μm → UF PES 10 kDa → NF NPO30P; H: MF 0.2 μm → UF PES 30 kDa → NF NPO10P; I: MF 0.2 μm → UF PES 30 kDa → NF NPO30P; J: MF 0.2 μm → UF C 10 kDa → NF NPO10P; K: MF 0.2 μm → UF C 10 kDa → NF NPO30P; L: MF 0.2 μm → UF C 30 kDa → NF NPO10P; M: MF 0.2 μm → UF C 30 kDa → NF NPO30P).

The application of the nanofiltration in the final stage of the solution purification process resulted in a significant improvement in its quality. This effect has been observed for all tested variants of the use of different types of membranes (both ultrafiltration and nanofiltration). The final quality of the permeate was determined by the combination of properties of sequentially applied membranes. The obtained values of DOC, BOD_5 and COD retention coefficients for both tested nanofiltration membranes show how much influence the type of NF membrane used had on the purified solution composition. It was easy to see that using a less compact membrane (NPO10P) resulted in the deterioration of the final permeate quality. For example, the NPO30P membrane allowed to achieve DOC, BOD_5 and COD retention coefficients of 58.9%, 63.0% and 58.9%, while the NPO10P membrane allowed to reduce DOC, BOD_5 and COD by 51.7%, 63.0% and 51.3% respectively (using a 10 kDa UF PES membrane prior to NF). Better separation properties of the NPO30P membrane could result from its lower cut-off and denser structure [90].

Based on the obtained results, it was found that the best results i.e., the best permeate quality in terms of organic substances content, were achieved by conducting sequential purification in the following variant: microfiltration (0.2 μm) → ultrafiltration (PES 10 membrane) → nanofiltration (NPO30P membrane) (G). Lack of literature sources on membrane separation of organics and water in HTC effluent makes direct comparison difficult. Urbanowska et al. [84] reported COD removal efficiency close to 30% for membrane purification of post HTC effluent, using a single ultrafiltration membrane. Owing to the use of membrane cascades in this study the efficiency was almost doubled (see G in Figures 4 and 5). Klein et al. [91] reported COD removal from effluent with similar organic load i.e., pyrogenic wastewater from oil shale retorting. COD of the samples ranges between 39,700 and 45,400 mg/dm^3 [91]. Air stripping resulted in COD removal of more than 30%, whereas ozonation was close to reach COD removal of 50% [91]. Fenton treatment resulted in COD removal ranging between almost 60% and 80%, depending on the COD/H_2O_2/Fe^{2+} *w/w/w* [91]. Only a combination of physical, chemical and biological methods allowed to achieve COD reduction of 98% [91]. However, the advantage of the membrane system is the fact that retentate could still be used as an energy source e.g., for production of biogas [34,81,92], which in turn could be used to produce heat necessary for the HTC process.

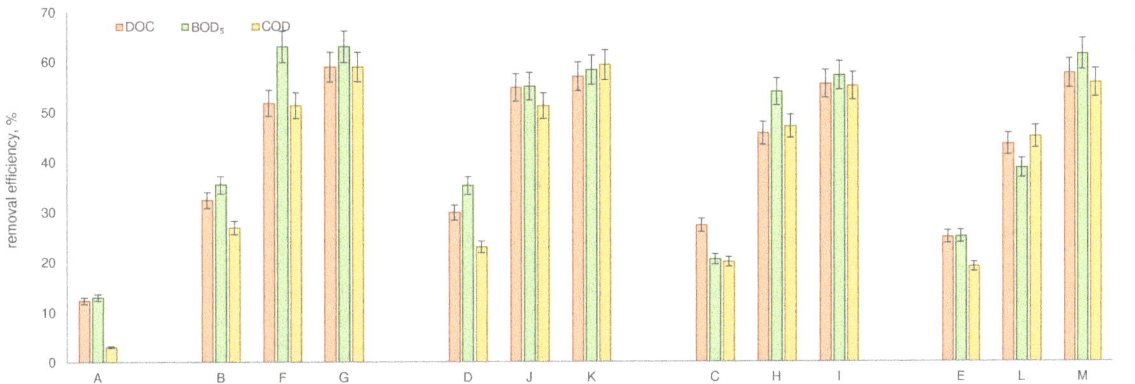

Figure 4. DOC, BOD$_5$ and COD removal efficiency in various membrane process configurations (TMP 0.4 MPa; A: MF 0.2 μm; B: MF 0.2 μm → UF PES 10 kDa; C: MF 0.2 μm → UF PES 30 kDa; D: MF 0.2 μm → UF C 10 kDa; E: MF 0.2 μm → UF C 30 kDa; F: MF 0.2 μm → UF PES 10 kDa → NF NPO10P; G: MF 0.2 μm → UF PES 10 kDa → NF NPO30P; H: MF 0.2 μm → UF PES 30 kDa → NF NPO10P; I: MF 0.2 μm → UF PES 30 kDa → NF NPO30P; J: MF 0.2 μm → UF C 10 kDa → NF NPO10P; K: MF 0.2 μm → UF C 10 kDa → NF NPO30P; L: MF 0.2 μm → UF C 30 kDa → NF NPO10P; M: MF 0.2 μm → UF C 30 kDa → NF NPO30P).

It was also advisable to assess the influence of the magnitude of the process driving force (in this case, transmembrane pressure) on the final quality of the effluent after HTC to be treated. It was found (Figure 5) that the transmembrane pressure value has no significant influence on the efficiency of removing organic compounds from the analysed liquid, both in the case of application of single (for MF) and sequentially successive membrane processes. In the analysed range of transmembrane pressures from 0.1 MPa to 0.4 MPa, the DOC, BOD$_5$ and COD retention coefficients were practically at the same level (differences did not exceed 10%).

The purpose of combining membrane processes is, in addition to increasing separation efficiency, to reduce the intensity of membrane fouling. This phenomenon is one of the main problems that occur during the operation of a membrane system, which is caused by the deposit of impurities from the solution to be purified on the membrane surface and in the membrane structure. This results in a decrease in the permeate flux (for processes run at a constant TMP) or an increase in the TMP (for processes run at a constant permeate flux). One method to reduce the intensity of membrane fouling is to purify solutions in integrated systems, such as using sequential membrane processes. As shown in Figure 6, the application of a sequence of the analysed membrane techniques has partially reduced the problem of membrane fouling. Comparison of the J/J$_0$ ratio, obtained in a separate UF process with the MF → UF and separate NF with MF → UF → NF, showed that the use of a pre-filtration improves the transport properties of the last membrane in the process chain. It was found that the application of MF 0.2 μm and UF PES 10 before any of the tested NF membranes reduced the problem of NF membrane fouling. MF applied before

UF removes some of the compounds blocking the subsequent membrane, and the use of a more compact UF membrane (10 kDa) made of both PES or C between MF and NF removes the remaining compounds deposited on the surfaces of NF membranes. The use of the 30 kDa cut-off UF membrane, which is characterised by larger pore size and thus lower efficiency of contaminants separation, does not as significantly improve the transport properties of NF membranes, as the fractions remaining in the solution (permeate) may penetrate the pores of the NF membranes or be deposited on its surface further contributing to their blocking. At the same time, it was observed that the type of membrane used in the NF is also crucial in terms of the membrane's susceptibility to blocking. It was found that after MF 0.2 μm and UF PES 10, the use of the NPO10P membrane resulted in the improvement of transport properties of this membrane. Relative membrane permeability (J/J_0) values were larger when using membrane NPO10P at the end of the sequence than when using membrane NPO30P, which may be due, among other things, to the more compact structure of the second membrane. In this case, its fouling is mainly due to the external fouling phenomenon.

Analysing the absolute values of permeate fluxes of nanofiltration membranes, it can be observed that for distilled water, the permeate flux of NPO10P membrane is 1.3272 m^3/m^2d and for NPO30P membrane: 0.2528 m^3/m^2d (at TMP 0.4 MPa). When these membranes were used to polish the solution after filtration with a sequence of MF 0.2 μm → UF PES 10 the permeate fluxes of these membranes decrease to 0.79 m^3/m^2d and 0.158 m^3/m^2d, respectively. This trend is in line with the literature reports [93]. It can be explained by analysing MWCO values and pore diameters of the membranes tested. According to [90], the cut-off of NPO10P membrane is greater and ranges between 1010 and 1400 Da (with pore diameter 0.80–1.29 nm), while for NPO30P membrane, it equals 500–700 Da (with pore diameter 0.57–0.93 nm). This can result in higher hydraulic resistance and thus lower permeability of NPO30P membranes.

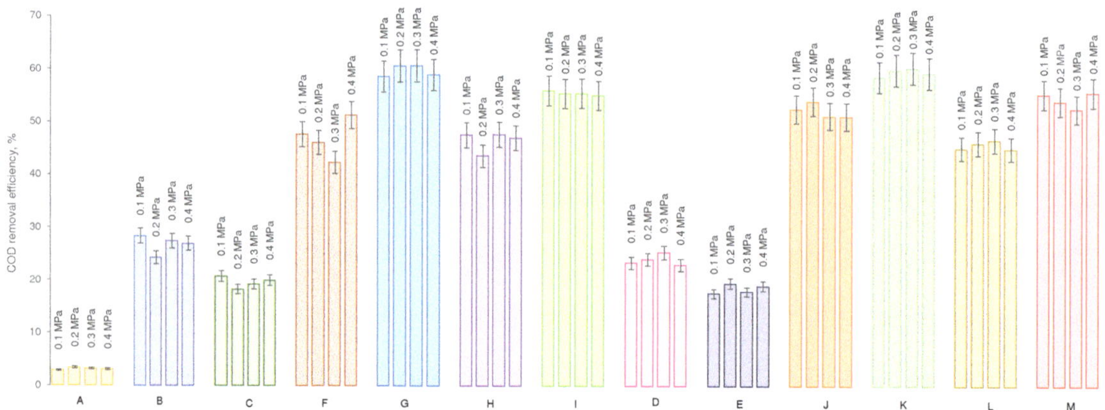

Figure 5. The influence of TMP on COD removal efficiency in various membrane processes configurations (A: MF 0.2 μm; B: MF 0.2 μm → UF PES 10 kDa; C: MF 0.2 μm → UF PES 30 kDa; D: MF 0.2 μm → UF C 10 kDa; E: MF 0.2 μm → UF C 30 kDa; F: MF 0.2 μm → UF PES 10 kDa → NF NPO10P; G: MF 0.2 μm → UF PES 10 kDa → NF NPO30P; H: MF 0.2 μm → UF PES 30 kDa → NF NPO10P; I: MF 0.2 μm → UF PES 30 kDa → NF NPO30P; J: MF 0.2 μm → UF C 10 kDa → NF NPO10P; K: MF 0.2 μm → UF C 10 kDa → NF NPO30P; L: MF 0.2 μm → UF C 30 kDa → NF NPO10P; M: MF 0.2 μm → UF C 30 kDa → NF NPO30P).

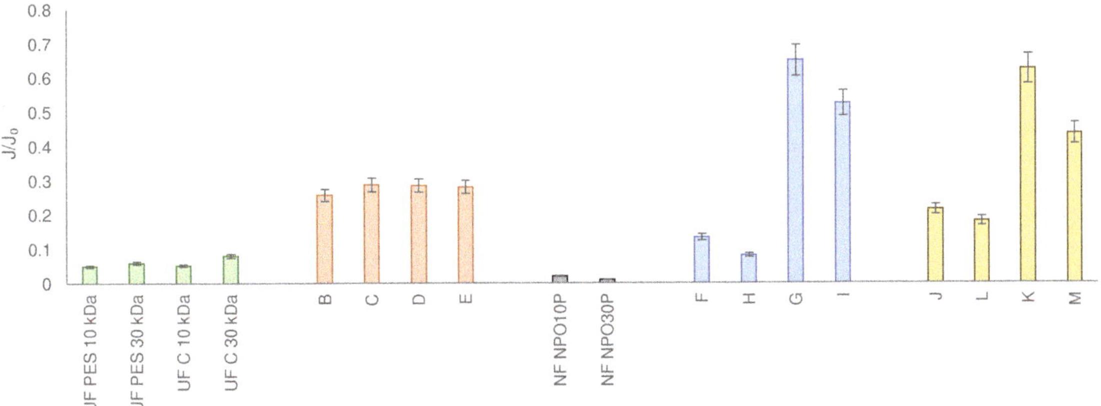

Figure 6. Comparison of relative membrane flux for various process configurations (TMP 0.4 MPa; (A: MF 0.2 μm; B: MF 0.2 μm → UF PES 10 kDa; C: MF 0.2 μm → UF PES 30 kDa; D: MF 0.2 μm → UF C 10 kDa; E: MF 0.2 μm → UF C 30 kDa; F: MF 0.2 μm → UF PES 10 kDa → NF NPO10P; G: MF 0.2 μm → UF PES 10 kDa → NF NPO30P; H: MF 0.2 μm → UF PES 30 kDa → NF NPO10P; I: MF 0.2 μm → UF PES 30 kDa → NF NPO30P; J: MF 0.2 μm → UF C 10 kDa → NF NPO10P; K: MF 0.2 μm → UF C 10 kDa → NF NPO30P; L: MF 0.2 μm → UF C 30 kDa → NF NPO10P; M: MF 0.2 μm → UF C 30 kDa → NF NPO30P).

Regarding the possibilities of recirculation of the organic fraction of HTC liquid by-products to AD reactor, the performance of an MF 0.2 μm → UF PES 10 kDa → NF NPO30P (G) cascade was the best among all of the investigated options: overall COD and BOD_5 removal efficiencies were close to the highest among all tested combinations (Figure 4). Overall, no trend can be found regarding the concentrations of the compounds found in the HTC effluent before and after different membranes and cascades (Table 4), which could be caused by the complexity of all the interactions between membranes, compounds as well as products of precipitation, produced during HTC.

GC-MS analysis of the raw solution of digestate liquid fraction after HTC samples and after the subsequent membrane filtration stages with different membrane sequences showed that for most low molecular weight organic compounds, the nanofiltration process's application does not allow for significant elimination of these substances.

The dominant compound in the HTC effluent, namely acetic acid, has not been detected in the permeate when these cascades were used (Table 4), which seems beneficial from the point of view of the use of retentate in the AD since volatile fatty acids are consumed during the acetogenesis [94], which is one of the main stages of AD process [95,96]. From the water-recovery perspective, especially in the context of its potential use in the agriculture, presence of acetic acid could be considered somewhat problematic as it is included on lists of pesticides, e.g., Pesticide Screening List for Luxembourg [97]. The dosage of pesticides is a subject of careful planning in agriculture, so a compound with pesticidal characteristics might be considered problematic when using reclaimed water containing acetic acid for watering the plants. The presence of 2,5-dimethyl pyrazine is also not desirable, as the compound is classified as harmful [98]. If acute toxicity for the compound, ranging between 1020 mg/kg and 1350 mg/kg of body mass (determined by LD50 tests) [99,100], is compared with the concentration presented in Table 4, it seems that significant exposure would be needed to achieve such doses. However, cautiousness is the foundation of present-day health and safety standards. Therefore, it seems sensible to advise an additional stage of treatment after the membrane cascade. Regarding the final concentration of 2,5-dimethyl pyrazine MF 0.2 μm → UF C 30 → NF NPO30P performed slightly better than MF 0.2 μm → UF PES 30 → NF NPO30P (Table 4). The presence of pyrazine is considered a health hazard by ECHA [101]. However, posing a problem of a much smaller magnitude [100]. Ethylpyrazine is a natural product of Maillard reaction [102], used as flavour [102,103], that does not seem problematic, especially in such low concentrations (Table 4).

Table 4. Results of the GC-MS analysis for the permeates.

Compound	Acetic Acid (MW 60.05 g/mol)	Pyrazine (MW 80.09 g/mol)	Pyrazine, 2,5-Dimethyl- (MW 108.14 g/mol)	Pyrazine, Ethyl- (MW 108.14 g/mol)	Propionic Acid (MW 74.08 g/mol)	Acetamide (MW 59.07 g/mol)	Cyclohexanecarboxylic Acid (MW 128.171 g/mol)	Hydrocinnamic Acid (MW 207.23 g/mol)
Sample/Unit	mg/dm^3							
HTC effluent	1080 ± 340	48 ± 9	31 ± 4	32 ± 3	110 ± 42	81 ± 14	18 ± 6	4 ± 3
(A) MF 0.2 µm	1210 ± 340	40 ± 6	36 ± 4	36 ± 3	91 ± 43	89 ± 14	24 ± 5	2 ± 3
(B) MF 0.2 µm → UF PES 10 kDa	960 ± 360	38 ± 6	34 ± 4	33 ± 3	155 ± 44	89 ± 14	20 ± 5	0.3 ± 3
(C) MF 0.2 µm → UF PES 30 kDa	1480 ± 350	42 ± 6	37 ± 4	38 ± 3	104 ± 45	94 ± 14	28 ± 5	2 ± 3
(D) MF 0.2 µm → UF C 10 kDa	1050 ± 350	38 ± 6	34 ± 4	34 ± 3	165 ± 45	87 ± 14	20 ± 5	4 ± 3
(E) MF 0.2 µm → UF C 30 kDa	1150 ± 340	40 ± 6	37 ± 4	37 ± 3	75 ± 42	90 ± 14	24 ± 5	1 ± 3
(F) MF 0.2 µm → UF PES 10 kDa → NF NPO10P	560 ± 340	40 ± 6	37 ± 4	33 ± 3	77 ± 42	97 ± 14	16 ± 5	n.d.
(G) MF 0.2 µm → UF PES 10 kDa → NF NPO30P	510 ± 340	37 ± 6	29 ± 4	30 ± 3	110 ± 42	86 ± 14	14 ± 5	n.d.
(H) MF 0.2 µm → UF PES 30 kDa → NF NPO10P	1650 ± 390	41 ± 7	33 ± 6	31 ± 6	104 ± 43	108 ± 15	28 ± 5	n.d.
(I) MF 0.2 µm → UF PES 30 kDa → NF NPO30P	n.d.	45 ± 6	36 ± 4	35 ± 3	15 ± 42	106 ± 14	12 ± 5	n.d.
(J) MF 0.2 µm → UF C 10 kDa → NF NPO10P	340 ± 350	40 ± 6	34 ± 4	33 ± 3	61 ± 43	96 ± 15	13 ± 6	n.d.
(K) MF 0.2 µm → UF C 10 kDa → NF NPO30P	590 ± 340	38 ± 6	31 ± 4	30 ± 3	128 ± 42	88 ± 14	15 ± 5	n.d.
(L) MF 0.2 µm → UF C 30 kDa → NF NPO10P	1500 ± 360	42 ± 6	37 ± 4	36 ± 3	94 ± 42	98 ± 14	25 ± 5	n.d.
(M) MF 0.2 µm → UF C 30 kDa → NF NPO30P	n.d.	40 ± 6	31 ± 4	30 ± 4	16 ± 42	103 ± 15	10 ± 5	n.d.

Propionic acid is merely an irritant with an unpleasant odour [104]. However, concentrations determined within the course of this study were higher than the advised threshold limit value (TLV) of 10 mg/dm^3 [104] for all membrane cascades (Table 4). Hydrocinnamic acid presence might not be desired as it is mentioned as an antifungal agent and plant metabolite [105], which cannot be unequivocally stated without further research. The presence of cyclohexanecarboxylic acid poses a severe problem, as the substance is considered a serious health hazard by ECHA [106]. The biggest problem is posed by acetamide's presence, as the substance is suspected of being carcinogenic to humans [107,108], based on rodent toxicity data and a common potentially genotoxic impurity in pharmaceutical manufacturing [108]. Acetamide is an impurity present in various medicines, and within the jurisdiction of pharmaceutical authorities, it falls under the limit of the threshold of toxicological concern (TTC), which is set as 1.5 µg/day for most known and all suspect carcinogens, unless experimental evidence can justify higher limits [108].

Due to these issues, it should be noted that the cascade membrane system cannot be considered as a stand-alone solution for water recovery from the effluent after HTC of agricultural digestate. Nonetheless, it should not be overlooked that such cascade systems make removal of such compounds significantly easier due to the significant reduction of COD in the permeates. All of the problematic compounds are hydrocarbons and can be oxidised. More research focused on the removal of the most problematic compounds from HTC effluent is advised.

4. Conclusions

In view of the insufficient treatment efficiency of the liquid fraction HTC effluent of the agricultural digestate in the independently used membrane processes, their sequential connections were used. The aim of such a procedure was both to improve the final quality of the treated digestate and to reduce the intensity of membrane fouling. The conducted research showed that the combination of sieving mechanisms of the examined membranes significantly increased the removal efficiency of organic compounds from the analysed HTC effluent. It was found that the final quality of the purified solution was determined by the membrane cut-off value, while the applied TMP and the ultrafiltration membrane material were of no significance. The best results were obtained by conducting sequential treatment of the solution in the following variant: MF 0.2 μm $\rightarrow$ UF PES 10 $\rightarrow$ NF NPO30P, which allowed reaching COD removal efficiency of almost 60%. All of the cascades outperformed a single membrane separation performed in the previous study regarding COD removal. Although the applied membrane sequence allows for very significant removal of organic compounds from the solution, there are still considerable amounts of low molecular weight substances remaining in the permeate. Overall, cascade system outperformed single membrane, reported in previous studies. Follow-up studies are advised; studies for further optimization of the cascade systems for separation of water and organics from liquid by-products of HTC of the agricultural digestate, as well as other types of biomass.

Author Contributions: Conceptualisation: A.U. and M.K.-K.; methodology: A.U., M.K.-K., M.W. and C.A.-B.; validation: A.U., M.W. and C.A.-B.; formal analysis: A.U., M.W. and C.A.-B.; investigation: A.U., M.C., M.B. and L.N.; resources: H.P.-K., P.S. and M.K.-K.; data curation: M.C. and M.B.; writing—original draft preparation: A.U., M.K.-K., L.N. and M.W.; writing—review and editing: A.U., M.K.-K., L.N., M.W. and P.S.; visualisation: A.U. and L.N.; supervision: H.P.-K., M.K.-K. and A.P.; project administration: H.P.-K., A.P., E.B. and G.B.; funding acquisition: H.P.-K., M.K.-K., E.B. and G.B. All authors have read and agreed to the published version of the manuscript.

Funding: The authors would like to thank the European Commission, the National Centre for Research and Development (Poland), Nederlandse Organisatie Voor Wetenschappelijk Onderzoek (Netherlands), and Swedish Research Council Formas for funding in the frame of the collaborative international consortium (RECOWATDIG) financed under the 2018 Joint call of the WaterWorks2017 ERA-NET Cofund. This ERA-NET is an integral part of the activities developed by the Water JPI. National Centre for Research and Development agreement number WATERWORKS2017/I/RECOWATDIG/01/2019.

Institutional Review Board Statement: Not applicable.

Informed Consent Statement: Not applicable.

Data Availability Statement: Not applicable.

Acknowledgments: The Authors would also like to thank Joanna Wolska and Katarzyna Smolińska-Kempisty for making microfiltration membranes available for research. The authors would like to thank Butor group for providing the digestate for this research.

Conflicts of Interest: The authors declare no conflict of interest. The funders had no role in the design of the study; in the collection, analyses, or interpretation of data; in the writing of the manuscript, or in the decision to publish the results.

References

1. Den Boer, J.; Obersteiner, G.; Gollnow, S.; den Boer, E.; Bodnárné Sándor, R. Enhancement of food waste management and its environmental consequences. *Energies* **2021**, *14*, 1790. [CrossRef]
2. Den Boer, J.; Kobel, P.; Dyjakon, A.; Urbańska, K.; Obersteiner, G.; Hrad, M.; Schmied, E.; den Boer, E. Food waste in Central Europe—Challenges and solutions. *E3S Web Conf.* **2017**, *22*, 00019. [CrossRef]
3. Piechota, G.; Igliński, B. Biomethane in Poland—Current Status, potential, perspective and development. *Energies* **2021**, *14*, 1517. [CrossRef]
4. Schmid, C.; Horschig, T.; Pfeiffer, A.; Szarka, N.; Thrän, D. Biogas upgrading: A review of national biomethane strategies and support policies in selected countries. *Energies* **2019**, *12*, 3803. [CrossRef]
5. Baltrénas, P.; Kolodynskij, V.; Zagorskis, A.; Baltrénaité, E. Research and analysis of biogas produced from sewage sludge using a batch bioreactor. *Environ. Technol.* **2018**, *39*, 3104–3112. [CrossRef]
6. Rasheed, T.; Anwar, M.T.; Ahmad, N.; Sher, F.; Khan, S.U.-D.; Ahmad, A.; Khan, R.; Wazeer, I. Valorisation and emerging perspective of biomass based waste-to-energy technologies and their socio-environmental impact: A review. *J. Environ. Manag.* **2021**, *287*, 112257. [CrossRef]
7. Plana, P.V.; Noche, B. A review of the current digestate distribution models: Storage and transport. In Proceedings of the 8 International Conference on Waste Management and The Environment (WM 2016), Valencia, Spain, 7–9 June 2016; Volume 202, pp. 345–357.
8. Monfet, E.; Aubry, G.; Ramirez, A.A. Nutrient removal and recovery from digestate: A review of the technology. *Biofuels* **2017**, *9*, 247–262. [CrossRef]
9. Daija, L.; Selberg, A.; Rikmann, E.; Zekker, I.; Tenno, T.; Tenno, T. The influence of lower temperature, influent fluctuations and long retention time on the performance of an upflow mode laboratory-scale septic tank. *Desalin. Water Treat.* **2016**, *57*, 18679–18687. [CrossRef]
10. Zekker, I.; Bhowmick, G.D.; Priks, H.; Nath, D.; Rikmann, E.; Jaagura, M.; Tenno, T.; Tämm, K.; Ghangrekar, M.M. ANAMMOX-denitrification biomass in microbial fuel cell to enhance the electricity generation and nitrogen removal efficiency. *Biodegradation* **2020**, *31*, 249–264. [CrossRef]
11. Tenno, T.; Uiga, K.; Mashirin, A.; Zekker, I.; Rikmann, E. Modeling closed equilibrium systems of H_2O-dissolved CO_2-solid $CaCO_3$. *J. Phys. Chem. A* **2017**, *121*, 3094–3100. [CrossRef]
12. Tenno, T.; Rikmann, E.; Uiga, K.; Zekker, I.; Mashirin, A.; Tenno, T. A novel proton transfer model of the closed equilibrium system H_2O-CO_2-$CaCO_3$-NHx. *Proc. Est. Acad. Sci.* **2018**, *67*, 260–270. [CrossRef]
13. Zekker, I.; Artemchuk, O.; Rikmann, E.; Ohimai, K.; Dhar Bhowmick, G.; Madhao Ghangrekar, M.; Burlakovs, J.; Tenno, T. Start-Up of Anammox SBR from non-specific inoculum and process acceleration methods by hydrazine. *Water* **2021**, *13*, 350. [CrossRef]
14. Zekker, I.; Rikmann, E.; Tenno, T.; Vabamäe, P.; Tomingas, M.; Menert, A.; Loorits, L.; Tenno, T. Anammox bacteria enrichment and phylogenic analysis in moving bed biofilm reactors. *Environ. Eng. Sci.* **2012**, *29*, 946–950. [CrossRef]
15. Zekker, I.; Raudkivi, M.; Artemchuk, O.; Rikmann, E.; Priks, H.; Jaagura, M.; Tenno, T. Mainstream-sidestream wastewater switching promotes anammox nitrogen removal rate in organic-rich, low-temperature streams. *Environ. Technol.* **2020**, *42*, 3073–3082. [CrossRef] [PubMed]
16. Vázquez-Rowe, I.; Golkowska, K.; Lebuf, V.; Vaneeckhaute, C.; Michels, E.; Meers, E.; Benetto, E.; Koster, D. Environmental assessment of digestate treatment technologies using LCA methodology. *Waste Manag.* **2015**, *43*, 442–459. [CrossRef] [PubMed]
17. Xiao, D.; Liu, D.L.; Feng, P.; Wang, B.; Waters, C.; Shen, Y.; Qi, Y.; Bai, H.; Tang, J. Future climate change impacts on grain yield and groundwater use under different cropping systems in the North China Plain. *Agric. Water Manag.* **2021**, *246*, 106685. [CrossRef]
18. Aghapour Sabbaghi, M.; Nazari, M.; Araghinejad, S.; Soufizadeh, S. Economic impacts of climate change on water resources and agriculture in Zayandehroud river basin in Iran. *Agric. Water Manag.* **2020**, *241*, 106323. [CrossRef]
19. Cortignani, R.; Dell'Unto, D.; Dono, G. Paths of adaptation to climate change in major Italian agricultural areas: Effectiveness and limits in supporting the profitability of farms. *Agric. Water Manag.* **2021**, *244*, 106433. [CrossRef]
20. Grusson, Y.; Wesström, I.; Svedberg, E.; Joel, A. Influence of climate change on water partitioning in agricultural watersheds: Examples from Sweden. *Agric. Water Manag.* **2021**, *249*, 106766. [CrossRef]
21. Angelidaki, I.; Ahring, B.K. Methods for increasing the biogas potential from the recalcitrant organic matter contained in manure. *Water Sci. Technol.* **2000**, *41*, 189–194. [CrossRef] [PubMed]
22. Żubrowska-Sudoł, M.; Podedworna, J.; Bisak, A.; Sytek-Szmeichel, K.; Krawczyk, P.; Garlicka, A. Intensification of anaerobic digestion efficiency with use of mechanical excess sludge disintegration in the context of increased energy production in wastewater treatment plants. *E3S Web Conf.* **2017**, *22*, 00208. [CrossRef]
23. Szatyłowicz, E.; Garlicka, A.; Żubrowska-Sudoł, M. The Effectiveness of the organic compounds released due to the hydrodynamic disintegration of sewage sludge. *Inżynieria Ekol.* **2017**, *18*, 47–55. [CrossRef]
24. Azman, S.; Milh, H.; Somers, M.H.; Zhang, H.; Huybrechts, I.; Meers, E.; Meesschaert, B.; Dewil, R.; Appels, L. Ultrasound-assisted digestate treatment of manure digestate for increased biogas production in small pilot scale anaerobic digesters. *Renew. Energy* **2020**, *152*, 664–673. [CrossRef]
25. Lippert, T.; Bandelin, J.; Xu, Y.; Liu, Y.C.; Robles, G.H.; Drewes, J.E.; Koch, K. From pre-treatment to co-treatment—How successful is ultrasonication of digested sewage sludge in continuously operated anaerobic digesters? *Renew. Energy* **2020**, *166*, 56–65. [CrossRef]

26. Villamil, J.A.; Mohedano, A.F.; Rodriguez, J.J.; De la Rubia, M.A. Anaerobic co-digestion of the aqueous phase from hydrothermally treated waste activated sludge with primary sewage sludge. A kinetic study. *J. Environ. Manag.* **2019**, *231*, 726–733. [CrossRef]
27. Seruga, P.; Krzywonos, M.; Seruga, A.; Niedzwiecki, L.; Pawlak-Kruczek, H.; Urbanowska, A. Anaerobic digestion performance: Separate collected vs. mechanical segregated organic fractions of municipal solid waste as feedstock. *Energies* **2020**, *13*, 3768. [CrossRef]
28. Seruga, P.; Krzywonos, M.; Pyżanowska, J.; Pawlak-Kruczek, H.; Urbanowska, A.; Niedzwiecki, L. Removal of ammonia from the municipal waste treatment effluents using natural minerals. *Molecules* **2019**, *24*, 3633. [CrossRef]
29. Pawlak-Kruczek, H.; Urbanowska, A.; Yang, W.; Brem, G.; Magdziarz, A.; Seruga, P.; Niedzwiecki, L.; Pozarlik, A.; Mlonka-Medrala, A.; Kabsch-Korbutowicz, M.; et al. Industrial process description for the recovery of agricultural water from digestate. *J. Energy Resour. Technol.* **2020**, *147*, 070917. [CrossRef]
30. Wilk, M.; Magdziarz, A.; Jayaraman, K.; Szymańska-Chargot, M.; Gökalp, I. Hydrothermal carbonization characteristics of sewage sludge and lignocellulosic biomass. A comparative study. *Biomass Bioenergy* **2019**, *120*, 166–175. [CrossRef]
31. Magdziarz, A.; Wilk, M.; Wądrzyk, M. Pyrolysis of hydrochar derived from biomass—Experimental investigation. *Fuel* **2020**, *267*, 117246. [CrossRef]
32. Smith, A.M.; Ekpo, U.; Ross, A.B. The influence of pH on the combustion properties of bio-coal following hydrothermal treatment of swine manure. *Energies* **2020**, *13*, 331. [CrossRef]
33. Sieradzka, M.; Rajca, P.; Zajemska, M.; Mlonka-Mędrala, A.; Magdziarz, A. Prediction of gaseous products from refuse derived fuel pyrolysis using chemical modelling software—Ansys Chemkin-Pro. *J. Clean. Prod.* **2020**, *248*, 119277. [CrossRef]
34. Aragón-Briceño, C.I.; Ross, A.B.; Camargo-Valero, M.A. Mass and energy integration study of hydrothermal carbonization with anaerobic digestion of sewage sludge. *Renew. Energy* **2021**, *167*, 473–483. [CrossRef]
35. Mihajlović, M.; Petrović, J.; Maletić, S.; Isakovski, M.K.; Stojanović, M.; Lopičić, Z.; Trifunović, S. Hydrothermal carbonization of Miscanthus × giganteus: Structural and fuel properties of hydrochars and organic profile with the ecotoxicological assessment of the liquid phase. *Energy Convers. Manag.* **2018**, *159*, 254–263. [CrossRef]
36. Gao, N.; Li, Z.; Quan, C.; Miskolczi, N.; Egedy, A. A new method combining hydrothermal carbonization and mechanical compression in-situ for sewage sludge dewatering: Bench-scale verification. *J. Anal. Appl. Pyrolysis* **2019**, *139*, 187–195. [CrossRef]
37. Jackowski, M.; Semba, D.; Trusek, A.; Wnukowski, M.; Niedzwiecki, L.; Baranowski, M.; Krochmalny, K.; Pawlak-Kruczek, H. Hydrothermal carbonization of brewery's spent grains for the production of solid biofuels. *Beverages* **2019**, *5*, 12. [CrossRef]
38. Wang, S.; Persson, H.; Yang, W.; Jönsson, P.G. Pyrolysis study of hydrothermal carbonization-treated digested sewage sludge using a Py-GC/MS and a bench-scale pyrolyzer. *Fuel* **2020**, *262*, 116335. [CrossRef]
39. Arauzo, P.; Olszewski, M.; Kruse, A. Hydrothermal carbonization Brewer's spent grains with the focus on improving the degradation of the feedstock. *Energies* **2018**, *11*, 3226. [CrossRef]
40. Reza, M.T.; Yan, W.; Uddin, M.H.; Lynam, J.G.; Hoekman, S.K.; Coronella, C.J.; Vásquez, V.R. Reaction kinetics of hydrothermal carbonization of loblolly pine. *Bioresour. Technol.* **2013**, *139*, 161–169. [CrossRef]
41. Román, S.; Libra, J.; Berge, N.; Sabio, E.; Ro, K.; Li, L.; Ledesma, B.; Alvarez, A.; Bae, S. Hydrothermal carbonization: Modeling, final properties design and applications: A review. *Energies* **2018**, *11*, 216. [CrossRef]
42. Kruse, A.; Zevaco, T.A. Properties of hydrochar as function of feedstock, reaction conditions and post-treatment. *Energies* **2018**, *11*, 674. [CrossRef]
43. Picone, A.; Volpe, M.; Messineo, A. Process water recirculation during hydrothermal carbonization of waste biomass: Current knowledge and challenges. *Energies* **2021**, *14*, 2962. [CrossRef]
44. Funke, A.; Ziegler, F. Hydrothermal carbonisation of biomass: A summary and discussion of chemical mechanisms for process engineering. *Biofuels Bioprod. Biorefin.* **2010**, *4*, 160–177. [CrossRef]
45. Reza, M.T.; Andert, J.; Wirth, B.; Busch, D.; Pielert, J.; Lynam, J.G.; Mumme, J. Hydrothermal carbonization of biomass for energy and crop production. *Appl. Bioenergy* **2014**, *1*, 11–29. [CrossRef]
46. Ameen, M.; Zamri, N.M.; May, S.T.; Azizan, M.T.; Aqsha, A.; Sabzoi, N.; Sher, F. Effect of acid catalysts on hydrothermal carbonization of Malaysian oil palm residues (leaves, fronds, and shells) for hydrochar production. *Biomass Convers. Biorefin.* **2021**, 1–12. [CrossRef]
47. Maniscalco, M.P.; Volpe, M.; Messineo, A. Hydrothermal carbonization as a valuable tool for energy and environmental applications: A review. *Energies* **2020**, *13*, 4098. [CrossRef]
48. Pauline, A.L.; Joseph, K. Hydrothermal carbonization of organic wastes to carbonaceous solid fuel—A review of mechanisms and process parameters. *Fuel* **2020**, *279*, 118472. [CrossRef]
49. Lucian, M.; Volpe, M.; Gao, L.; Piro, G.; Goldfarb, J.L.; Fiori, L. Impact of hydrothermal carbonization conditions on the formation of hydrochars and secondary chars from the organic fraction of municipal solid waste. *Fuel* **2018**, *233*, 257–268. [CrossRef]
50. Shafie, S.A.; Al-attab, K.A.; Zainal, Z.A. Effect of hydrothermal and vapothermal carbonization of wet biomass waste on bound moisture removal and combustion characteristics. *Appl. Therm. Eng.* **2018**, *139*, 187–195. [CrossRef]
51. Acharjee, T.C.; Coronella, C.J.; Vasquez, V.R. Effect of thermal pretreatment on equilibrium moisture content of lignocellulosic biomass. *Bioresour. Technol.* **2011**, *102*, 4849–4854. [CrossRef] [PubMed]

52. Ahmed, M.; Andreottola, G.; Elagroudy, S.; Negm, M.S.; Fiori, L. Coupling hydrothermal carbonization and anaerobic digestion for sewage digestate management: Influence of hydrothermal treatment time on dewaterability and bio-methane production. *J. Environ. Manag.* **2021**, *281*, 111910. [CrossRef]

53. Szufa, S.; Piersa, P.; Adrian, Ł.; Czerwińska, J.; Lewandowski, A.; Lewandowska, W.; Sielski, J.; Dzikuć, M.; Wróbel, M.; Jewiarz, M.; et al. Sustainable drying and torrefaction processes of miscanthus for use as a pelletized solid biofuel and biocarbon-carrier for fertilizers. *Molecules* **2021**, *26*, 1014. [CrossRef]

54. Wang, L.; Zhang, L.; Li, A. Hydrothermal treatment coupled with mechanical expression at increased temperature for excess sludge dewatering: Influence of operating conditions and the process energetics. *Water Res.* **2014**, *65*, 85–97. [CrossRef]

55. Wang, L.F.; Qian, C.; Jiang, J.K.; Ye, X.D.; Yu, H.Q. Response of extracellular polymeric substances to thermal treatment in sludge dewatering process. *Environ. Pollut.* **2017**, *231*, 1388–1392. [CrossRef]

56. Louwes, A.C.; Halfwerk, R.B.; Bramer, E.A.; Brem, G. Experimental study on fast pyrolysis of raw and torrefied woody Biomass. *Energy Technol.* **2020**, *8*, 1900799. [CrossRef]

57. Louwes, A.C.; Basile, L.; Yukananto, R.; Bhagwandas, J.C.; Bramer, E.A.; Brem, G. Torrefied biomass as feed for fast pyrolysis: An experimental study and chain analysis. *Biomass Bioenergy* **2017**, *105*, 116–126. [CrossRef]

58. Kaczor, Z.; Buliński, Z.; Werle, S. Modelling approaches to waste biomass pyrolysis: A review. *Renew. Energy* **2020**, *159*, 427–443. [CrossRef]

59. Sobek, S.; Werle, S. Solar pyrolysis of waste biomass: Part 1 reactor design. *Renew. Energy* **2019**, *143*, 1939–1948. [CrossRef]

60. Kantorek, M.; Jesionek, K.; Polesek-Karczewska, S.; Ziółkowski, P.; Stajnke, M.; Badur, J. Thermal utilization of meat-and-bone meal using the rotary kiln pyrolyzer and the fluidized bed boiler—The performance of pilot-scale installation. *Renew. Energy* **2021**, *164*, 1447–1456. [CrossRef]

61. Volpe, M.; Wüst, D.; Merzari, F.; Lucian, M.; Andreottola, G.; Kruse, A.; Fiori, L. One stage olive mill waste streams valorisation via hydrothermal carbonisation. *Waste Manag.* **2018**, *80*, 224–234. [CrossRef]

62. Procentese, A.; Russo, M.E.; Di Somma, I.; Marzocchella, A. Kinetic characterization of enzymatic hydrolysis of apple pomace as feedstock for a sugar-based biorefinery. *Energies* **2020**, *13*, 1051. [CrossRef]

63. Steinbach, D.; Kruse, A.; Sauer, J.; Vetter, P. Sucrose is a promising feedstock for the synthesis of the platform chemical hydroxymethylfurfural. *Energies* **2018**, *11*, 645. [CrossRef]

64. Lühmann, T.; Wirth, B. Sewage sludge valorization via hydrothermal carbonization: Optimizing dewaterability and phosphorus release. *Energies* **2020**, *13*, 4417. [CrossRef]

65. Meisel, K.; Clemens, A.; Führner, C.; Breulmann, M.; Majer, S.; Thrän, D. Comparative life cycle assessment of HTC concepts valorizing sewage sludge for energetic and agricultural use. *Energies* **2019**, *12*, 786. [CrossRef]

66. Paul, S.; Dutta, A.; Defersha, F. Mechanical and alkaline hydrothermal treated corn residue conversion in to bioenergy and biofertilizer: A resource recovery concept. *Energies* **2018**, *11*, 516. [CrossRef]

67. Aragon-Briceño, C.I.; Pozarlik, A.K.; Bramer, E.A.; Niedzwiecki, L.; Pawlak-Kruczek, H.; Brem, G. Hydrothermal carbonization of wet biomass from nitrogen and phosphorus approach: A review. *Renew. Energy* **2021**, *171*, 401–415. [CrossRef]

68. Ismail, H.Y.; Shirazian, S.; Skoretska, I.; Mynko, O.; Ghanim, B.; Leahy, J.J.; Walker, G.M.; Kwapinski, W. ANN-Kriging hybrid model for predicting carbon and inorganic phosphorus recovery in hydrothermal carbonization. *Waste Manag.* **2019**, *85*, 242–252. [CrossRef]

69. Wilk, M.; Magdziarz, A.; Kalemba-Rec, I.; Szymańska-Chargot, M. Upgrading of green waste into carbon-rich solid biofuel by hydrothermal carbonization: The effect of process parameters on hydrochar derived from acacia. *Energy* **2020**, *202*, 117717. [CrossRef]

70. Atallah, E.; Zeaiter, J.; Ahmad, M.N.; Kwapinska, M.; Leahy, J.J.; Kwapinski, W. The effect of temperature, residence time, and water-sludge ratio on hydrothermal carbonization of DAF dairy sludge. *J. Environ. Chem. Eng.* **2020**, *8*, 103599. [CrossRef]

71. Atallah, E.; Kwapinski, W.; Ahmad, M.N.; Leahy, J.J.; Zeaiter, J. Effect of water-sludge ratio and reaction time on the hydrothermal carbonization of olive oil mill wastewater treatment: Hydrochar characterization. *J. Water Process. Eng.* **2019**, *31*, 100813. [CrossRef]

72. Ghanim, B.M.; Kwapinski, W.; Leahy, J.J. Hydrothermal carbonisation of poultry litter: Effects of initial pH on yields and chemical properties of hydrochars. *Bioresour. Technol.* **2017**, *238*, 78–85. [CrossRef] [PubMed]

73. Aragón-Briceño, C.I.; Grasham, O.; Ross, A.B.; Dupont, V.; Camargo-Valero, M.A. Hydrothermal carbonization of sewage digestate at wastewater treatment works: Influence of solid loading on characteristics of hydrochar, process water and plant energetics. *Renew. Energy* **2020**, *157*, 959–973. [CrossRef]

74. Śliz, M.; Wilk, M. A comprehensive investigation of hydrothermal carbonization: Energy potential of hydrochar derived from Virginia mallow. *Renew. Energy* **2020**, *156*, 942–950. [CrossRef]

75. Surup, G.R.; Leahy, J.J.; Timko, M.T.; Trubetskaya, A. Hydrothermal carbonization of olive wastes to produce renewable, binder-free pellets for use as metallurgical reducing agents. *Renew. Energy* **2020**, *155*, 347–357. [CrossRef]

76. Ghanim, B.M.; Pandey, D.S.; Kwapinski, W.; Leahy, J.J. Hydrothermal carbonisation of poultry litter: Effects of treatment temperature and residence time on yields and chemical properties of hydrochars. *Bioresour. Technol.* **2016**, *216*, 373–380. [CrossRef]

77. Shrestha, A.; Acharya, B.; Farooque, A.A. Study of hydrochar and process water from hydrothermal carbonization of sea lettuce. *Renew. Energy* **2021**, *163*, 589–598. [CrossRef]

78. Atallah, E.; Zeaiter, J.; Ahmad, M.N.; Leahy, J.J.; Kwapinski, W. Hydrothermal carbonization of spent mushroom compost waste compared against torrefaction and pyrolysis. *Fuel Process. Technol.* **2021**, *216*, 106795. [CrossRef]
79. Atallah, E.; Kwapinski, W.; Ahmad, M.N.; Leahy, J.J.; Al-Muhtaseb, A.H.; Zeaiter, J. Hydrothermal carbonization of olive mill wastewater: Liquid phase product analysis. *J. Environ. Chem. Eng.* **2019**, *7*, 102833. [CrossRef]
80. Parmar, K.R.; Ross, A.B. Integration of hydrothermal carbonisation with anaerobic digestion; Opportunities for valorisation of digestate. *Energies* **2019**, *12*, 1586. [CrossRef]
81. Ferrentino, R.; Merzari, F.; Fiori, L.; Andreottola, G. Coupling hydrothermal carbonization with anaerobic digestion for sewage sludge treatment: Influence of HTC liquor and hydrochar on biomethane production. *Energies* **2020**, *13*, 6262. [CrossRef]
82. Musa, B.; Arauzo, P.J.; Olszewski, M.P.; Kruse, A. Electricity generation in microbial fuel cell from wet torrefaction wastewater and locally developed corncob electrodes. *Fuel Cells* **2021**, *21*, 182–194. [CrossRef]
83. Román, F.; Adolph, J.; Hensel, O. Hydrothermal treatment of biogas digestate as a pretreatment to reduce fouling in membrane filtration. *Bioresour. Technol. Rep.* **2021**, *13*, 100638. [CrossRef]
84. Urbanowska, A.; Kabsch-Korbutowicz, M.; Wnukowski, M.; Seruga, P.; Baranowski, M.; Pawlak-Kruczek, H.; Serafin-Tkaczuk, M.; Krochmalny, K.; Niedzwiecki, L. Treatment of liquid by-products of hydrothermal carbonization (HTC) of agricultural digestate using membrane separation. *Energies* **2020**, *13*, 262. [CrossRef]
85. Cao, Z.; Hülsemann, B.; Wüst, D.; Illi, L.; Oechsner, H.; Kruse, A. Valorization of maize silage digestate from two-stage anaerobic digestion by hydrothermal carbonization. *Energy Convers. Manag.* **2020**, *222*, 113218. [CrossRef]
86. Cao, Z.; Jung, D.; Olszewski, M.P.; Arauzo, P.J.; Kruse, A. Hydrothermal carbonization of biogas digestate: Effect of digestate origin and process conditions. *Waste Manag.* **2019**, *100*, 138–150. [CrossRef] [PubMed]
87. MICRODYN-NADIR. Flat Sheet Membrane Data Sheets—MICRODYN-NADIR. Available online: https://www.microdyn-nadir.com/flat-sheet-membrane-data-sheets/ (accessed on 15 February 2021).
88. Kabsch-Korbutowicz, M. Ultrafiltration as a method of separation of natural organic matter from water. *Mater. Sci. Pol.* **2008**, *26*, 459–467.
89. Rice, E.W.; Baird, R.B.; Eaton, A.D. (Eds.) *Standard Methods for the Examination of Water and Wastewater*, 23rd ed.; American Public Health Association: Washington, DC, USA; American Water Works Association: Denver, CO, USA; Water Environment Federation: Alexandria, VA, USA, 2017.
90. Kovacs, Z.; Samhaber, W. Characterization of nanofiltration membranes with uncharged solutes. *Membrantechnika* **2008**, *12*, 22–36.
91. Klein, K.; Kattel, E.; Goi, A.; Kivi, A.; Dulova, N.; Saluste, A.; Zekker, I.; Trapido, M.; Tenno, T. Combined treatment of pyrogenic wastewater from oil shale retorting. *Oil Shale* **2017**, *34*, 82–96. [CrossRef]
92. Lucian, M.; Volpe, M.; Merzari, F.; Wüst, D.; Kruse, A.; Andreottola, G.; Fiori, L. Hydrothermal carbonization coupled with anaerobic digestion for the valorization of the organic fraction of municipal solid waste. *Bioresour. Technol.* **2020**, *314*, 123734. [CrossRef]
93. Córdova, A.; Astudillo, C.; Giorno, L.; Guerrero, C.; Conidi, C.; Illanes, A.; Cassano, A. Nanofiltration potential for the purification of highly concentrated enzymatically produced oligosaccharides. *Food Bioprod. Process.* **2016**, *98*, 50–61. [CrossRef]
94. Reza, M.T.; Werner, M.; Pohl, M.; Mumme, J. Evaluation of Integrated Anaerobic Digestion and Hydrothermal Carbonization for Bioenergy Production. *J. Vis. Exp.* **2014**, *88*, e51734. [CrossRef]
95. Kozłowski, K.; Dach, J.; Lewicki, A.; Malińska, K.; Do Carmo, I.E.P.; Czekała, W. Potential of biogas production from animal manure in Poland. *Arch. Environ. Prot.* **2019**, *45*, 99–108. [CrossRef]
96. Seruga, P.; Krzywonos, M.; Wilk, M. Thermophilic co-digestion of the organic fraction of municipal solid wastes—The influence of food industry wastes addition on biogas production in full-scale operation. *Molecules* **2018**, *23*, 3146. [CrossRef]
97. Krier, J. S69 | LUXPEST | Pesticide Screening List for Luxembourg. *Zenodo* **2020**, *23*, 3146. [CrossRef]
98. Substance Information ECHA. 2,5-Dimethylpyrazine. Available online: https://echa.europa.eu/substance-information/-/substanceinfo/100.004.200 (accessed on 19 February 2021).
99. Moran, E.J.; Easterday, O.D.; Oser, B.L. Acute oral toxicity of selected flavor chemicals. *Drug Chem. Toxicol.* **1980**, *3*, 249–258. [CrossRef] [PubMed]
100. Nishie, K.; Waiss, A.C.; Keyl, A.C. Pharmacology of alkyl and hydroxyalkylpyrazines. *Toxicol. Appl. Pharmacol.* **1970**, *17*, 244–249. [CrossRef]
101. Substance Information ECHA. Pyrazine. Available online: https://echa.europa.eu/substance-information/-/substanceinfo/100.005.480 (accessed on 19 February 2021).
102. PubChem. 2-Ethylpyrazine | C6H8N2. Available online: https://pubchem.ncbi.nlm.nih.gov/compound/26331 (accessed on 19 February 2021).
103. FEMA. 2-Ethylpyrazine. Available online: https://www.femaflavor.org/flavor-library/2-ethylpyrazine (accessed on 19 February 2021).
104. Amoore, J.E.; Hautala, E. Odor as an aid to chemical safety: Odor thresholds compared with threshold limit values and volatilities for 214 industrial chemicals in air and water dilution. *J. Appl. Toxicol.* **1983**, *3*, 272–290. [CrossRef]
105. PubChem. 3-Phenylpropionic acid C9H10O2. Available online: https://pubchem.ncbi.nlm.nih.gov/compound/107 (accessed on 19 February 2021).
106. Substance Information ECHA. Cyclohexanecarboxylic Acid. Available online: https://echa.europa.eu/substance-information/-/substanceinfo/100.002.465 (accessed on 19 February 2021).

107. Substance Information ECHA. Acetamide. Available online: https://echa.europa.eu/substance-information/-/substanceinfo/10 0.000.430 (accessed on 19 February 2021).
108. Székely, G.; Fritz, E.; Bandarra, J.; Heggie, W.; Sellergren, B. Removal of potentially genotoxic acetamide and arylsulfonate impurities from crude drugs by molecular imprinting. *J. Chromatogr. A* **2012**, *1240*, 52–58. [CrossRef] [PubMed]

Article

Linear and Non-Linear Regression Analysis for the Adsorption Kinetics of SO$_2$ in a Fixed Carbon Bed Reactor—A Case Study

Anna M. Kisiela-Czajka [1,*] and **Bartosz Dziejarski** [2]

[1] Faculty of Mechanical and Power Engineering, Wroclaw University of Science and Technology, 50-370 Wroclaw, Poland

[2] Faculty of Environmental Engineering, Wroclaw University of Science and Technology, 50-370 Wroclaw, Poland; bartosz.dziejarski@pwr.edu.pl

* Correspondence: anna.kisiela-czajka@pwr.edu.pl; Tel.: +48-713-204-485

Abstract: Here, we determined the kinetic parameters of SO$_2$ adsorption on unburned carbons from lignite fly ash and activated carbons based on hard coal dust. The model studies were performed using the linear and non-linear regression method for the following models: pseudo first and second order, intraparticle diffusion, and chemisorption on a heterogeneous surface. The quality of the fitting of a given model to empirical data was assessed based on: R^2, R, Δq, SSE, ARE, χ^2, HYBRID, MPSD, EABS, and SNE. It was clearly shown that the linear regression more accurately reflects the behaviour of the adsorption system, which is consistent with the first-order kinetic reaction—for activated carbons (SO$_2$ + Ar) or chemisorption on a heterogeneous surface—for unburned carbons (SO$_2$ + Ar and SO$_2$ + Ar + H$_2$O$_{(g)}$ + O$_2$) and activated carbons (SO$_2$ + Ar + H$_2$O$_{(g)}$ + O$_2$). Importantly, usually, each of the approaches (linear/non-linear) indicated a different mechanism of the studied phenomenon. A certain universality of the χ^2 and HYBRID functions has been proved, the minimization of which repeatedly led to the lowest SNE values for the indicated models. Fitting data by any of the non-linear equations based on the R or R^2 functions only cannot be treated as evidence/prerequisite of the existence of a given adsorption mechanism.

Keywords: unburned carbon; fly ash; activated carbon; adsorption kinetics; statistical regression

Citation: Kisiela-Czajka, A.M.; Dziejarski, B. Linear and Non-Linear Regression Analysis for the Adsorption Kinetics of SO$_2$ in a Fixed Carbon Bed Reactor—A Case Study. *Energies* **2022**, *15*, 633. https://doi.org/10.3390/en15020633

Academic Editors: Robert Król, Witold Kawalec, Izabela Sówka and Krzysztof M. Czajka

Received: 27 November 2021
Accepted: 14 January 2022
Published: 17 January 2022

Publisher's Note: MDPI stays neutral with regard to jurisdictional claims in published maps and institutional affiliations.

1. Introduction

The structure of fuel consumption in Poland, based on hard coal and lignite, makes the energy sector one of the main sources of pollutants emitted into the air. According to the information presented in the report of the National Center for Balancing and Emission Management in Warsaw, in 2015–2017 the commercial power industry was responsible for 43–52% of the national SO$_2$ emissions [1]. In EU countries, on the other hand, the emission of sulfur oxides (total) from the sector of thermal power plants and other combustion installations, in 2014 accounted for 66.9% of the total emissions from all installations covered by the provisions of the Directive on the Establishment of the European Pollutant Release and Transfer Register (E-PRTR) [2].

Due to the fast and unlimited spread of pollutants and direct impact on the natural environment, a significant tightening of emission standards for air pollutants is observed. Pursuant to EU regulations, emission limits of up to 200 mg SO$_2$·Nm^{-3} have been in force since 2016, and, according to the projections developed in 2019, the national commitment to reduce emissions in the period 2020–2029 and from 2030 was set at 59% and 70%, respectively, compared to the emissions recorded in 2005 [3].

In the light of the information presented in the literature, the least invasive method that does not interfere with the combustion process is the capture of pollutants after the combustion process (i.e., post-combustion capture of pollutants). One of the solutions presented in the literature is an innovative technology for the use of unburned carbon from fly ash for flue gas cleaning [4–9]. Meanwhile, the attempts to re-utilize unburned carbon in this way may not only reduce the emission of pollutants but also enable an increase in the efficiency of electricity generation, minimize additional costs related to the storage of high-calorific waste (considering the legalization of the recovery process), and increase the commercial attractiveness of valorized fly ash.

The correctness of the method of adsorptive desulfurization of flue gas with the use of porous carbon materials is based on the knowledge of the adsorption mechanism and the state of adsorbate molecules in the pores of the adsorbent. According to literature reports on the methods of reducing SO_2 emissions in installations in the energy production and transformation sector, adsorption on the surface of the carbon adsorbent turns out to be one of the most frequently analyzed solutions [10]. Despite the quite extensive variety of methods for removing sulfur dioxide from boiler flue gases [11,12], the practical significance of most of them is limited, and the research does not go beyond laboratory work.

The research on the kinetics and dynamics of adsorption is used to understand the interaction between the adsorbent and the adsorbate. While in the case of preparation on a laboratory scale testing the reaction rate is not necessary, it is imperative if one wants to adapt a given reaction on a technical scale. In light of the information presented in the literature, adsorption on a heterogeneous surface is most often described by the following models: pseudo first-order and pseudo second-order kinetic models (PFO and PSO), intraparticle diffusion model (Weber–Morris), and chemisorption on a heterogeneous surface (Elovich) [13–16]. However, a commonly used tool for the analysis of empirical data is linear regression, and the classic method of least squares is used to determine the optimal values of unknown parameters [17–21]. Nevertheless, the greatest disadvantage of the above method is the undefined distribution of empirical data errors when determining the parameters of a given model, as a result of transforming kinetic equations into linearized forms. This may affect its variance (a measure of the accuracy of fitting to experimental data) and cause a misinterpretation of kinetic parameters, ultimately leading to an incorrect indication of the optimal model and the form of its equation [22–25].

This makes non-linear regression or non-linear fit analysis worth considering, as it provides a mathematically rigorous method for determining the kinetic parameters and adsorption dynamics while using the basic form of the equation, which offers the most accurate fit of the model curve function to the experimental data [26,27]. It is closely related to the minimization of the value of the error function distribution between the experimental data and the predicted model value obtained based on the convergence criteria, i.e., the ability of a given model to "lead" towards the empirical result [27,28].

In view of the above requirements, it is necessary to identify and explain the usefulness of linear and non-linear regression in various adsorption systems. Interpretation of the values of individual error functions enables the selection of the most convenient and precise optimization criteria in the kinetics and dynamics of adsorption.

Referring to the above, this study aims to determine the parameters of SO_2 adsorption kinetics by the method of linear and non-linear regression for the following models: pseudo first-order and pseudo second-order kinetic model, intraparticle diffusion, and chemisorption on a heterogeneous surface. The quality of the fitting of a given model to empirical data was assessed based on the following: determination coefficient (R^2), correlation coefficient (R), relative standard deviation (Δq), sum squared error (SSE), average relative error (ARE), chi-square test (χ^2), hybrid fractional error function (HYBRID), Marquardt's percent standard deviation (MPSD), the sum of absolute errors (EABS), and the sum of normalized errors (SNE). The subject of research is selected fractions of unburned carbon recovered from lignite fly ash, created as a result of the nominal operation of the pulverized carbon boiler of a Polish power unit. Selected commercial activated carbons dedicated to industrial gas purification processes and traded on the domestic and foreign markets were used as reference materials.

2. Experimental Section

2.1. Materials

The subjects of the research presented in this paper are selected fractions of unburned carbon recovered from lignite fly ash, resulting from the nominal operation of the pulverized carbon boiler BB-1150 in Bełchatów Power Plant (370 MW unit). Unburned carbon along with fly ash was collected with the use of demonstration installation from the ash hoppers located under the second pas chamber and rotary air heater (more in [29]). The combustible parts have been separated by a mechanical classification system with a capacity of 500 kg·h^{-1} into three grain classes: ~0.8 mm and 57.3% (marked UnCarb_HAsh), ~1.0 mm and 44.6% (marked UnCarb_MAsh), and ~1.5 mm and 12.8% (marked UnCarb_LAsh). The commercial activated carbons AKP-5 and AKP-5/A were used as reference materials, manufactured and distributed by GRYFSKAND Sp. z o.o. (Gryfino, Poland), Hajnówka Branch, active carbon production plant (more in [30]). Both products were developed for the treatment of industrial gases, boiler flue gases in power plants, or waste incineration plants, including sulfur dioxide, nitrogen oxides, hydrogen chloride or dioxins, and furans.

The characterization of the carbon substance structure of the test material, including the analysis of the degree of pore expansion and the identification of surface oxygen functional groups, was presented in an earlier work by one of the authors [30].

2.2. Experimental Studies

The model tests were carried out on the results of laboratory tests for SO$_2$ adsorption on a fixed carbon bed, which were the subject of one of the author's earlier works, published in [30]. The experiments were carried out at a temperature of 120 °C, in the presence of gas mixtures flowing linearly through 1.73×10^{-4} m^3 of the bed and with the following composition (in volume concentration):

1. 5% of sulfur dioxide and 95% of argon (as carrier gas) and a volumetric flow rate of 2×10^{-3} m^3·min^{-1};
2. 2.5% of sulfur dioxide, 11% of water vapor, 20% of oxygen and 66.5% of argon (as carrier gas) and a volumetric flow rate of 2.05×10^{-3} m^3·min^{-1}.

Measurements were made on a fixed-bed reactor (Figure 1), which enabled the assessment of both the degree and dynamics of the adsorption process. The water vapor was generated using Ar from a bubbling container that was bathed in 60.5 ± 0.1 °C water, and the relative humidity was controlled using the Ar flow based on the water vapor Antonio equation. The gas flow line to the reactor was maintained at an elevated temperature (120 °C) to prevent condensation. The final concentration of sulfur in the solid phase was used to assess the effectiveness of sulfur dioxide adsorption, which was carried out in accordance with the PN-EN 04584:2001 standard while correcting this value by the share of the so-called fuel sulfur:

$$m_{S,\,t} = m_{S,\infty} - m_{S,0} \tag{1}$$

where $m_{S,t}$ is the mass of adsorbed sulfur, mg; $m_{S,\infty}$ is the total mass of sulfur in the sample after the adsorption process, mg; $m_{S,0}$ represents the mass of sulfur in the sample before the adsorption process, mg.

Due to the possibility of adsorption of various forms of sulfur dioxide and the occurrence of indirect chemical reactions, as a consequence of the presence of O$_2$ and H$_2$O$_{(g)}$ in the reaction system, no comparative analyzes were performed for the participation of sulfur dioxide in the solid phase.

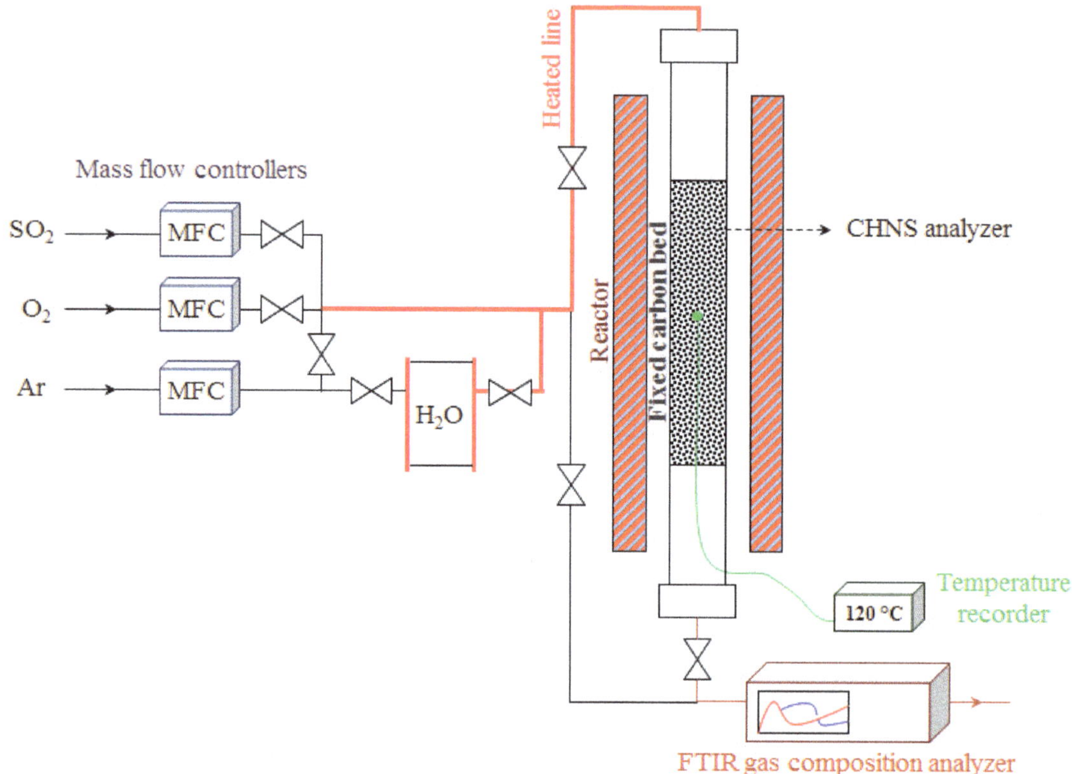

Figure 1. Scheme of the measuring system for SO$_2$ adsorption.

2.3. Modelling Studies

2.3.1. Reaction Kinetics Models

Processes carried out in the environment of SO$_2$ + Ar gases (UnCarb_HAsh, Un-Carb_MAsh, UnCarb_LAsh, AKP-5, and AKP-5/A samples) and SO$_2$ + O$_2$ + H$_2$O$_{(g)}$ + Ar (UnCarb_LAsh and AKP-5/A samples) were subjected to model tests. For this study, four models were chosen [31–35], i.e.,:

- pseudo first-order kinetic model developed by Legergren,
- pseudo second-order kinetic model developed by Ho i McKaya,
- Weber-Morris intraparticle diffusion model, and
- chemisorption on a heterogeneous surface called the Elovich or Roginski-Zeldowicz model,

which were verified by means of linear regression determined with the use of the least squares method and non-linear regression determined with the use of a numerical algorithm solved by means of the Solver in MS Excel.

Pseudo First-Order Kinetic Model (PFO)

The pseudo-first-order kinetic model, hereinafter referred to as model 1, makes the adsorption rate of sulfur dioxide/oxidized forms of sulfur dioxide ($dm_{S,t} \cdot dt^{-1}$, $g \cdot kg^{-1} min^{-1}$) dependent on the reaction rate constant k_1 (min^{-1}) and the difference in adsorbate mass after time t ($m_{S,t}$, $g \cdot kg^{-1}$) and ∞ ($m_{S,\infty}$, $g \cdot kg^{-1}$), according to the relationship:

$$\frac{dm_{S,t}}{dt} = k_1(m_{S,\infty} - m_{S,t}) \tag{2}$$

The $m_{S,\infty}$ value was determined experimentally by washing the adsorbent bed with the gas mixture for 1, 5, 15, and 30 min. In order to determine the rate constant k_1 (min^{-1}), the relationship (2) was integrated with the range from 0 to $m_{S,\infty}$, obtaining a linear equation:

$$\ln(m_{S,\infty} - m_{S,t}) = \ln(m_{S,\infty}) - k_1 t \tag{3}$$

which was then presented in semi-logarithmic coordinates $(t, \ln(m_{S,\infty} - m_{S,t}))$ so that the parameter k_1 corresponds to the slope a, according to the relationship $a = -k_1$. Integrating the differential Equation (2) with the above boundary conditions also gave a non-linearized function:

$$m_{S,t} = m_{S,\infty}[1 - \exp(-k_1 t)] \tag{4}$$

Pseudo Second-Order Kinetic Model (PSO)

The pseudo second-order kinetic model hereinafter referred to as model 2, assumes that the adsorption rate changes depending on the constant k_2 ($kg \cdot g^{-1} \cdot min^{-1}$) and the square of the adsorbate mass difference over time t and ∞, according to the equation:

$$\frac{dm_{S,t}}{dt} = k_2 \cdot (m_{S,\infty} - m_{S,t})^2 \tag{5}$$

the integration of which in the range from 0 (for t = 0) to $m_{S,\infty}$ (for t = t), allowed to obtain the relationship:

$$\frac{t}{m_{S,t}} = \frac{1}{k_2 m_{S,\infty}{}^2} + \frac{t}{m_{S,\infty}} \tag{6}$$

The value of the total adsorbate mass (after time ∞) $m_{S,\infty}$, was not determined experimentally (as was the case for model 1), but it was determined together with the rate constant k_2, based on the slope of the line (6) and the intercept in the system coordinates with a linear scale $(t, t \cdot m_{S,t}{}^{-1})$. Integrating the differential Equation (5) with the above boundary conditions also gave a non-linearized function:

$$m_{S,t} = \frac{m_{S,\infty} k_2 t}{1 + m_{S,\infty} k_2 t} \tag{7}$$

Model of Intraparticle Diffusion

The intraparticle diffusion model, hereinafter referred to as model 3, assumes that the amount of adsorbed sulfur dioxide/oxidized forms of sulfur dioxide at time t can be written by a simple equation:

$$m_{S,t} = k_{id} \cdot t^{0,5} + C \tag{8}$$

where the k_{id} coefficient is called the intraparticle diffusion rate constant ($g \cdot kg^{-1} \cdot min^{-0.5}$), and C ($g \cdot kg^{-1}$) is the thickness of the layer, called the thickness. If the only factor determining the speed of the process is intramolecular diffusion, then the linear relationship of q(t) to time $t^{1/2}$ should be a straight line with a slope coefficient k_{id} and going through the zero intercept, i.e., C = 0. However, the deviation from linearity indicates the existence of other factors limiting the rate of the adsorption process, such as: surface diffusion, diffusion of the boundary layer, gradual adsorption in the adsorbent pores, and adsorption on the active sites of the adsorbent [26].

Model of Chemisorption on a Heterogeneous Surface

The last of the applied models (model 4) was developed to describe the chemisorption on a heterogeneous surface. According to the Elovich equation, the adsorption rate of sulfur dioxide/oxidized forms of sulfur dioxide is described by the relationship:

$$\frac{dm_{S,t}}{dt} = \alpha \exp(-\beta \cdot m_{S,t}) \tag{9}$$

the integration of which in the range from 0 (for t = 0) to $m_{S,\infty}$ (for t = t) allows for the obtainment of the relationship:

$$m_{S,t} = \frac{\ln(t)}{\beta} + \frac{\ln(\alpha\beta)}{\beta} \tag{10}$$

where α is the initial adsorption rate (g·kg^{-1}min^{-1}), and β is the Elovich constant, reflecting the degree of surface coverage and activation energy for chemisorption (kg·g^{-1}). Presenting it in the system of semi-logarithmic coordinates ($\ln(t)$, $m_{S,t}$) makes it possible to determine the parameters α and β based on the slope of the straight line and the intercept. Integrating the differential Equation (9) with the above boundary conditions also gave a non-linearized function:

$$m_{S,t} = \frac{1}{\beta} \ln(\alpha\beta t) \tag{11}$$

2.3.2. Linear vs. Non-Linear Approach

In order to determine the linear kinetic parameters, the equations presented in Chapter 2.3.1 were used, i.e., Equation (3) for model 1, Equation (6) for model 2, Equation (8) for model 3, and Equation (10) for model 4. The determined kinetic parameters made it possible to determine the curve which shows the course of the reaction as a function of time. On the basis of these curves, a model amount of adsorbed components was determined and compared with the values measured experimentally. The discrepancies between the model and experimental data were analyzed by comparing 9 statistical criteria (summarized in Table 1), i.e., the determination coefficient (R^2), the correlation coefficient (R), the relative standard deviation (Δq), sum squared error (SSE), average relative error (ARE), chi-square test (χ^2), hybrid fractional error function (HYBRID), Marquardt's percent standard deviation (MPSD), and the sum of absolute errors (EABS):

Table 1. Statistic error functions [36,37].

Function	Equation	
Determination coefficient (R^2)	$R^2 = \dfrac{\sum_{i=1}^{n}\left(m_{S,t,mod} - \overline{m_{S,t,exp}}\right)^2}{\sum_{i=1}^{n}\left(m_{S,t,mod} - \overline{m_{S,t,exp}}\right)^2 + \sum_{i=1}^{n}\left(m_{S,t,mod} - m_{S,t,exp}\right)^2}$	(12)
Correlation coefficient (R)	$\sqrt{R^2} = R$	(13)
Relative standard deviation (Δq)	$\Delta q = \sqrt{\dfrac{\sum_{i=1}^{n}\left(\frac{m_{S,t,exp} - m_{S,t,mod}}{m_{S,\,t,exp}}\right)^2}{N-1}}$	(14)
Sum of squared deviations (SSE)	$SSE = \sum_{i=1}^{n}\left(m_{S,t,exp} - m_{S,t,mod}\right)^2$	(15)
Average Relative Error (ARE)	$ARE = \dfrac{100}{N}\sum_{i=1}^{n}\left|\dfrac{m_{S,t,exp} - m_{S,t,mod}}{m_{S,t,exp}}\right|$	(16)
Chi-square test (χ^2)	$\chi^2 = \sum_{i=1}^{n}\dfrac{\left(m_{S,t,exp} - m_{S,t,mod}\right)^2}{m_{S,t,exp}}$	(17)
Hybrid fractional error function (HYBRID)	$HYBRID = \dfrac{100}{N-p}\sum_{i=1}^{n}\dfrac{\left(m_{S,t,exp} - m_{S,t,mod}\right)^2}{m_{S,t,exp}}$	(18)
Marquardt's percent standard deviation (MPSD)	$MPSD = 100\sqrt{\dfrac{1}{N-p}\sum_{i=1}^{n}\left(\dfrac{m_{S,t,exp} - m_{S,t,mod}}{m_{S,t,exp}}\right)^2}$	(19)
Sum of absolute errors (EABS)	$EABS = \sum_{i=1}^{n}\left|m_{S,t,exp} - m_{S,t,mod}\right|$	(20)

where: $m_{S,t,mod}$ is the model amount of adsorbate adsorbed by the adsorbent mass as a function of time (g·kg^{-1}), $m_{S,t,exp}$ is the experimental amount of adsorbate adsorbed by the adsorbent mass as a function of time (g·kg^{-1}), N is the number of experimental points and p is the number of parameters in a given mathematical model. The high data convergence is evidenced by the lowest possible value of the criteria, Δq, SSE, ARE, χ^2, HYBRID, MPSD, and EABS, and the highest possible values for the criteria, R^2 and R (Figure 2 and Table 2).

In order to determine the kinetic parameters via the linear method, the equations presented in Section 2.3.1 were used, i.e., Equation (4) for model 1, Equation (7) for model 2, Equation (8) for model 3, and Equation (11) for model 4. For each data series, these equations were solved in 9 different variants, assuming the minimization of individual statistical criteria (collected in Table 1). To select the optimal variant for the best convergence of the model and experimental results, the criterion of the sum of normalized errors (SNE) was applied, which took into account the values of each statistical error, in accordance with the method described in [38,39]. The variant with minimal SNE error was considered to be the optimal non-linear variant. In order to compare the effectiveness of the linear and non-linear approach, Section 3.3 compares the values of 9 statistical error functions and model curves for the best linear variant with the selected optimal non-linear variant (determined based on the lowest SNE value).

3. Results and Discussion

A detailed analysis of the adsorption capacity of unburned carbon from lignite fly ash and activated carbons based on hard coal dust in relation to SO_2 was presented in the previous work by one of the authors [30]. This work also includes the characterization of the porous structure and the quantitative and qualitative analysis of surface oxygen functional groups, the key to the efficiency of the sulphur dioxide binding process. Therefore, this paper focuses on the mathematical description, which enables a deeper understanding of the mechanism of the observed reactions and to identify the optimum way to predict the behaviour of unburned carbons.

3.1. Linear Regression

The results of the model tests for linear regression are shown in Figure 2. As shown by the test results, the highest sorption capacity against sulfur dioxide is shown by unburned carbons UnCarb_MAsh and UnCarb_LAsh (Figure 2b,c). By mass, these materials adsorbed 28.90 and 28.95 g of S per kg of adsorbent, respectively. Among the selected materials, the lowest concentration of the active agent is characteristic of commercial activated carbons formed on the basis of hard coal dust. The mass of adsorbed sulfur dioxide for the AKP-5 and AKP-5/A samples is 41 and 32% lower than the least adsorbing unburned carbon (UnCarb_HAsh), for which 25.15 g S per kg of adsorbent was demonstrated. Additionally, due to the presence of oxygen and water vapor in the measurement system, the sorption capacity of the samples increased. The percentage of sulfur in the solid phase after the process increased 1.6 times for the UnCarb_LAsh material, while for commercial materials this value did not exceed 1.3 (Figure 2f,g).

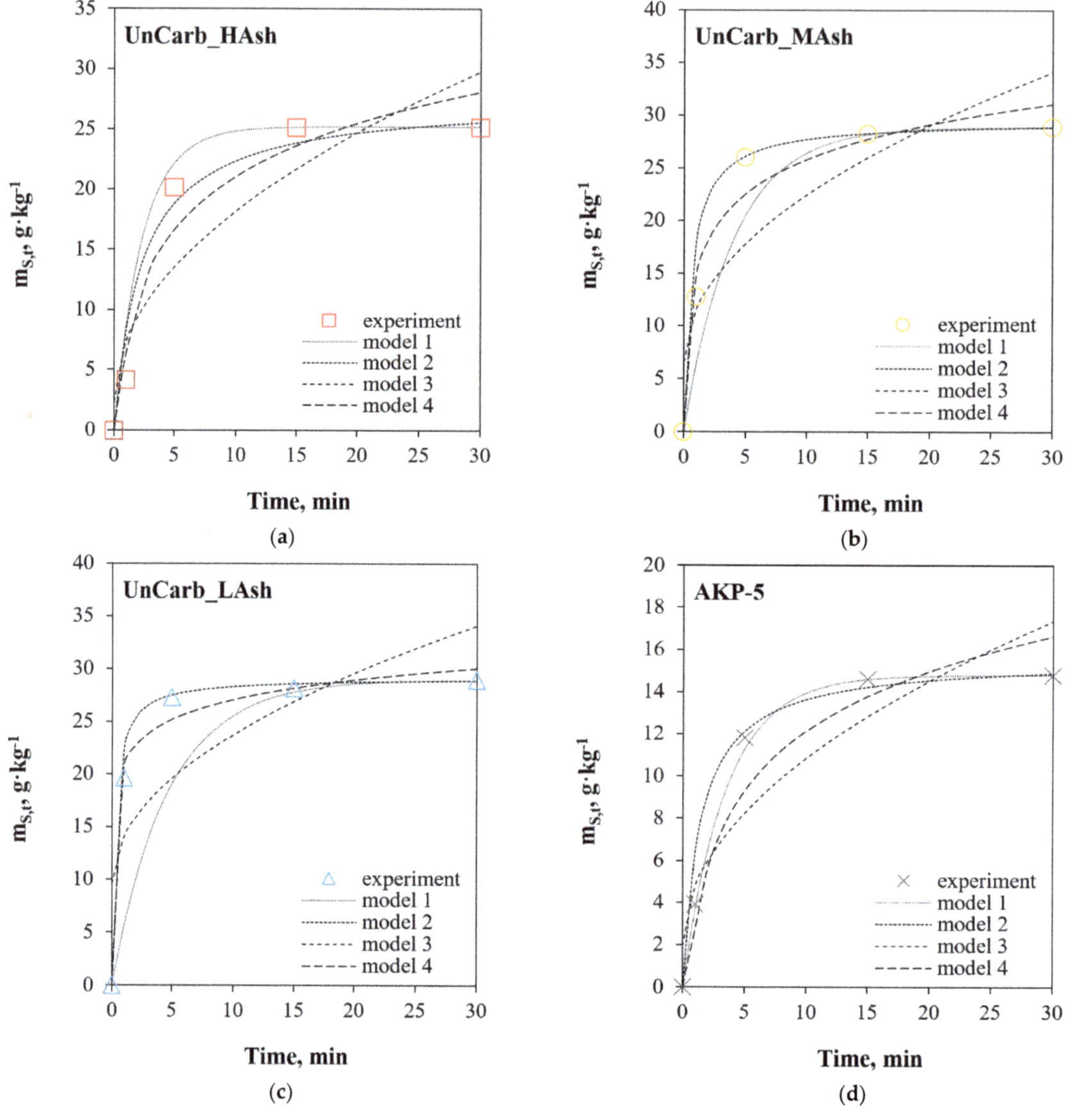

Figure 2. *Cont.*

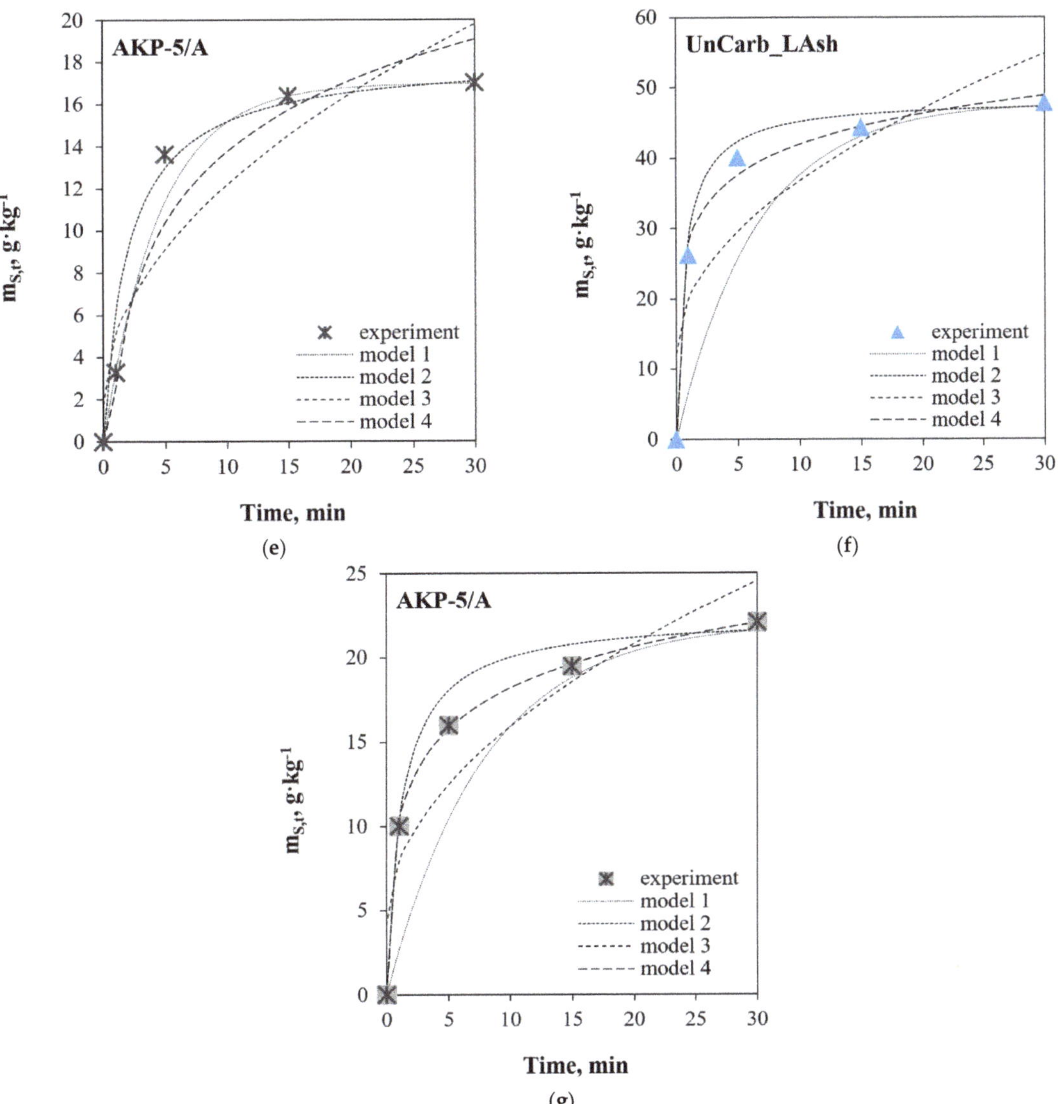

Figure 2. Summary of model and experimental curves for linear regression for the following mixtures: (**a–e**) SO_2 + Ar, (**f,g**) SO_2 + O_2 + $H_2O_{(g)}$ + Ar.

As can be observed, the reaction rate constants determined during the tests range from 0.123 min^{-1} (AKP-5/A, SO_2 + Ar + $H_2O_{(g)}$ + O_2) to 0.423 min^{-1} (UnCarb_HAsh, SO_2 + Ar) for model 1 and from 0.0156 kg·g^{-1}·min^{-1} (UnCarb_HAsh, SO_2 + Ar) up to 0.114 kg·g^{-1}·min^{-1} (UnCarb_LAsh, SO_2 + Ar) for model 2 (Table 2). According to the theory, for both models, materials that quickly bind the adsorbate should be characterized by high reaction rates. However, in practice, the correlation between the values of k_1 and k_2 has not been confirmed. Interestingly, the calculations made for model 1 show a reduction in the rate of the adsorption process in the presence of $H_2O_{(g)}$ and O_2 (0.123–0.155 min^{-1} under SO_2 + Ar + $H_2O_{(g)}$ + O_2 vs. 0.214–0.423 min^{-1} under SO_2 + Ar).

Table 2. Kinetic parameters determined by the method of linear regression.

Sample	Model 1	Model 2		Model 3		Model 4	
	k_1 $\mathrm{min^{-1}}$	k_2 $\mathrm{kg \cdot g^{-1} \cdot min^{-1}}$	$m_{S,\infty}$ $\mathrm{g \cdot kg^{-1}}$	k_{id} $\mathrm{g \cdot kg^{-1} \cdot min^{-0.5}}$	C $\mathrm{g \cdot kg^{-1}}$	α $\mathrm{g \cdot kg^{-1} min^{-1}}$	β $\mathrm{kg \cdot g^{-1}}$
$SO_2 + Ar$							
UnCarb_HAsh	0.423	0.0156	27.5	5.01	2.33	17.2	0.156
UnCarb_MAsh	0.247	0.0527	29.5	5.03	6.56	107	0.210
UnCarb_LAsh	0.214	0.114	29.2	4.47	9.56	5917	0.370
AKP-5	0.286	0.0449	15.6	2.81	1.94	7.92	0.244
AKP-5/A	0.222	0.0273	18.2	3.30	1.76	8.24	0.206
$SO_2 + O_2 + H_2O_{(g)} + Ar$							
UnCarb_LAsh	0.155	0.0293	48.3	7.77	12.1	515	0.160
AKP-5/A	0.123	0.0363	22.5	3.73	4.13	62.6	0.285

A model parameter of great practical importance is the amount of adsorbate related to the equilibrium conditions $m_{S,\infty}$ (for unlimited contact time). It is interesting that this coefficient, determined on the basis of model 2, reaches a value similar to that obtained experimentally (for a contact time of 30 min), and the discrepancies (averaged for all analyzes) do not exceed 3.5% (Figure 3).

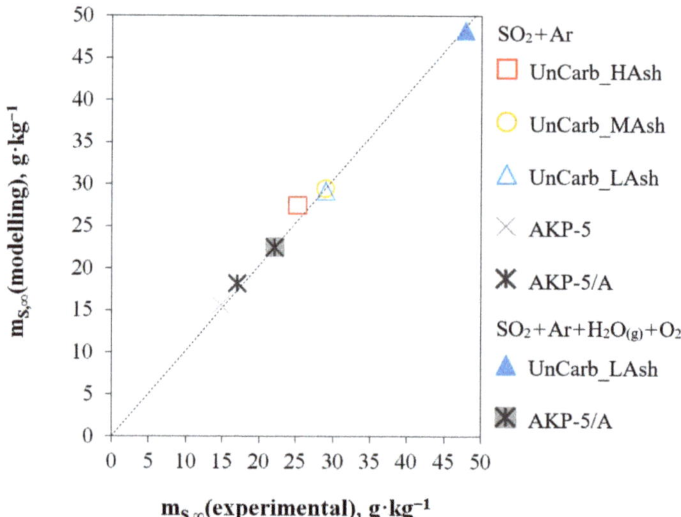

Figure 3. Relationship between the model and experimental amount of bound adsorbate.

The calculations made for model 3 show that the values of the k_{id} coefficient range from 2.81 AKP-5/A, $SO_2 + Ar$) to 7.77 $\mathrm{g \cdot kg^{-1} \cdot min^{-0.5}}$ (UnCarb_LAsh, $SO_2 + O_2 + H_2O_{(g)} + Ar$), while parameter C varies from 1.76 (AKP-5/A, $SO_2 + Ar$) to 12.1 $\mathrm{g \cdot kg^{-1}}$ (UnCarb_LAsh, $SO_2 + O_2 + H_2O_{(g)} + Ar$) (Table 2). In view of the information from [40], high C values and the low k_{id} would indicate a role that the diffusion-controlled boundary layer could play. The reverse configuration of the discussed parameters would prove that the speed-limiting stage of the process was diffusion inside the pores of the solid phase surface. Nevertheless, as shown in Figure 2, the described model does not faithfully reflect the course of the reaction, which to some extent confirms the kinetic nature of the experiments performed. However, it is interesting that the addition of $H_2O_{(g)}$ and O_2 to the gas mixture significantly increased the C value (4.13 and 12.1 $\mathrm{g \cdot kg^{-1}}$ vs. 1.76 and 9.56 $\mathrm{g \cdot kg^{-1}}$, respectively, for the

AKP-5/A and UnCarb_LAsh tests). Considering the information presented above, it should be assumed that under the conditions of the SO_2 + Ar + $H_2O_{(g)}$ + O_2 mixture, the boundary layer effect in the SO_2 adsorption process will be greater.

The kinetic parameters determined for model 4 are theoretical and physicochemical interpretation is difficult. Moreover, as far as the author is aware, the literature lacks studies on the kinetics of SO_2 adsorption on unburned carbons, which would make it possible to compare the obtained results. Interestingly, the registered change in kinetic parameters for the addition of $H_2O_{(g)}$ and O_2 to the gas mixture would indicate a change in the kinetics of SO_2 adsorption. In the case of the UnCarb_LAsh test, a decrease was noted in both the value of the reaction rate constant α and the degree of surface coverage with the β adsorbate (0.515 g·kg^{-1}min^{-1} and 0.160 kg·g^{-1} vs. 5917 g·kg^{-1}min^{-1} and 0.370 kg·g^{-1}); for the AKP-5/A sample, the intensification of each of them (62.6 g·kg^{-1}min^{-1} and 0.285 kg·g^{-1} against 8.24 g·kg^{-1}min^{-1} and 0.206 kg·g^{-1}) (Table 2).

Table 3 presents the analysis of statistical errors in kinetic models solved by the linear regression method. The highlighted data (in colours and bold) indicate the most appropriate values for a given sample out of the four analyzed models.

Table 3. Error analysis for kinetic models solved by linear regression method.

Sample	R^2	R	Δq	SSE	ARE	χ^2	HYBRID	MPSD	EABS
				Model 1 [1]					
UnCarb_HAsh	0.955	0.977	54.9	24.6	23.9	5.16	172	63.4	**6.55**
UnCarb_MAsh	0.906	0.952	27.6	73.6	14.5	4.51	150	31.8	12.2
UnCarb_LAsh	0.745	0.863	38.9	268	20.8	12.6	421	45.0	22.9
AKP-5	**0.998**	**0.999**	**3.72**	**0.346**	**2.10**	**0.0383**	**1.28**	**4.29**	0.776
AKP-5/A	**0.979**	**0.989**	**8.39**	5.05	3.98	0.373	**12.4**	9.68	2.38
				Model 2 [1]					
UnCarb_HAsh	**0.958**	**0.979**	49.9	**20.7**	22.6	4.27	142	57.6	7.15
UnCarb_MAsh	0.958	0.979	19.9	26.2	8.06	2.04	67.9	23.0	**5.23**
UnCarb_LAsh	0.987	0.994	7.04	7.77	**3.28**	0.392	13.1	8.13	**3.42**
AKP-5	0.961	0.980	31.1	6.52	14.0	1.62	54.1	37.0	3.27
AKP-5/A	0.962	0.981	42.4	8.24	18.4	2.38	79.5	48.9	3.84
				Model 3 [1]					
UnCarb_HAsh	0.840	0.917	43.2	92.0	28.3	5.91	197	49.9	20.1
UnCarb_MAsh	0.771	0.878	19.3	144.8	13.5	3.85	128	22.2	23.5
UnCarb_LAsh	0.647	0.804	22.1	211	15.9	4.77	159	25.5	29.4
AKP-5	0.851	0.922	21.3	26.8	16.2	1.91	63.8	24.6	10.7
AKP-5/A	0.847	0.920	33.3	38.1	23.1	3.14	105	38.5	12.7
				Model 4 [1]					
UnCarb_HAsh	0.952	0.976	**28.3**	27.8	**17.4**	**2.17**	**72.5**	**32.7**	10.1
UnCarb_MAsh	**0.966**	**0.983**	**11.0**	**21.5**	**7.70**	**0.961**	**32.0**	**12.7**	8.24
UnCarb_LAsh	**0.988**	**0.994**	**5.20**	**6.99**	3.50	**0.272**	**9.07**	**6.01**	4.40
AKP-5	0.945	0.972	20.0	11.8	14.0	1.18	39.3	23.1	6.36
AKP-5/A	0.944	0.972	17.4	16.0	12.4	1.22	40.6	20.1	6.74
				Model 1 [2]					
UnCarb_LAsh	0.776	0.881	41.1	587.4	22.7	19.6	653	47.5	35.6
AKP-5/A	0.824	0.908	40.8	86.3	22.8	7.42	247	47.1	14.1
				Model 2 [2]					
UnCarb_LAsh	0.992	0.996	5.17	12.4	3.69	0.355	11.8	5.97	6.59
AKP-5/A	0.982	0.991	7.33	6.14	4.46	0.361	12.0	8.46	3.90

Table 3. *Cont.*

Sample	R^2	R	Δq	SSE	ARE	χ^2	HYBRID	MPSD	EABS
Model 3 [2]									
UnCarb_LAsh	0.978	0.876	19.5	354	14.0	5.48	183	22.6	38.2
AKP-5/A	0.867	0.931	16.6	41.1	11.9	1.57	52.3	19.2	13.2
Model 4 [2]									
UnCarb_LAsh	0.994	0.997	4.14	9.15	2.75	0.245	8.17	4.78	5.03
AKP-5/A	1.00	1.00	0.898	0.0787	0.614	0.00494	0.165	1.04	0.471

[1] $SO_2 + Ar$; [2] $SO_2 + O_2 + H_2O_{(g)} + Ar$.

In the case of the SO_2 + Ar mixture, for commercial samples of activated carbons, regardless of the statistical error function, the quality of the results suggests that SO_2 adsorption is a first-order kinetic reaction. However, bearing in mind the considerations of Płaziński and Rudziński in [41,42], we should be cautious to hypothesize about a specific physical model of adsorption in the case of Equation (3). There is a belief that the indicated equation is not able to reflect changes in the mechanism controlling the adsorption kinetics, and the adjustment of the model data to the experimental data, especially in the case of systems close to the equilibrium state, results rather from mathematical foundations.

In the case of the UnCarb_HAsh trial, inconsistency in the indication of error values was obtained. It is highly likely related to the heterogeneity of the sample (ash content 57.3% for UnCarb_HAsh, 44.6% for UnCarb_MAsh, 12.8% for the UnCarb_LAsh [30]). Nevertheless, as evidenced in Table 3, 5 (Δq, ARE, χ^2, HYBRID, MPSD) out of 9 functions indicate that model 4 reflects the empirical data most accurately. The determination (R^2) and correlation (R) coefficients, as well as the sum squared error (SSE) indicate model 2; and the sum of absolute errors (EABS)—model 1. However, bearing in mind the information that in the case of the first and second-order models (models 1 and 2), the ability to fit data may result only from the mathematical properties of Equations (3) and (6), and not from specific physical assumptions, the compliance of adsorption with the kinetic mechanism of chemisorption on a heterogeneous surface was adopted for further comparative analyzes (according to model 4).

In the case of the UnCarb_MAsh and UnCarb_LAsh trials, greater consistency of the statistical error values was obtained, and their quality indicates the importance of the chemisorption phenomenon. This confirms the observations described in [30] that even in the absence of molecular oxygen in the gas mixture, the interaction between the adsorbate molecules and the carbon material occurs both due to relatively weak intermolecular van der Waals forces (corresponding to physical adsorption), as well as the chemical binding of sulfur dioxide.

The change of the atmosphere into $SO_2 + O_2 + H_2O_{(g)} + Ar$ indicates that the reliability of the analyzed models changes towards model 1 < model 3 < model 2 < model 4. These data, in line with the results of experimental research [30], also prove the formation of strong chemical bonds between the adsorbent and the adsorbate in the presence of oxygen and water vapor, thus indicating a strong inhomogeneity of the adsorbent surface.

3.2. Non-Linear Regression

In the case of describing sulfur dioxide adsorption by non-linear regression, the so-called sum of normalized errors (SNE) method was applied, allowing to select the most appropriate error function used to optimize kinetic parameters. This method makes it possible to estimate the values that are not burdened with the error resulting from the use of only one type of function and enables the selection of the model that best describes the adsorption process.

Figure 4 shows the distribution of the parameter of the sum of normalized errors for all tested samples. As can be seen, the SNE value determined for one data series varies greatly. Within a given model, it may even decrease twofold (e.g., for the UnCarb_HAsh trial and model 3: 8.81 in the case of minimizing the R^2 criterion and 4.39 in the case of minimizing the EABS criterion). Especially in the case of models 3 and 4, there is a correlation that minimization of the determination coefficient (R^2) and correlation (R) leads to high SNE values. This observation does not confirm the commonly used assumption that the models with $R^2 > 0.7$ describe the studied phenomena reliably [43,44]. It is therefore clear that fitting data by any of the non-linear equations based on the R or R^2 functions only, cannot be treated as evidence or prerequisite of the existence of a mechanism that determines the kinetics or dynamics of adsorption in a given system. Notwithstanding the fact that it is quite common in the literature to use them as a basis for the assessment of the quality of fitting kinetic data to experimental data [45–47]. Interestingly, the analyses were performed to prove a certain universality of the χ^2 and HYBRID functions. As noted, in 15 out of 28 cases the minimization of these functions led to the lowest SNE values for individual models (Table 4). For example, for the AKP-5 sample, HYBRID values in the range 5.60–6.28 were recorded—the lowest for models 1, 2, and 4; in the case of the AKP-5/A sample ($SO_2 + Ar + H_2O_{(g)} + O_2$), the noted values of χ^2 were in the range 4.27–8.09—the lowest for models 2, 3, and 4.

Table 4. SNE error analysis for kinetic models solved by non-linear regression method—the most appropriate values.

Sample	Model 1	Model 2	Model 3	Model 4
	$SO_2 + Ar$			
UnCarb_HAsh	**HYBRID** **5.40**	**HYBRID** **4.84**	EABS 4.39	χ^2 5.06
UnCarb_MAsh	SSE 8.16	χ^2 7.13	χ^2 5.37	Δq 5.00
UnCarb_LAsh	R^2 8.97	EABS 8.43	**MPSD** **3.16**	EABS 5.59
AKP-5	HYBRID 6.08	HYBRID 6.28	EABS 6.94	HYBRID 5.60
AKP-5/A	HYBRID 5.42	χ^2 4.97	χ^2 6.77	**MPSD** **4.84**
	$SO_2 + Ar + H_2O_{(g)} + O_2$			
UnCarb_LAsh	SSE 8.92	R 7.55	χ^2 4.04	EABS 5.52
AKP-5/A	SSE 7.91	χ^2 8.09	χ^2 4.27	χ^2 6.18

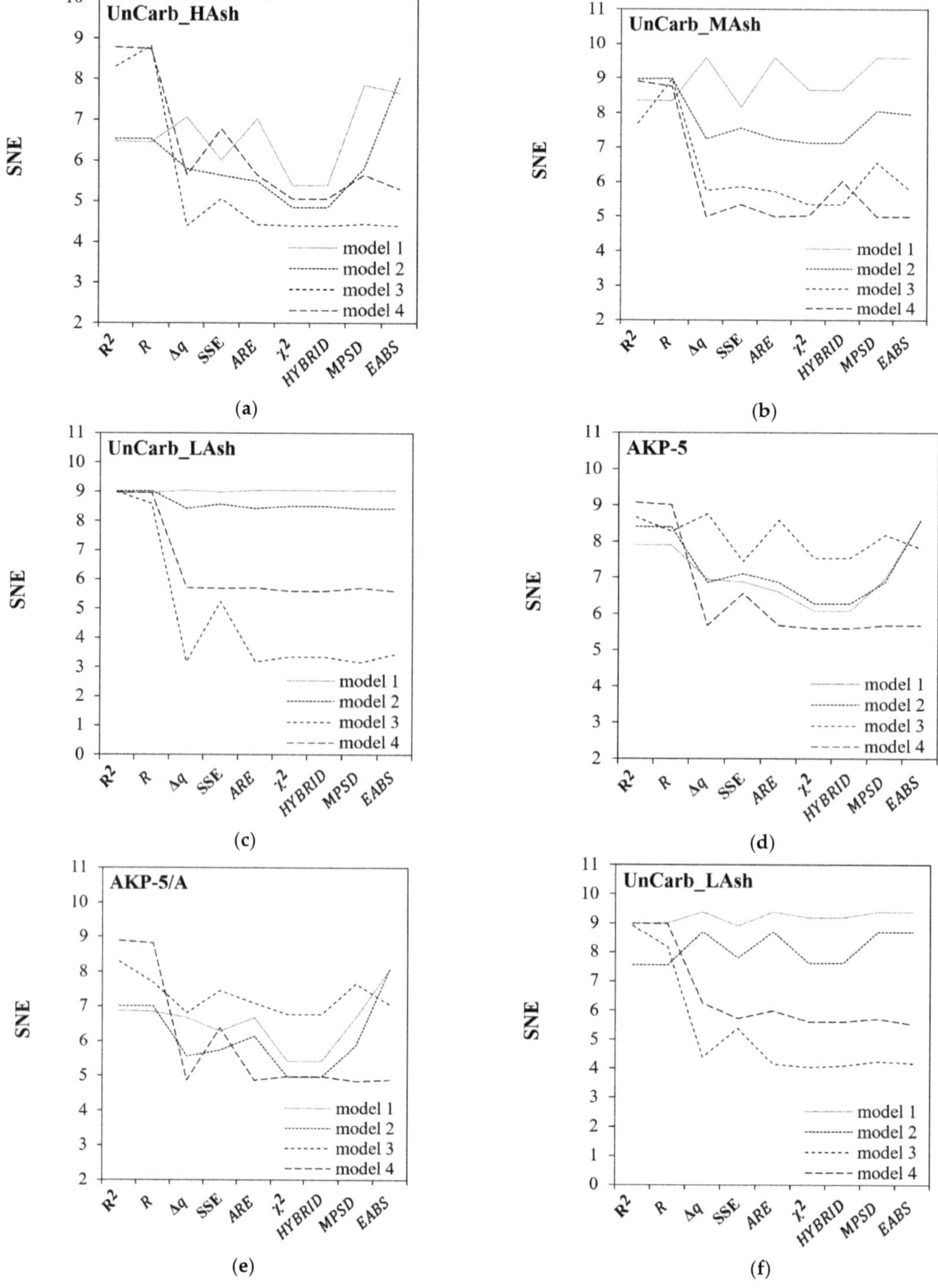

Figure 4. *Cont.*

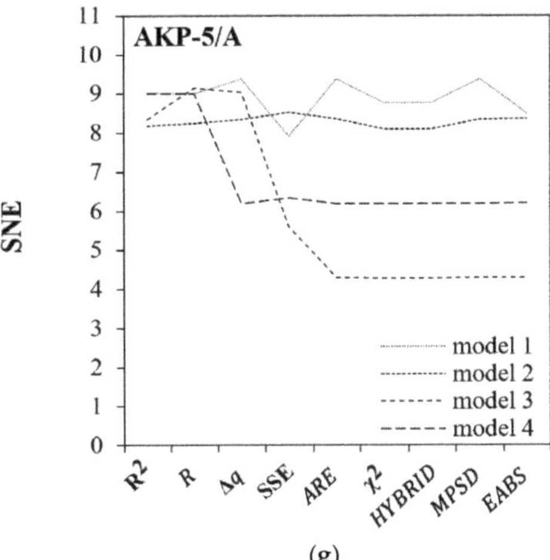

(g)

Figure 4. SNE error analysis for kinetic models solved by non-linear regression method for the following mixtures: **(a–e)** SO_2 + Ar, **(f,g)** SO_2 + O_2 + $H_2O_{(g)}$ + Ar.

Table 4 distinguishes the error functions used for non-linear regression (out of 9), for which the most appropriate values of the SNE function were obtained. These values served as a criterion for selecting an appropriate mathematical model for the discussed adsorption case. As can be seen, regardless of the tested sample and process conditions, in the case of models 1 and 2, the lowest SNE values were obtained by minimizing the complex fractional error function (HYBRID), and for models 3 and 4, by Marquardt's percentage standard deviation (MPSD). Interestingly, all the indicated values correspond to the SO_2 + Ar mixture. As a result of wetting and oxygenating the gas mixture, the functions of 9 statistical errors for each model generated higher SNE values.

A detailed analysis of the nonlinear fit and SNE values (Tables 4 and 5), at the level of the tested samples and process conditions, clearly indicates that under the conditions of the SO_2 + Ar mixture, in the case of commercial activated carbons and the unburned activated carbon UnCarb_MAsh sample, permanent bonding of sulfur dioxide could have occurred. Compatibility of adsorption with the Elovich equation (model 4) shows that the adsorption sites increased exponentially with the course of the process, which resulted in multilayer adsorption. Interestingly, for the UnCarb_HAsh and UnCarb_LAsh (SO_2 + Ar and SO_2 + Ar + $H_2O_{(g)}$ + O_2) and AKP-5/A (SO_2 + Ar + $H_2O_{(g)}$ + O_2) samples, diffusion in boundary layers or inside the pores of adsorbents (model 3) could have been the stage limiting the adsorption rate. Considering the high values of parameter C (od 8.17 do 24.3 $g \cdot kg^{-1}$) (Table 5), it can be indicated that in the case of the UnCarb_LAsh and AKP-5/A samples, internal diffusion of sulfur dioxide dominated over the general adsorption kinetics. The phenomenon of external diffusion should rather be noted for the UnCarb_HAsh sample (C = 0) (Table 5), similar to the case [48].

Table 5. Kinetic parameters determined by the method of linear regression.

Sample	Model 1	Model 2		Model 3		Model 4	
	k_1 min^{-1}	k_2 $kg \cdot g^{-1} \cdot min^{-1}$	$m_{S,\infty}$ $g \cdot kg^{-1}$	k_{id} $g \cdot kg^{-1} \cdot min^{-0.5}$	C $g \cdot kg^{-1}$	α $g \cdot kg^{-1} min^{-1}$	β $kg \cdot g^{-1}$
$SO_2 + Ar$							
UnCarb_HAsh	0.224	6.57E-03	31.8	**4.59**	**0**	13.6	0.123
UnCarb_MAsh	0.581	2.41E-02	31.0	5.38	7.48	**54.4**	**0.175**
UnCarb_LAsh	1.19	6.85E-02	29.4	**1.97**	**18.2**	3955	0.353
AKP-5	0.312	1.97E-02	17.1	3.45	0.460	**11.8**	**0.280**
AKP-5/A	0.252	1.19E-02	20.6	3.52	0.775	**9.44**	**0.213**
$SO_2 + O_2 + H_2O_{(g)} + Ar$							
UnCarb_LAsh	0.865	2.59E-02	47.5	**4.80**	**24.3**	390	0.153
AKP-5/A	0.578	3.71E-02	21.5	**2.74**	**8.17**	61.7	0.284

It is worth emphasizing, disregarding the values of the SNE function, that as a consequence of the addition of $H_2O_{(g)}$ and O_2 to the gas mixture, the kinetic parameters of SO_2 adsorption have changed (Table 5). Analogously to the linear regression method (Table 2), for both samples (UnCarb_LAsh and AKP-5/A) an increase in the boundary layer effect was noted (in accordance with C). Moreover, for the AKP-5/A test, the rate of SO_2 adsorption under the $SO_2 + Ar + H_2O_{(g)} + O_2$ mixture was intensified (in accordance k_1, k_2 and α).

3.3. Comparative Analysis of Linear and Non-Linear Regression

To assess the validity of the description of the kinetics and dynamics of adsorption by means of linear or nonlinear regression, the values of statistical errors and model curves were compared for the models for which the smallest deviations from empirical data were recorded (Figure 5, Table 6). As can be seen, for 6 out of 7 tested trials, the research clearly proves that it is the linear regression that more accurately reflects the behaviour of the adsorption system (regardless of the process conditions). What is particularly interesting, only for the UnCarb_MAsh sample, the method of linear and nonlinear fitting indicates the same mechanism of the studied phenomenon (model 4). Depending on the applied statistical error, the linear and nonlinear approaches may differ even several dozen times. For example, for the AKP-5/A ($SO_2 + Ar + H_2O_{(g)} + O_2$) sample it was noted that the HYBRID error reached the value of 0.2 with linear regression and as much as 56 times more with non-linear regression (11.2).

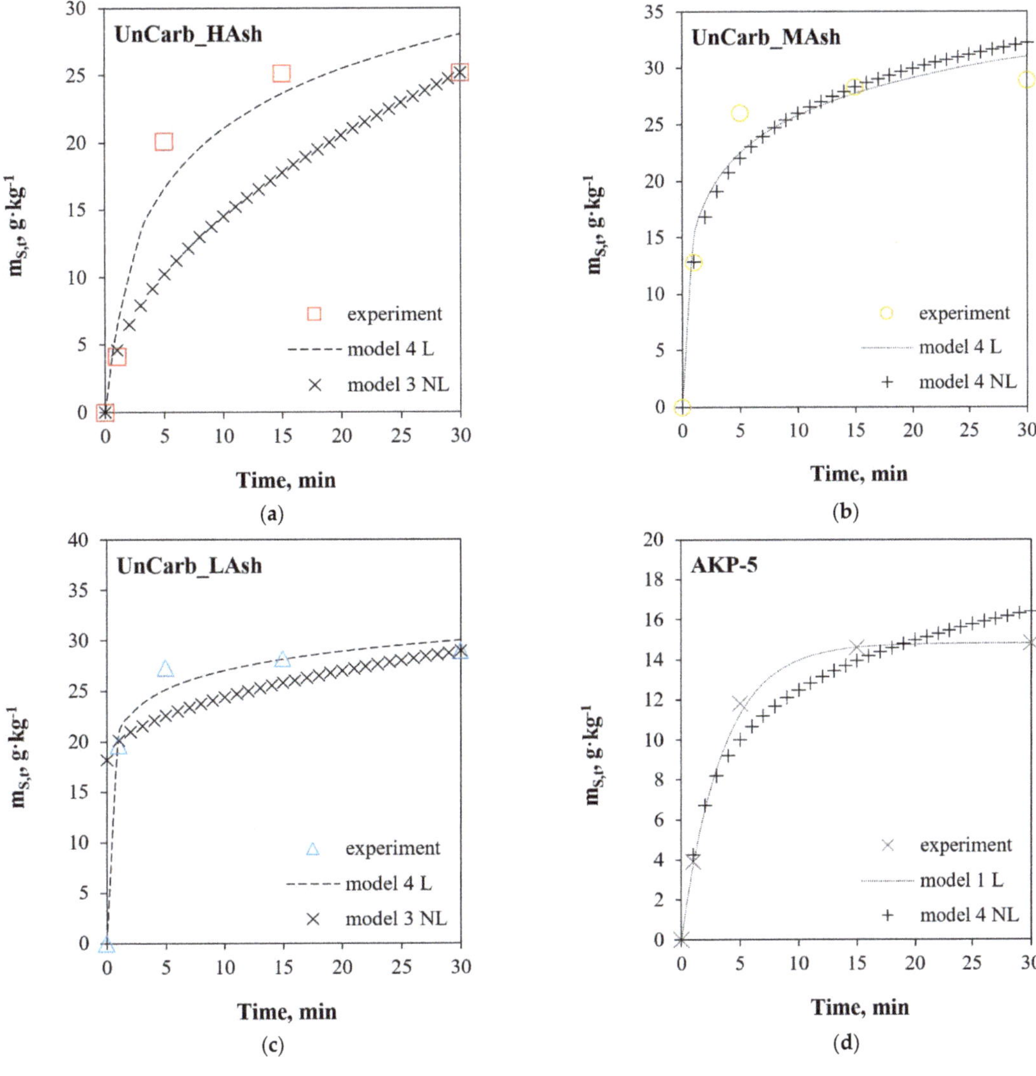

Figure 5. *Cont.*

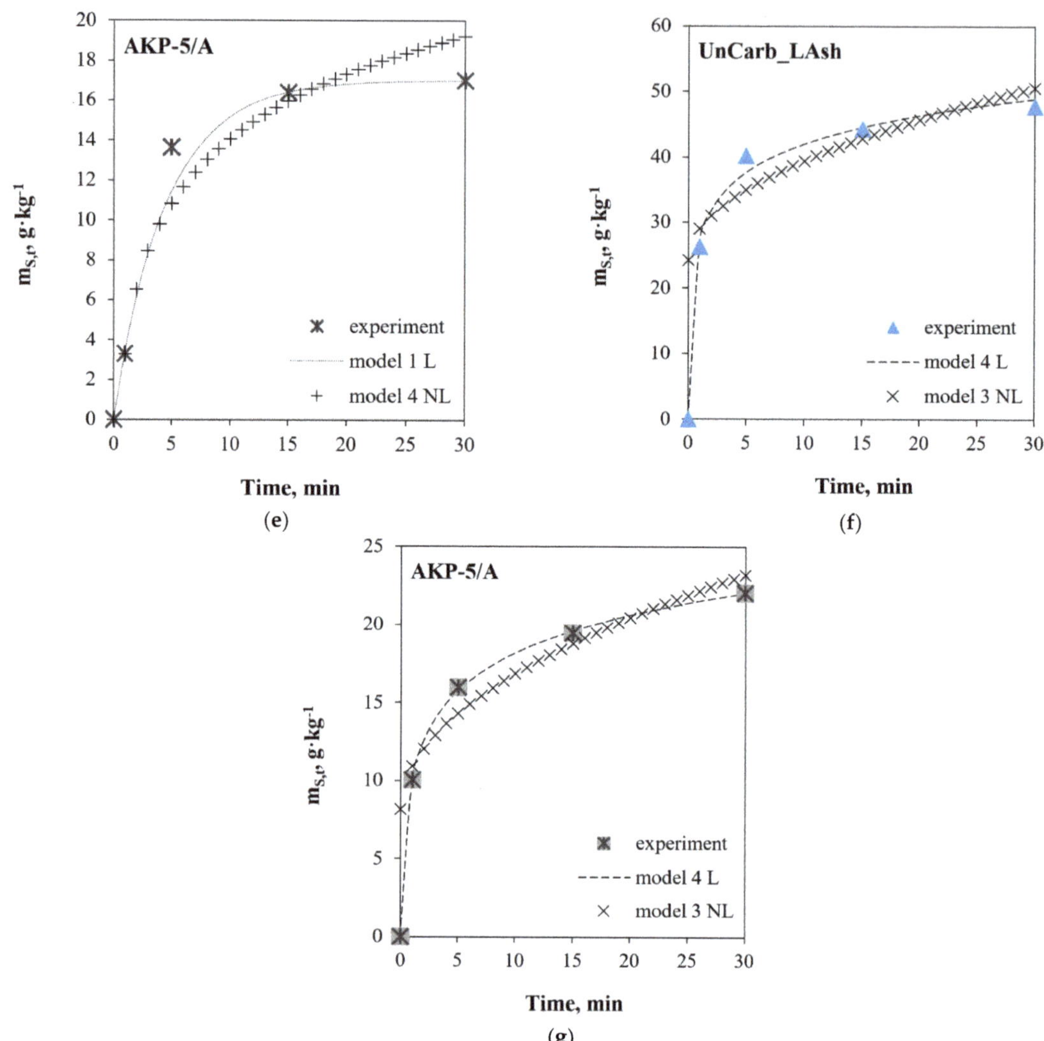

Figure 5. Summary of model and experimental curves for linear and non-linear regression for the following mixtures: (**a–e**) SO_2 + Ar, (**f**,**g**) SO_2 + O_2 + $H_2O_{(g)}$ + Ar.

Table 6. Error analysis for kinetic models solved by linear and non-linear regression methods.

Model	R^2	R	Δq	SSE	ARE	χ^2	HYBRID	MPSD	EABS
UnCarb_HAsh [1]									
Model 4 L	**0.952**	**0.976**	**28.3**	**27.8**	**17.4**	**2.2**	**72.5**	**32.7**	**10.1**
Model 3 NL	0.752	0.867	29.1	152.9	17.8	7.1	235.9	33.6	17.7
UnCarb_Mash [1]									
Model 4L	**0.966**	**0.983**	11.0	**21.5**	7.7	**0.96**	**32.0**	12.7	8.24
Model 4NL	0.961	0.980	**9.6**	27.4	**5.4**	1.01	33.6	**11.1**	**7.4**

Table 6. *Cont.*

Model	R^2	R	Δq	SSE	ARE	χ^2	HYBRID	MPSD	EABS
				UnCarb_Lash [1]					
Model 4L	0.988	0.994	5.2	7.0	3.5	0.3	9.1	6.0	4.4
Model 3NL	0.220	0.469	9.7	358.9	5.7	1.0	34.4	11.2	25.8
				AKP-5 [1]					
Model 1L	0.998	0.999	3.7	0.3	2.1	0.04	1.3	4.3	0.8
Model 4NL	0.967	0.983	10.6	6.4	7.9	0.5	17.0	12.2	4.4
				AKP-5/A [1]					
Model 1L	0.979	0.989	8.4	5.1	4.0	0.4	12.4	9.7	2.4
Model 4NL	0.953	0.976	12.3	13.1	7.3	0.9	29.6	14.2	5.5
				UnCarb_Lash [2]					
Model 4L	0.994	0.997	4.1	9.2	2.8	0.2	8.2	4.8	5.0
Model 3 NL	0.465	0.682	9.0	633.3	6.5	1.2	38.9	10.4	36.5
				AKP-5/A [2]					
Model 4L	1.00	1.00	0.9	0.08	0.6	0.005	0.2	1.0	0.5
Model 3NL	0.686	0.828	7.5	72.1	5.6	0.3	11.2	8.7	12.5

[1] SO_2 + Ar; [2] SO_2 + O_2 + $H_2O_{(g)}$ + Ar.

What is also noteworthy, comparing the kinetic parameters from Table 2 for the linear regression method with the parameters from Table 5 for the non-linear regression method, it can be seen that the differences between them can be over 100%. As can be seen, the k_{id} rate constant for the UnCarb_LAsh trial for the linear fit is 1.97 $g \cdot kg^{-1} \cdot min^{-0.5}$, and for the non-linear fit it is as much as 4.47 $g \cdot kg^{-1} \cdot min^{-0.5}$ (the difference is 227%).

4. Conclusions

The aim of this article was to determine the parameters of the kinetics and dynamics of adsorption by linear and non-linear regression for the following models: the pseudo first-order (model 1) and pseudo second-order (model 2) models, intraparticle diffusion (model 3), and chemisorption on a heterogeneous surface (model 4). The quality of fitting the model data to the experimental data was analyzed based on 9 statistical error functions (R, R^2, Δq, SSE, ARE, χ^2, HYBRID, MPSD, EABS) and, in the case of non-linear regression, the normalized error sum (SNE) method. The performed measurements and analyzes lead to the conclusion that:

- confronting 9 statistical error functions for the models was the most reliable for linear and non-linear regression, respectively, leading to an unequivocal conclusion that it is the linear regression that more accurately reflects the behaviour of the adsorption system (regardless of the process conditions);
- in the case of the SO_2+Ar mixture, for commercial samples of activated carbons AKP-5 and AKP-5/A, regardless of the statistical error function, the quality of the results suggests that SO_2 adsorption is a first-order kinetic reaction (model 1). However, it should be noted that fitting model data to experimental data for the systems close to the equilibrium state can only result from the mathematical foundations of model 1;
- in the case of unburned carbons samples (UnCarb_HAsh, UnCarb_MAsh, UnCarb_LAsh), regardless of the process conditions, and the AKP-5/A (SO_2 + Ar + $H_2O_{(g)}$ + O_2) sample, the quality of the results shows that the adsorption is compatible with the kinetic mechanism of chemisorption on the heterogeneous surface (according to model 4);
- the sum of normalized errors, regardless of the tested sample and process conditions, reaches the lowest values for models 1 and 2 by minimizing the hybrid fractional error function (HYBRID), and for models 3 and 4 by the Marquardt's percentage standard deviation (MPSD);
- minimization of the determination coefficient (R^2) and correlation (R) leads to high SNE values. Fitting data by any of the non-linear equations based on the R or R^2

functions only cannot be treated as evidence or a prerequisite of the existence of a given mechanism determining the kinetics or dynamics of adsorption in a given system.

- only in 1 case (UnCarb_MAsh) out of 7 possible, both linear and non-linear regression indicate the same mechanism of the adsorption phenomenon—identical to chemisorption on a heterogeneous surface (according to model 4).

The analysis presented above proves that linear methods generally enable the determination of kinetic parameters that reflect the character of adsorption more reliably than non-linear methods, although it is puzzling that usually each of the approaches indicates a different mechanism of the phenomenon. Hence, in order to determine the optimal set of kinetic pairs as faithfully reproducing the course of the analyzed processes as possible, it is recommended to perform both linear and non-linear regression, in accordance with the methodology presented in this paper. Moreover, the assessment of the mechanism of the adsorption reaction based solely on the accuracy of the kinetic model may be misleading and, in the opinion of the authors, requires additional discussion supported by experimental studies, as in the case of [30]. Considering the limited amount of data in the literature on SO_2 adsorption on unburned carbon from lignite fly ash, the indicated work may be the first attempt at a thorough analysis of the chemical kinetics of this process, constituting the basis for considering the industrial application of the adsorption reaction.

Author Contributions: Conceptualization, A.M.K.-C.; methodology A.M.K.-C. and B.D.; validation, A.M.K.-C.; formal analysis, A.M.K.-C. and B.D.; investigation, A.M.K.-C.; resources, A.M.K.-C.; data curation, A.M.K.-C. and B.D.; writing—original draft preparation, A.M.K.-C.; writing—review and editing, A.M.K.-C.; visualization, A.M.K.-C., supervision, A.M.K.-C. All authors have read and agreed to the published version of the manuscript.

Funding: Not applicable.

Institutional Review Board Statement: Not applicable.

Informed Consent Statement: Not applicable.

Data Availability Statement: Not applicable.

Conflicts of Interest: The authors declare no conflict of interest.

References

1. Institute of Environmental Protection—The National Centre for Emissions Management (KOBiZE). National Emission Balance of SO2, NOx, CO, NH3, NMVOC, Dust, Heavy Metals and POPs in the SNAP and NFR Classification System. Warsaw, 2015. Available online: https://www.kobize.pl/uploads/materialy/materialy_do_pobrania/krajowa_inwentaryzacja_emisji/Bilans_emisji_za_2017.pdf (accessed on 7 July 2021).
2. Lecomte, T.; de la Fuente, J.F.F.; Neuwahl, F.; Canova, M.; Pinasseau, A.; Jankov, I.; Brinkmann, T.; Roudier, S.; Sancho, L.D. *Best Available Techniques (BAT) Reference Document for Large Combustion Plants*; EUR 28836 EN; Publications Office of the European Union: Luxembourg, 2017.
3. Common Format for a National Air Pollution Control Program under Art. 6 of Directive (EU) 2016/2284. Available online: https://ec.europa.eu/environment/air/pdf/reduction_napcp/PL%20final%20NAPCP%2027Jun19%20annexed%20report.pdf (accessed on 14 July 2021).
4. Bartonova, L. Unburned carbon from coal combustion ash: An overview. *Fuel Process. Technol.* **2015**, *134*, 136–158. [CrossRef]
5. Hower, J.C.; Groppo, J.G.; Graham, U.M.; Ward, C.R.; Kostova, I.J.; Maroto-Valer, M.M.; Dai, S. Coal-derived unburned carbons in fly ash: A review. *Int. J. Coal Geol.* **2017**, *179*, 11–27. [CrossRef]
6. Musyoka, N.M.; Wdowin, M.; Rambau, K.M.; Franus, W.; Panek, R.; Madej, J.; Czarna-Juszkiewicz, D. Synthesis of activated carbon from high-carbon coal fly ash and its hydrogen storage application. *Renew. Energy* **2020**, *155*, 1264–1271. [CrossRef]
7. Kisiela-Czajka, A.M.; Hull, S.; Albiniak, A. Investigation of the activity of unburned carbon as a catalyst in the decomposition of NO and NH_3. *Fuel* **2022**, *309*, 122170. [CrossRef]
8. Kisiela, A.M.; Czajka, K.M.; Moroń, W.; Rybak, W.; Andryjowicz, C. Unburned carbon from lignite fly ash as an adsorbent for SO_2 removal. *Energy* **2016**, *116*, 1454–1463. [CrossRef]
9. Rubio, B.; Izquierdo, M.T.; Mayoral, M.C.; Bona, M.T.; Andres, J.M. Unburnt carbon from coal fly ashes as a precursor of activated carbon for nitric oxide removal. *J. Hazard. Mater.* **2007**, *143*, 561–566. [CrossRef]
10. Bansal, R.C.; Goyal, M. *Activated Carbon Adsorption*; CRC Press: Boca Raton FL, USA, 2005.

11. Roy, P.; Sardar, A. SO$_2$ emission control and finding a way to produce sulphuric acid from industrial SO$_2$ emission. *J. Chem. Eng. Process Technol.* **2015**, *6*, 1000230. [CrossRef]
12. Moroń, W.; Ferens, W.; Czajka, K.M. Explosion of different ranks coal dust in oxy-fuel atmosphere. *Fuel Process. Technol.* **2016**, *148*, 388–394. [CrossRef]
13. Wang, J.; Yi, H.; Tang, X.; Zhao, S.; Gao, F. Simultaneous removal of SO$_2$ and NOx by catalytic adsorption using γ-Al$_2$O$_3$ under the irradiation of non-thermal plasma: Competitiveness, kinetic, and equilibrium. *Chem. Eng. J.* **2020**, *384*, 123334. [CrossRef]
14. Li, B.; Zhang, Q.; Ma, C. Adsorption kinetics of SO$_2$ on powder activated carbon. *IOP Conf. Ser. Earth Environ. Sci.* **2018**, *121*, 022019. [CrossRef]
15. Zhangm, S.; Li, Z.; Yang, Y.; Jian, W.; Ma, D.; Jia, F. Kinetics and thermodynamics of SO$_2$ adsorption on metal-loaded multiwalled carbon nanotubes. *Open Phys.* **2020**, *18*, 1201–1214. [CrossRef]
16. Wang, H.; Duan, Y.; Ying, Z.; Xue, Y. Effects of SO$_2$ on Hg adsorption by activated carbon in O$_2$/CO$_2$ conditions. Part 1: Experimental and kinetic study. *Energy Fuels* **2018**, *32*, 10773–10778. [CrossRef]
17. Silalahi, D.D.; Midi, H.; Arasan, J.; Mustafa, M.S.; Caliman, J.-P. Kernel partial least square regression with high resistance to multiple outliers and bad leverage points on near-infrared spectral data analysis. *Symmetry* **2021**, *13*, 547. [CrossRef]
18. Fehintola, E.O.; Amoko, J.S.; Obijole, O.A.; Oke, I.A. Pseudo second order kinetics model of adsorption of Pb^{2+} onto powdered corn cobs: Comparison of linear regression methods. *Direct Res. J. Chem. Mater. Sci.* **2015**, *3*, 1–10.
19. Adnan, F.; Thanasupsin, S.P. Kinetic studies using a linear regression analysis for a sorption phenomenon of 17a-methyltestosterone by Salvinia cucullata in an active plant reactor. *Environ. Eng. Res.* **2016**, *21*, 384–392. [CrossRef]
20. Moroń, W.; Czajka, K.M.; Ferens, W.; Babul, K.; Szydełko, A.; Rybak, W. NOx and SO$_2$ emission during oxy-coal combustion. *Chem. Process Eng.* **2013**, *34*, 337–346. [CrossRef]
21. Pilatau, A.; Czajka, K.M.; Filho, G.P.; Medeiros, H.S.; Kisiela, A.M. Evaluation criteria for the assessment of the influence of additives (AlCl$_3$ and ZnCl$_2$) on pyrolysis of sunflower oil cake. *Waste Biomass Valorization* **2017**, *8*, 2595–2607. [CrossRef]
22. Hanna, O.T.; Sandall, O.C. *Computational Methods in Chemical Engineering*; Prentice Hall International Series in the Physical and Chemical Engineering Sciences; Prentice Hall: Hoboken, NJ, USA, 1995.
23. Ngakou, C.S.; Anagho, G.S.; Ngomo, H.M. Non-linear regression analysis for the adsorption kinetics and equilibrium isotherm of phenacetin onto activated carbons. *Curr. J. Appl. Sci. Technol.* **2019**, *36*, 1–18. [CrossRef]
24. Lin, J.; Wang, L. Comparison between linear and non-linear forms of pseudo-first-order and pseudo-second-order adsorption kinetic models for the removal of methylene blue by activated carbon. *Front. Environ. Sci. Eng.* **2009**, *3*, 320–324. [CrossRef]
25. Kumar, K.V. Comparative analysis of linear and non-linear method of estimating the sorption isotherm parameters for malachite green onto activated carbon. *J. Hazard. Mater.* **2006**, *136*, 197–202. [CrossRef] [PubMed]
26. Lataye, D.H.; Mishra, I.M.; Mall, I.D. Adsorption of 2-picoline onto bagasse fly ash from aqueous solution. *Chem. Eng. J.* **2008**, *138*, 35–46. [CrossRef]
27. Kumar, K.V.; Porkodi, K.; Rocha, F. Isotherms and thermodynamics by linear and non-linear regression analysis for the sorption of methylene blue onto activated carbon: Comparison of various error functions. *J. Hazard. Mater.* **2008**, *151*, 794–804. [CrossRef] [PubMed]
28. Edgar, T.F.; Himmelblau, D.M. *Optimization of Chemical Processes*; McGraw-Hill College: New York, NY, USA, 1988.
29. Rybak, W.; Moroń, W.; Czajka, K.M.; Kisiela, A.M.; Ferens, W.; Jodkowski, W.; Andryjowicz, C. Co-combustion of unburned carbon separated from lignite fly ash. *Energy Procedia* **2017**, *120*, 197–205. [CrossRef]
30. Kisiela-Czajka, A.M. Adsorption behaviour of SO$_2$ molecules on unburned carbon from lignite fly ash in the context of developing commercially applicable environmental carbon adsorbent. *Preprint (SSRN/Elsevier)*. [CrossRef]
31. Lagergren, S. About the theory of so-called adsorption of soluble substances. *K. Sven. Vetensk. Akad. Handl.* **1989**, *241*, 1–39.
32. Ho, Y.S.; McKay, G. Pseudo-second order model for sorption processes. *Process Biochem.* **1999**, *34*, 451–465. [CrossRef]
33. Weber, W.J.; Morris, J.C. Kinetics of adsorption on carbon from solutions. *J. Sanit. Eng. Div.* **1963**, *89*, 31–60. [CrossRef]
34. Low, M.D. Kinetics of chemisorption of gases on solids. *Chem. Rev.* **1960**, *60*, 267–312. [CrossRef]
35. Shafiq, M.; Alazba, A.A.; Amin, M.T. Kinetic and isotherm studies of Ni^{2+} and Pb^{2+} adsorption from synthetic wastewater using *Eucalyptus camdulensis*–derived biochar. *Sustainability* **2021**, *13*, 3785. [CrossRef]
36. Ghaffari, H.R.; Pasalari, H.; Tajvar, A.; Dindarloo, K.; Goudarzi, B.; Alipour, V.; Ghanbarneajd, A. Linear and non-linear two-parameter adsorption isotherm modeling: A case-study. *Int. J. Eng. Sci.* **2017**, *6*, 1–11. [CrossRef]
37. Sivarajasekar, N.; Baskar, R. Adsorption of basic magenta II onto H$_2$SO$_4$ activated immature Gossypium hirsutum seeds: Kinetics, isotherms, mass transfer, thermodynamics and process design. *Arab. J. Chem.* **2019**, *12*, 1322–1337. [CrossRef]
38. Ho, Y.S.; Porter, J.F.; McKay, G. Equilibrium isotherm studies for the sorption of divalent metal ions onto peat: Copper, nickel and lead single component systems. *Water Air Soil Pollut.* **2002**, *141*, 1–33. [CrossRef]
39. Anirudhan, T.S.; Radhakrishnan, P.G. Kinetic and equilibrium modelling of Cadmium (II) ions sorption onto polymerized tamarind fruit shell. *Desalination* **2009**, *249*, 1298–1307. [CrossRef]
40. Soco, E.; Kalembkiewicz, J. Effect of chemical modification of the coal fly ash onto adsorption of lead (II) ions in the presence of cadmium (II) ions in a single- and bi-component system. *Eng. Prot. Environ.* **2016**, *19*, 81–95. [CrossRef]
41. Płaziński, W.; Rudziński, W.; Płazińska, A. Theoretical models of sorption kinetics including a surface mechanism: A review. *Adv. Colloid Interface Sci.* **2009**, *152*, 2–13. [CrossRef]

42. Płaziński, W.; Rudziński, W. Adsorption kinetics at solid/solution interfaces. The meaning of the pseudo-first and pseudo-second-order equations. *Wiadomości Chem.* **2011**, *65*, 1055–1067.

43. Czajka, K.M.; Modliński, N.; Kisiela-Czajka, A.M.; Naidoo, R.; Peta, S.; Nyangwa, B. Volatile matter release from coal at different heating rates—Experimental study and kinetic modeling. *J. Anal. Appl. Pyrolysis* **2019**, *139*, 282–290. [CrossRef]

44. Moore, D.S.; Notz, W.I.; Flinger, M.A. *The Basic Practise of Statistics*, 3rd ed.; W. H. Freeman and Company: New York, NY, USA, 2013.

45. Hashem, A.; Badawy, S.M.; Farag, S.; Mohamed, L.A.; Fletcher, A.J.; Taha, G.M. Non-linear adsorption characteristics of modified pine wood sawdust optimized for adsorption of Cd(II) from aqueous systems. *J. Environ. Chem. Eng.* **2020**, *8*, 103966. [CrossRef]

46. Saravanan, A.; Karishma, S.; Jeevanantham, S.; Jeyarsi, S.; Kiruthika, A.R.; Senthil Kumar, P.; Yaashikaa, P.R. Optimization and modeling of reactive yellow adsorption by surface modified Delonix regia seed: Study on nonlinear isotherm and kinetic parameters. *Surf. Interfaces* **2020**, *20*, 100520. [CrossRef]

47. Mohammed, N.A.S.; Abu-Zurayk, R.A.; Hamadneh, I.; Al-Dujaili, A.H. Phenol adsorption on biochar prepared from the pine fruit shells: Equilibrium, kinetic and thermodynamics studies. *J. Environ. Manag.* **2018**, *226*, 377–385. [CrossRef]

48. Módenes, A.N.; Scheufele, F.B.; Espinoza-Quiñones, F.R.; de Souza, P.S.C.; Cripa, C.R.B.; dos Santos, J.; Steffen, V.; Kroumov, A.D. Adsorption of direct of yellow ARLE dye by activated carbon of shell of coconut palm: Diffusional effects on kinetics and equilibrium states. *Int. J. Bioautom.* **2015**, *19*, 187–206.

Review

Non-Destructive Diagnostic Methods for Fire-Side Corrosion Risk Assessment of Industrial Scale Boilers, Burning Low Quality Solid Biofuels—A Mini Review

Tomasz Hardy [1,*], Amit Arora [2], Halina Pawlak-Kruczek [1], Wojciech Rafajłowicz [3], Jerzy Wietrzych [3], Łukasz Niedźwiecki [1], Vishwajeet [1] and Krzysztof Mościcki [1]

[1] Department of Energy Conversion Engineering, Wroclaw University of Science and Technology, Wybrzeże Wyspiańskiego 27, 50-370 Wrocław, Poland; halina.pawlak@pwr.edu.pl (H.P.-K.); lukasz.niedzwiecki@pwr.edu.pl (Ł.N.); vishwajeet@pwr.edu.pl (V.); krzysztof.moscicki@pwr.edu.pl (K.M.)

[2] Department of Chemical Engineering, Shaheed Bhagat Singh State University, Ferozepur 152004, India; amitarora@sbsstc.ac.in

[3] Department of Automation, Mechatronics and Control Systems, Wroclaw University of Science and Technology, Wybrzeże Wyspiańskiego 27, 50-370 Wrocław, Poland; wojciech.rafajlowicz@pwr.wroc.pl (W.R.); jerzy.wietrzych@pwr.edu.pl (J.W.)

* Correspondence: tomasz.hardy@pwr.edu.pl

Citation: Hardy, T.; Arora, A.; Pawlak-Kruczek, H.; Rafajłowicz, W.; Wietrzych, J.; Niedźwiecki, Ł.; Vishwajeet; Mościcki, K. Non-Destructive Diagnostic Methods for Fire-Side Corrosion Risk Assessment of Industrial Scale Boilers, Burning Low Quality Solid Biofuels—A Mini Review. *Energies* **2021**, *14*, 7132. https://doi.org/10.3390/en14217132

Academic Editors: Robert Król, Witold Kawalec, Izabela Sówka and Krzysztof M. Czajka

Received: 28 September 2021
Accepted: 27 October 2021
Published: 1 November 2021

Abstract: The use of low-emission combustion technologies in power boilers has contributed to a significant increase in the rate of high-temperature corrosion in boilers and increased risk of failure. The use of low quality biomass and waste, caused by the current policies pressing on the decarbonization of the energy generation sector, might exacerbate this problem. Additionally, all of the effects of the valorization techniques on the inorganic fraction of the solid fuel have become an additional uncertainty. As a result, fast and reliable corrosion diagnostic techniques are slowly becoming a necessity to maintain the security of the energy supply for the power grid. Non-destructive testing methods (NDT) are helpful in detecting these threats. The most important NDT methods, which can be used to assess the degree of corrosion of boiler tubes, detection of the tubes' surface roughness and the internal structural defects, have been presented in the paper. The idea of the use of optical techniques in the initial diagnosis of boiler evaporators' surface conditions has also been presented.

Keywords: fire-side corrosion; boiler tube wastage; diagnostics; industrial-scale boilers; non-destructive inspection; pipe inspection; wall thickness measurement

1. Introduction

The use of low-emission combustion techniques in pulverized coal-fired boilers has contributed, to a large extent, to intensifying the high-temperature corrosion processes, consequently causing the relatively quick wastage of wall tubes [1,2]. Fire-side corrosion in coal-fired boilers has been well investigated [3–8]. In general, fire-side corrosion can be subdivided into two distinct types, namely:

- water wall (evaporator) corrosion
- corrosion of the superheaters

The main causes of the boiler tubes high-temperature are: impurities in the fuel, such as sulphur alkali metals and chlorine; the lack of control of the combustion process resulting in a reducing gaseous environment at the tube surface; impingement of flames; and the temperature of tube metal.

The erosion-corrosion boiler tube-thick losses increase the exploitation expenses by increasing the frequency of the necessary full and partial screen replacements. They also strongly negatively affect the reliability and availability of the whole power unit. In general, the problem of the severe wastage of the wall tubes in boilers emerged a long time ago, for boilers fired with hard coal, after the modernization of their furnace systems; i.e., after staged combustion had been introduced. Reducing atmosphere can be detrimental to high-chromium heat-resisting steels with ferritic matrix, working at elevated or high temperature, making them liable to brittleness growth, as mentioned by Golanski and Lachowicz [9]. For temperatures as high as 720 °C, the presence of a regular grid or Sigma phase residue at the ferrite grain boundaries can be observed [9].

In the near future, industrial size power units will remain vitally important as power sources due to their flexible ability to cover demand when intermittent power sources are incapable of supplying the energy to the grid [10]. The ability to utilize biomass in these units will allow them to decrease their carbon footprint while maintaining the energy supply security for the grid. However, due to the inherently high variability of biomass, in terms of its properties, even the use of novel valorization techniques might not alleviate the problem completely. Therefore, there is a need for quick and reliable corrosion diagnostics techniques that will allow for optimizing the maintenance of power plant boilers. Nowadays, diagnostics are performed using visual inspection, followed by sampling sections of piping for laboratory testing. This is time-consuming as removed sections need to be replaced before the startup of the boiler after the inspection. Furthermore, any left-over tensile stress can lead to the formation of cracks, as observed by Duarte et al. [11].

2. The Problem of the Fire-Side Corrosion

Biomass is an energy source that is considered neutral in terms of CO_2 emissions [12]. That statement is based upon an oversimplification. Of course, carbon dioxide is always released as a product of combustion when carbon is being burned. However, biomass absorbs carbon present in CO_2 particles during the photosynthesis process. Carbon, along with oxygen and hydrogen, is one of the three main elements present in biomass, namely in three main carbohydrate compounds (cellulose, hemicelluloses and lignin) that form the orthotropic, composite organic structure of plants.

Two main types of biomass that may serve as fuel are woody biomass and herbaceous biomass [13–17]. Woody biomass originates from the forest, and it is often called lingo-cellulosic biomass because trees typically contain quite significant amounts of lignin [14,18]. Herbaceous biomass consists mainly of hemicellulose and cellulose and contains only a minor part of lignin. It is mostly represented by perennial plants, which in many cases are grown as crops [14,19,20]. One of the major problems that is especially related to the combustion of herbaceous biomass is high alkali (K, P) and chlorine content [21–25]. Similarly, this problem is important in combustion in waste-to-energy plants [26,27]. Moreover, similar behaviour can be observed in the deficiency of oxygen, e.g., during gasification [28]. Due to the relatively low evaporation temperature, compounds that consist of the aforementioned elements may evaporate in the combustion chamber (Figure 1) and later on condense on the superheater tubes. The problem consists of two different mechanisms and could be detrimental for heat exchanging surfaces, as has been confirmed for a wide range of alloys [29,30].

The first mechanism is called fouling, and it introduces problems with heat exchange and flue gas flow within the heat exchanger area. It is caused by deposits of previously mentioned compounds and also unburned char and ash that stick to the molten deposits and make them grow by volume. The second mechanism is fire-side (high temperature) corrosion (Figure 1). This mechanism is of chemical nature but works in between the tube surface and the deposit. It weakens the alloys that tubes are made of, which is dangerous in terms of pressurized devices. Moreover, it decreases the total lifetime of the device. It is known that the reducing environment (with mineral matter, sulfur and chlorine present in the flue gas) inside the furnace leads to high corrosion rates [4,31].

The analyses of the composition of the gas boundary layer and that of the wall tube deposits have shown that the corrosion process is mainly due to combustion with air deficiency, i.e., with the occurrence of reducing zones and the presence of hydrogen sulfide [32]. The corrosion wastage of the tubes is compounded by the presence of sodium and potassium. A strongly correlated relationship between the reducing conditions and the corrosion process has been established. The investigations of the deposits showed that sulfur usually occurs in the second deposit layer, and its content reaches a high value, and the quantity of deposits does not decide the rate of corrosion. Moreover, the presence of K_2SO_4 is an additional factor that could influence the behavior of the eutectics [33].

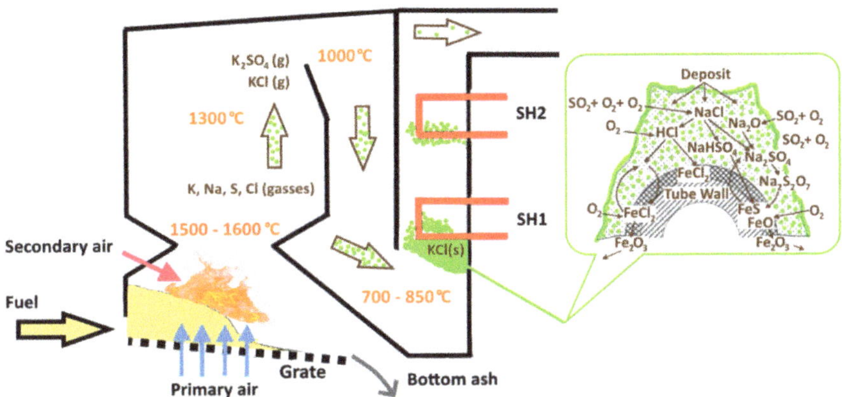

Figure 1. Fouling by KCl in a fixed bed straw boiler (**left**), with chloride and sulfide corrosion mechanism (**right**)—adapted, based on [34,35] (SH1 and 2—sections of the superheaters in the boiler).

It is possible to mitigate the corrosion potential of the fuel by applying carefully selected fuel mixtures [36] or fuel valorization techniques. Some of them, e.g., washing [37], are simple but require large quantities of water or alternatively are dependent on atmospheric conditions (rain). Other processes, such as hydrothermal carbonization, can influence the composition of the inorganic fraction of the material [38–43]. However, this influence can be both positive and negative, depending on the process conditions (e.g., Figure 2). These techniques are getting increased attention as they are especially suitable for types of biomass with high moisture content [44–49]. In some of the cases, mixing of different fuels can be utilized effectively, e.g., co-firing of problematic biomass with sewage sludge [50,51] and peat [52].

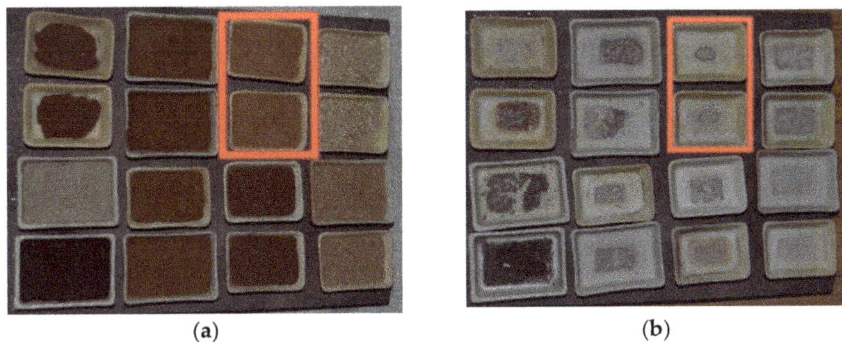

(a)   (b)

Figure 2. Miscanthus after wet torrefaction in different process conditions (samples after wet torrefaction in 200 °C, 10 min, water: biomass ratio 12:1): (**a**) raw samples; (**b**) samples after 3 h of ashing in the furnace at 850 °C.

In addition, using additives, as an injection of ammonia sulfate [52] or adding mineral additives [53,54] (kaolin, zeolites), could be feasible in terms of reducing the risk of deposition and corrosion. However, due to required changes in the feeding system and the availability of such supplementary fuels in close proximity to the plant, fuel logistics become increasingly difficult in both of these cases. The work on improvements in the materials for the boilers is also intensive at the moment [55]. Moreover, the pace of the development of different coatings is also considerable [56]. Technical means are also used to remove these deposits, e.g., by using soot blowers [57]. However, improvements in material science will not make diagnostics obsolete, only less intensive.

Many researchers have established the relationship between high corrosion rates and the strongly reducing atmosphere near the wall. It is well proven that the reducing environment inside the furnace [58,59] and mineral matter of the ash composition, such as the presence of sulfur, chlorine, and potassium leads to high corrosion rates. Pronobis and Litka [60] showed that the corrosion rate depends on the temperature of the external tube surface and on the distribution of the local CO concentration in the boundary layer in the furnace.

Two systems of online monitoring of the boundary layer gas composition have been developed independently and tested in some boilers [61]. The results of working with such a system might be helpful, e.g., in making decisions related to protective-air systems and anti-corrosion coatings implementation. Modelling of ash deposition by computational fluid dynamics (CFD) has turned out to be helpful in indicating the areas most susceptible to corrosion [62–64]. The results obtained may be useful in optimizing the boiler operation and modernizing burners and the boiler furnace. It was observed by Modliński and Hardy that CFD simulation was one of the best ways to detect eventual problems that could potentially be caused by the unfavorable distribution of CO and O_2 in a pulverized coal boiler [65–67]. On the other hand, CFD simulation could be a good way of assessing improvement offered by designs impeding the corrosion process, e.g., slot wall jet [68].

Despite the fact that such techniques are constantly being improved, a lot of damage caused by the boiler tube-thick losses (Figure 3) results in the intensification of interest in and need of simple and reliable non-destructive testing methods. The methods are expected to be sufficiently precise and easy to utilize in the conditions one experiences during work inside a boiler combustion chamber. Non-destructive testing methods (NDT) that aim at inspecting the state of heating steam boiler surfaces can be used as a fast screening technique for corrosion effects. Utilizing such methods during periodic inspections contributes to taking preventive actions and avoiding emergency situations caused by high-temperature corrosion. These methods may also be used in studies of the effectiveness of functioning of the new protective coatings types that are used more and more in our energy economics. Some of the methods give a chance to automatize the measurement process.

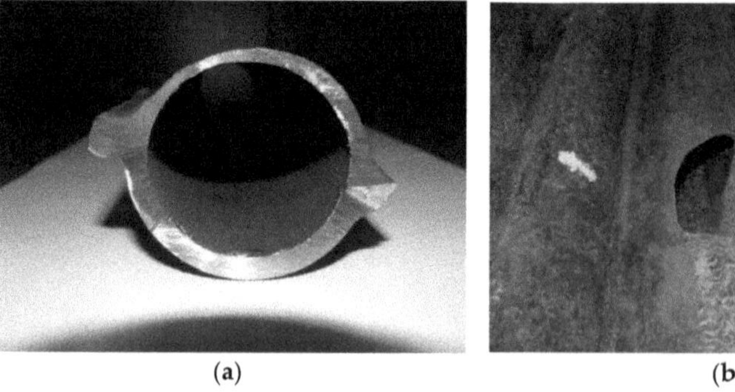

(a)   (b)

Figure 3. Visible darkening of the tube wall and its effect on the boiler OP430 damage: (**a**) a fragment of corroded pipe, cut-out during the maintenance; (**b**) evaporator pipe ruptured in the boiler (photo by authors).

3. Non-Destructive Testing Methods for Fire-Side Corrosion Testing

Non-destructive testing methods (NDT) pertain to non-invasive measurement techniques that allow observers both to study the structure of the construction material and to take quantitative measurement of certain characteristics of the sample studied, for example, of the material thickness. As opposed to destructive testing methods, NDT serves to assess the material's state without causing damage to the sample studied [69,70].

The NDT methods are used in many industrial fields as well as in nearly every stage of the manufacturing processes for a wide range of products [71,72]. They play a crucial role in ensuring low operating costs, security, and reliability of the whole industrial plant's functioning. The techniques of non-destructive weld testing are widely used in many industrial fields, including in energy economics as well [73].

Ultrasonographic techniques, as well as the ones using eddy currents, usually exploit small surface probes, which restrict the whole area only to control a tiny bit of the boiler water wall. Moreover, methods using, for example, ionizing radiation and infrared radiation are also being developed.

One of the interesting techniques is in situ metallographic testing, which can be performed during periodic boiler inspections [74]. The method gives additional knowledge regarding the structural changes that occurred in the inspected sections during the exploitation of the boiler. However, it should be noted that such inspections should be performed by well-skilled staff to avoid misinterpretation of the metallographic data [74]. The time and resources needed for advanced diagnostic methods could be saved by initial identification of areas particularly exposed to high corrosion rates, using simple methods.

3.1. Ultrasonographic Methods

Ultrasonographic methods utilize phenomena that result from differences pertaining to the speed of mechanical vibration movement in the materials studied and their surroundings [70,75,76]. The differences trigger the reflection of waves at the media boundaries. The ultrasonographic measurements are realized by sending an ultrasonic impulse to the sample studied. The impulse bounces off the sample edge, comes back to the receptive head and then is presented to the operator. By measuring the reflected wave movement times and by knowing the speed with which the wave moves in a given medium, it is possible to estimate the thickness of the measured sample as well as the localization of the potential unevenness and defect in its structure.

The ultrasonographic technique is of two types:

- the classic one, which uses the transmitter head generating vibrations;
- the EMAT Method, in which the mechanical vibrations are evoked in the material studied.

In the first method, one head works as a transmitter, the second one as a receiver. The ultrasonic wave is most frequently generated and converted into electric impulses by means of the piezoelectric phenomenon. Techniques of various technological advancement levels can be distinguished here; however, the classic method (manual measurement with the use of light and hand-held measuring devices) is of the utmost importance in boiler tubes studies.

The classic method requires the appropriate preparation of the sample studied—that is, cleansing its surface as well as using the surface, which transmits the transmitter head vibrations to the sample studied. As for the studies of the boiler tubes' thickness, they refer to the necessity of the appropriate preparation of the tube surface where the measuring head is located—that is, the deposit has to be removed, and the surface is expected to be polished (Figure 4).

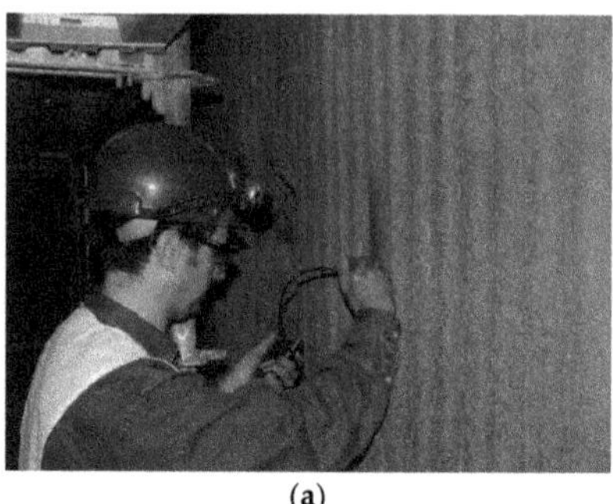 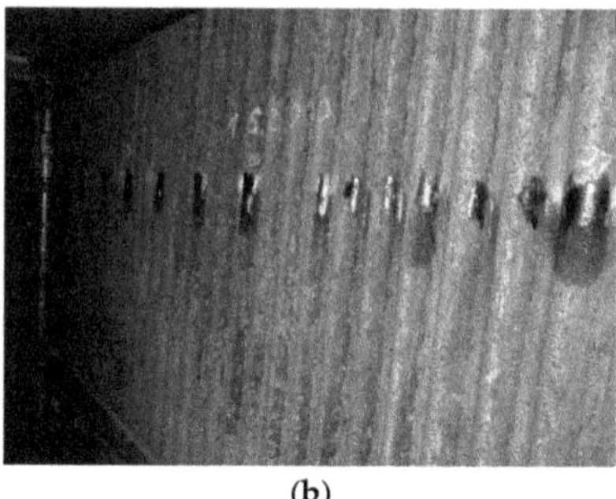

(**a**)   (**b**)

Figure 4. Measuring the tube thickness in the boiler with the aid of the ultrasonic method: (**a**) the measurement process; (**b**) the measurement points.

The electromagnetic acoustic transducer method (EMAT) assumes the actuation of vibrations being studied by using the magnetic field. The vibrations inside the sample studied are actuated through the overlapping of two kinds of magnetic fields: constant and variable. These are incited by means of the winding powered by a generator. The material thickness measurement itself is conducted in a similar way to that in which the classic ultrasonic method with the piezoelectric head is conducted. The measurement is carried out using a special head with a transducer not requiring direct contact with the tested material; therefore, there is no need for mechanical removal of the deposit (oxide layer on the fire-side of the water wall or superheater tubes), and measurements can be carried out very quickly without the need for a liquid couplant (as is the case with the ultrasonic method) [70,77]. Such a diagnostic system is suitable for quick wall thickness measurements in power boilers because it enables a detailed examination of the entire evaporator surface and obtains a clear visualization of the control results (color maps).

There are solutions for carrying out the inspection of large surfaces operating at high temperatures [78]. However, measurements of the wall tubes' thickness are still carried out during the boiler outages because cleaning of the surface is usually necessary before the inspection inside the furnace chamber (especially during biomass combustion, which is accompanied by increased fouling of the heating surface).

However, the automation of measurements (use of robots) would be beneficial, as it would allow measurements to start earlier and reduce the time of the entire diagnostic process.

Undoubtedly, the EMAT Method, in comparison to the classic ultrasonic method, boasts, as its greatest advantage, the possibility of contactless measurement as well as less sensitivity to the state of the surface of the sample studied. This method, nevertheless, has some limitations. One of them is the accuracy of the measurement. Another is the influence of the accuracy of the distance of the head from the test object, and the effect of chemical composition and structure of ash deposits on the signal disturbance [79].

3.2. Methods Employing the Magnetic and Electromagnetic Fields

Because of the possibility of using the study materials utilized in the construction of industrial-scale boilers, one may apply two, out of many, methods of measurement employing magnetic and electromagnetic fields: the magnetic flux leakage method (MFL) and the saturated low-frequency eddy current method (SLOFEC).

The MFL method employs the magnetization of the material studied to the saturation level as well as the measurement of magnetic field corona in places in which material losses and other anomalies are located [71,72,75]. Detecting defects by means of this technique has been known for a long time; new studies concerning this issue refer to methods of establishing the structure of detected defects. The constant magnetic field applied to the material studied is generated by an adequately strong magnet or electromagnet constituting an element of the measuring probe. The field closes inside the sample studied and should trigger saturation of the sample. In the places where losses, material defects, or contractions are present, the magnetic field corona occurs, which is captured by the measuring probe. In one of the recommended solutions, detecting corrosion inside the tubes constitutes the original function of the measuring device. Prior to the test proper, calibration tests are necessary, whose task is to adjust the measurement system to the specifics of the tested object. In these studies, various methods of simulation of defects characteristic of the material studied are employed. High-resolution devices using the MFL Method gather the data roughly every 2 mm along the tube axis. The measurements are processed by the specialized software, which, on the basis of the whole set of data received, assesses the present state and makes a prediction concerning the corrosion progress. The measurement accuracy constitutes the limitation of the method—the material losses can indeed be identified but without the information on which side of the sample studied the loss is located.

The SLOFEC Method uses eddy current along with the constant magnetic field [80]. The classic method employing eddy current is not quite suitable for ferromagnetic materials studies because of the low ability of the current to deeply penetrate materials of this kind. Applying the constant magnetic field to the sample studied causes improvement of the eddy current penetration depth. In the case of a defect, the magnetic field lines have a higher density in the remaining wall thickness where the fault is located. This, as a consequence, changes the medium magnetic permeability and changes the arrangement of the eddy current field lines. Alterations in the arrangement of the eddy current field lines undergo analysis in reference to calibrations with regards to differences in amplitude and the signal phase. The possibility of observation of the phase, amplitude, and shape of the received return signal in the SLOFEC Method allows observers to assess the tube wall thickness loss and distinguishing defects at both sides of the wall [80]. In addition, the SLOFEC Method is not an absolute wall thickness measurement technique, and calibration by a comparison technique, like the ultrasonographic method, is required. Although some similarity in using the magnetic field between the SLOFEC (the constant and variable magnetic field) and EMAT methods can be observed, a fundamental difference between the two exists, too. In the EMAT method, the issue under investigation is the signal resulting from mechanical vibrations in the sample studied. On the other hand, in the SLOFEC method, the induced current is observed.

Another non-contact inspection method that could be used for the purpose of the fire-side corrosion method is eddy current testing (ECT). The method can potentially be used for measuring the wall thickness through insulation and cladding [81]. Pulsed eddy current (PEC) testing is still a relatively new technique used for the inspection of electrically conducting and ferrous materials with corrosion under insulation; it provides considerably more defect information when compared to conventional eddy current techniques due to the use of an excitation pulse that contains broadband frequencies [82]. Pulsed eddy current testing (PECT) uses broadband excitation for sufficient electromagnetic penetration into test objects [81,83]. The PECT signal changes over time along with the penetration of eddy current, and therefore it might be possible to characterize a test object by depth from the analysis of the time-varying signal. The short duty cycle pulses consume lower power in comparison to those of its sinusoidal counterpart. With the same large excitation current, higher excitation pulses can be employed to strengthen induced eddy currents. Overall, the PECT method can be applied to wall-thinning measurement when the pipe wall and cladding sheet's properties are fully understood. It is possible to apply this approach to other insulated or difficult-to-access test objects, where no contact with the probe is required [81]. Other methods, such as RFID (radio frequency identification) [75], as well as microwave-based sensors [75,84], have attracted interest as novel diagnostic techniques for online monitoring of structural materials' integrity in the process industry, especially in the context of the development of so-called industry 4.0 [75].

3.3. Methods Employing Ionizing Radiation

Utilizing ionizing radiation may present another method of tube thickness alteration analysis. In this case, however, the tube must be located between the source and receiver of radiation, the latter of which constitutes a specialized scintillator-multiplier phototube system. The electronics in this construction consist of a CCD camera equivalent. In addition, software analyzing the received radiograph results plays a crucial role in the measuring system.

An image from the camera is sent to a computer, where it undergoes an analysis aimed at establishing the thickness of the wall of the x-rayed tube. Methods utilizing ionizing radiation are, in practice, used for tube wall thickness measurement. However, they require synchronized radiation source placement and a receiver head at both sides of the analyzed tube. They detect corrosion and erosion defects both sides on the outer and inner tube surface, as well as the insulated and deposit-covered tubes [85–87]. In the case of field studies in boilers with airtight screens, this may cause some difficulties, especially when additional angular exposure is required. This method type, nevertheless, allows for the measurement process automatization with minimal costs destined for the preparation of the surfaces for analysis.

3.4. Infrared Radiation-Based Methods

Methods employing infrared radiation concern observation of the heat corona of the material studied heated by the local heat impulse. The observation is assisted with the camera working in the infrared mode. The pace at which the material regions with different thicknesses give up the heat varies. This allows for distinguishing and detecting the material regions. Moreover, on the basis of temperature change dynamics, one can roughly assess the material thickness changes as well as other defects influencing the way heat is conducted in the material. In this method, the infrared camera moves (as one device) with a linear heat source in such a way as to situate in the camera view the material region before and after heating it. The temperature recorded from the unheated region contains information about the initial temperature schema in the sample studied. One of the advantages this method offers is the easiness with which the measuring process is automated as well as the possibility to gain the map of the thickness of large-surface screens after the scan process is over. Unfortunately, the results are subordinate to the material heat capacity as well as heat conductivity both of the tube material and the deposited layer, the latter one being, in most cases, heterogeneous. That is why it is necessary, prior to the analysis, to carefully cleanse the surface [88]. An example of using this measurement technique for the study of the boiler screen is the Line Scanning Thermography System (LST) produced by the Mistras Company [88].

3.5. Other Methods

Other methods of non-destructive testing could likely find their way to corrosion detecting systems in boilers. Al-Mayouf and Al-Shalwi [89] investigated a simple galvanic sensor that was useful for determining the start of iron corrosion under situations similar to those used for separating layers of iron oxide magnetite from the steam boiler with the 5% HCL solution at 60 °C. The time at which the inclusion of corrosion inhibitors is needed to safeguard the base metal can be signaled by this sensor [89]. The fiber brag grating (FBG) system employs the use of photoacoustic sensors [75]. With a photoacoustic sensor, two types of ultrasonic waves are generated: body waves (i.e., p-waves and s-waves) and surface waves (fundamental mode of Rayleigh waves), which have higher amplitude and are more sensitive to corrosion [75]. Recently, PARC Company has developed a technique to allow optical fiber sensors to detect the wavelength deviation with a 50 femtometer resolution [75]. The idea is to use a distributed array of high precision photosensors. Brillouin distributed fiber is another sensor used for corrosion monitoring since it measures the strain very precisely [75].

4. Available Automatic Systems with Visual Data Processing

In order to determine the boiler tubes' thickness, it is necessary to put scaffolds in the boiler and to carefully cleanse the surface with a water stream. It is essential to study an appropriately large screen surface region in as short a time as possible. Thus, it seems profitable to automate such a study by utilizing special robots intended to relocate measuring probes. Vakhguelt et al. [76] reported a conceptual design of a climbing robot for corrosion measurements in power plant scale boilers, using EMA sensors [76].

The Bhabha Atomic Research Centre (BARC) is one of the companies that has attempted to construct such a robot, which was intended for vertical movement along the boiler wall tubes. The robot was adapted to relocate the EMAT measuring head and was able to carry weight amounting to 10 kg at speed up to 120 mm/s [90]. Such a solution allows for scanning the region of a large wall, detecting all of the critical wall tubes regions, and, subsequently, making meticulous measurements of the thickness of the tubes by means of the classic ultrasonic method in the very places that were detected.

The Vertiscan™ System by the Russell NDE Systems Company from Canada uses, on the other hand, a magnetic robot—TubeCAT™—capable of self-moving along the vertical tubes of the boiler. The robot, which uses the electromagnetic measurement method, has a scanner (known as E-PIT) able to analyze five tubes at a time [91]. The system is calibrated by using ultrasonic measurements.

The Alstom Company has been working for a few years on the development of mobile robots. It is in possession of a few solutions which may be successfully used in the boiler combustion chamber conditions playing the role of an automatized platform, for example, in non-destructive studies (ultrasonography, eddy current, thermography, and radiation measurements), in the visual inspection of the boiler (a video camera, 3D scanning), and in the heating surface cleansing [92].

Using the visual systems may constitute technique supplementation, complementing the tube thickness diagnostics, and may aid the initial evaluation of the heating boilers' surface state. Systems of this kind utilize the tube surfaces photographs taken in visible light. A prototype of such a computer vision system intended for evaluating the state of the boiler evaporators' surface was designed in the Department of Automation, Mechatronics and Control Systems at the Wrocław University of Science and Technology.

The system equipment consists of a digital camera, LED illuminator, and devices that aid in directing towards and indicating the current position in which the camera was located while taking a photograph. The device may be moved by an operator (Figure 5), although plans exist for the construction of some sort of mobile platform capable of moving along the boiler walls by itself (as a robot).

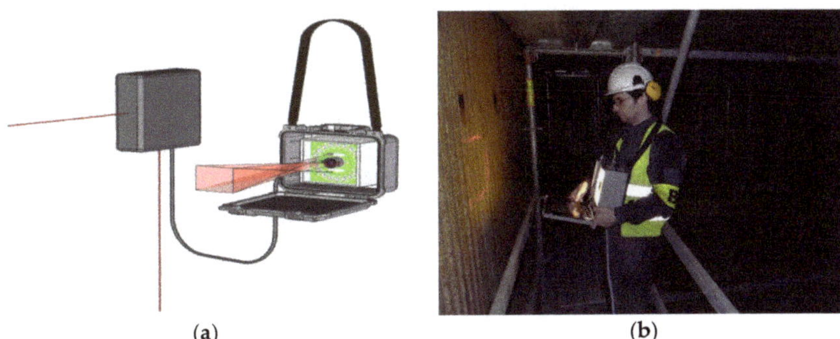

(a)   (b)

Figure 5. Schematic construction of the device intended for taking photographs of the boiler walls and illustration demonstrating the OP-230 boiler firebox during a test: (**a**) a sketch of the device; (**b**) the measurement process.

The system software performs the photograph archiving and registers the data indicating the position as well as further analysis of what is visible in the photographs [93]. The information is saved in the database, which enables one to visualize the whole wall or a selected fragment bigger than a single photograph in the form of a photomosaic (Figure 6). The possible use of machine vision systems depends on what has been photographed and what kind of functions are implemented in the software. In the case of self-propelled devices, the software will also aid the device movement control.

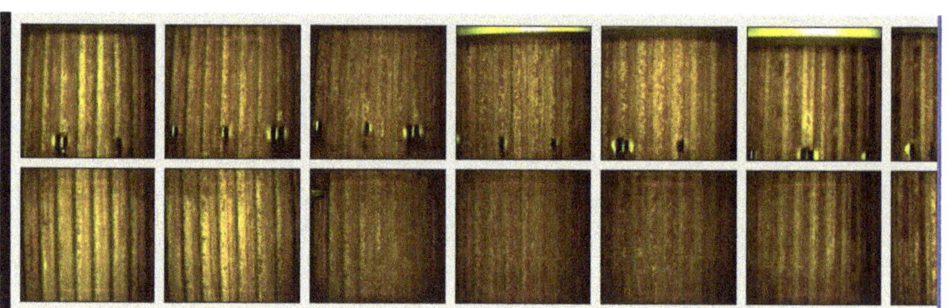

Figure 6. The example of a screenshot from the program aimed at processing and analyzing the evaporator photographs [93].

The field of application of vision systems is determined by what is intended for photographing as well as the functions performed by the software that analyzes the collected photographs. One of the possible applications of the vision methods is the analysis of deposits that have accumulated on the surface of the tube. In such a situation, the photographs are taken prior to getting rid of the deposit from the evaporator surface. By comparing photographs of selected regions that were taken in various time periods, one may, for example, assess the effectiveness of deposit blower work and the influence of variable fuel properties on deposit formation.

After partial ash deposit removal, it is possible to search out, by means of picture analysis methods, the places where the deposit underwent melting. Drawing conclusions about the state of the boiler walls on the basis of the appearance of the deposit accumulated on the walls is risky and requires the collection of the data from manifold measurements as well as search for correlations by means of mathematical statistics methods. In the designed software, the possibility of carrying out such analyses was provided.

The vision analysis may also concern the surface of the metal of which the tubes have been made; nonetheless, it requires, in the beginning, the removal of the deposit that has accumulated on this surface. The photographs taken in such a situation may be analyzed through the picture detection methods, which are color irregularity measurement oriented. Based on this, one may draw conclusions about the corrosion pitting on the metal. Generally, images are obtained in the form of RGB components (with the possible Bayer matrix at the sensor level). For this kind of analysis, a form of hue and saturation, the HSV value is considered to be better. The size of the color space is roughly the same since we also have three parameters. The hue parameter controls color but in the form of a color wheel; therefore, similar values have similar colors (if we accept that 0 and 255 are close). This allows for much simpler processing methods.

Descriptions of the measuring devices offered by various producers usually contain information about the necessity to calibrate the device to particular applications. This requirement concerns carrying out the test measurements, processing the collected data, and, on the basis thereof, setting the parameters used during the measurement proper. Data collected during the measurements is also compared to the model measurements database, and, on this basis, one makes the final evaluation of the interpretation of the collected material.

Generally, the received measurement data constitutes data emerging as the sum of responses of fragments studied from many samples. It causes some difficulties in determining the unambiguous location and size of the detected defect in the structure of the material studied. Usually, the ultrasonographic study is exemplified as the model method of tube thickness measurement.

5. Conclusions

As far as the matter of the measurement methods used in not-destroying-metals testing is concerned, one should focus on two issues:

- a physical phenomenon used for collecting measurement data;
- processing the collected data by means of computer techniques.

The methods that have the best chance of being widely used are those that allow for relatively meticulous determination of the tube walls' thickness without the need to build a scaffold in the boiler firebox. The experience gained so far with the automation of wall tube thickness measurement using the EMAT method shows its great potential and the possibility of making quick measurements without time-consuming surface cleaning, unlike the most commonly used method today—the ultrasonic method. If robots carrying the measuring head are used (automatized measurement), it will be possible to test the entire wall without the need to erect scaffolding.

Additionally, the vision methods will serve as aids; they will, in turn, be aided by the appropriate computer software intended for processing and analyzing pictures. In this case, it will be possible to use drones to take photos of the water wall surfaces in the boiler.

Non-destructive testing methods for fire-side corrosion testing are well proven for the corrosion testing of water walls of the boiler. Straightforward utilization for inspection of the superheaters might prove challenging due to difficult access. Further work is required on this topic.

Author Contributions: Conceptualization: T.H., H.P.-K., W.R. and J.W.; investigation: T.H., H.P.-K., W.R., J.W., K.M., V. and Ł.N.; resources, T.H. and H.P.-K.; writing—original draft preparation, T.H., A.A., H.P.-K., W.R., J.W., K.M., V. and Ł.N.; writing—review and editing, T.H., H.P.-K., A.A., K.M. and Ł.N.; visualization, T.H., K.M. and Ł.N.; supervision, T.H. and H.P.-K.; project administration, T.H. and H.P.-K.; funding acquisition, T.H. and H.P.-K. All authors have read and agreed to the published version of the manuscript.

Funding: Support from the project "Corrosion and Emission Reduction of Utility Boilers through Intelligent Systems" CERUBIS (RFCR-CT-2014-00008), funded by the European Research Fund for Coal and Steel, is gratefully acknowledged.

Institutional Review Board Statement: Not applicable.

Informed Consent Statement: Not applicable.

Data Availability Statement: Not applicable.

Conflicts of Interest: The authors declare no conflict of interest.

References

1. Pronobis, M. Evaluation of the influence of biomass co-combustion on boiler furnace slagging by means of fusibility correlations. *Biomass Bioenergy* **2005**, *28*, 375–383. [CrossRef]
2. Pronobis, M. *Modernizacja Kotłów Energetycznych*; WNT: Warszawa, Poland, 2009; ISBN 9788379261352.
3. Hein, K.R.G. *Operational Problems, Trace Emissions and By-Product Management for Industrial Biomass Co-Combustion*; CORDIS: Brussels, Belgium, 2000.
4. Harb, J.; Smith, E. Fireside corrosion in pc-fired boilers. *Prog. Energy Combust. Sci.* **1990**, *16*, 169–190. [CrossRef]
5. Li, G.; Li, S.; Huang, Q.; Yao, Q. Fine particulate formation and ash deposition during pulverized coal combustion of high-sodium lignite in a down-fired furnace. *Fuel* **2015**, *143*, 430–437. [CrossRef]
6. Zheng, S.; Zeng, X.; Qi, C.; Zhou, H. Mathematical modeling and experimental validation of ash deposition in a pulverized-coal boiler. *Appl. Eng.* **2017**, *110*, 720–729. [CrossRef]

7. Wang, X.; Xu, Z.; Wei, B.; Zhang, L.; Tan, H.; Yang, T.; Mikulčić, H.; Duić, N. The ash deposition mechanism in boilers burning Zhundong coal with high contents of sodium and calcium: A study from ash evaporating to condensing. *Appl. Eng.* **2015**, *80*, 150–159. [CrossRef]
8. Gao, Q.; Li, S.; Yang, M.; Biswas, P.; Yao, Q. Measurement and numerical simulation of ultrafine particle size distribution in the early stage of high-sodium lignite combustion. *Proc. Combust. Inst.* **2017**, *36*, 2083–2090. [CrossRef]
9. Golański, G.; Lachowicz, M.M. Failure cause analysis for the suspension element of boiler superheater. *Eng. Fail. Anal.* **2019**, *105*, 490–495. [CrossRef]
10. Pawlak-Kruczek, H.; Niedźwiecki, Ł.; Ostrycharczyk, M.; Czerep, M.; Plutecki, Z. Potential and methods for increasing the flexibility and efficiency of the lignite fired power unit, using integrated lignite drying. *Energy* **2019**, *181*, 1142–1151. [CrossRef]
11. Duarte, C.A.; Espejo, E.; Martinez, J.C. Failure analysis of the wall tubes of a water-tube boiler. *Eng. Fail. Anal.* **2017**, *79*, 704–713. [CrossRef]
12. Sims, R. *The Brilliance of Bioenergy: In Business and Practice*; Earthscan Publications Ltd.: London, UK, 2002; ISBN 978-1902916286.
13. Sher, F.; Iqbal, S.Z.; Liu, H.; Imran, M.; Snape, C.E. Thermal and kinetic analysis of diverse biomass fuels under different reaction environment: A way forward to renewable energy sources. *Energy Convers. Manag.* **2020**, *203*, 112266. [CrossRef]
14. Tzelepi, V.; Zeneli, M.; Kourkoumpas, D.S.; Karampinis, E.; Gypakis, A.; Nikolopoulos, N.; Grammelis, P. Biomass availability in europe as an alternative fuel for full conversion of lignite power plants: A critical review. *Energies* **2020**, *13*, 3390. [CrossRef]
15. Jewiarz, M.; Wróbel, M.; Mudryk, K.; Szufa, S. Impact of the drying temperature and grinding technique on biomass grindability. *Energies* **2020**, *13*, 3392. [CrossRef]
16. Romanowska-Duda, Z.; Piotrowski, K.; Wolska, B.; Debowski, M.; Zielinski, M.; Dziugan, P.; Szufa, S. *Stimulating Effect of Ash from Sorghum on the Growth of Lemnaceae—A New Source of Energy Biomass*; Springer International Publishing: 2020; pp. 341–349, ISBN 9783030138882. Available online: https://link.springer.com/chapter/10.1007/978-3-030-13888-2_34 (accessed on 1 September 2021).
17. Szufa, S.; Wielgosiński, G.; Piersa, P.; Czerwińska, J.; Dzikuć, M.; Adrian, Ł.; Lewandowska, W.; Marczak, M. Torrefaction of Straw from Oats and Maize for Use as a Fuel and Additive to Organic Fertilizers—TGA Analysis, Kinetics as Products for Agricultural Purposes. *Energies* **2020**, *13*, 2064. [CrossRef]
18. Luo, H.; Lu, Z.; Jensen, P.A.; Glarborg, P.; Lin, W.; Dam-Johansen, K.; Wu, H. Experimental and modelling study on the influence of wood type, density, water content, and temperature on wood devolatilization. *Fuel* **2020**, *260*, 116410. [CrossRef]
19. Tran, K.Q.; Werle, S.; Trinh, T.T.; Magdziarz, A.; Sobek, S.; Pogrzeba, M. Fuel characterization and thermal degradation kinetics of biomass from phytoremediation plants. *Biomass Bioenergy* **2020**, *134*, 105469. [CrossRef]
20. Polesek-Karczewska, S.; Turzyński, T.; Kardaś, D.; Heda, Ł. Front velocity in the combustion of blends of poultry litter with straw. *Fuel Process. Technol.* **2018**, *176*, 307–315. [CrossRef]
21. Shao, Y.; Wang, J.; Preto, F.; Zhu, J.; Xu, C. Ash deposition in biomass combustion or co-firing for power/heat generation. *Energies* **2012**, *5*, 5171–5189. [CrossRef]
22. Magdziarz, A.; Gajek, M.; Nowak-Woźny, D.; Wilk, M. Mineral phase transformation of biomass ashes—Experimental and thermochemical calculations. *Renew. Energy* **2018**, *128*, 446–459. [CrossRef]
23. Zuwała, J.; Lasek, J. Co-combustion of low-rank coals with biomass. In *Low-Rank Coals for Power Generation, Fuel and Chemical Production*; Woodhead Publishing: Sawston, UK, 2017; pp. 125–158. ISBN 9780081008959.
24. Mlonka-Mędrala, A.; Magdziarz, A.; Gajek, M.; Nowińska, K.; Nowak, W. Alkali metals association in biomass and their impact on ash melting behaviour. *Fuel* **2020**, *261*, 116421. [CrossRef]
25. Mlonka-Mędrala, A.; Magdziarz, A.; Dziok, T.; Sieradzka, M.; Nowak, W. Laboratory studies on the influence of biomass particle size on pyrolysis and combustion using TG GC/MS. *Fuel* **2019**, *252*, 635–645. [CrossRef]
26. Phongphiphat, A.; Ryu, C.; Yang, Y.B.; Finney, K.N.; Leyland, A.; Sharifi, V.N.; Swithenbank, J. Investigation into high-temperature corrosion in a large-scale municipal waste-to-energy plant. *Corros. Sci.* **2010**, *52*, 3861–3874. [CrossRef]
27. Viklund, P.; Hjörnhede, A.; Henderson, P.; Stålenheim, A.; Pettersson, R. Corrosion of superheater materials in a waste-to-energy plant. *Fuel Process. Technol.* **2013**, *105*, 106–112. [CrossRef]
28. Wan, W.; Engvall, K.; Yang, W. Model investigation of condensation behaviors of alkalis during syngas treatment of pressurized biomass gasification. *Chem. Eng. Process.-Process. Intensif.* **2018**, *129*, 28–36. [CrossRef]
29. Mlonka-Mędrala, A.; Gołombek, K.; Buk, P.; Cieślik, E.; Nowak, W. The influence of KCl on biomass ash melting behaviour and high-temperature corrosion of low-alloy steel. *Energy* **2019**, *188*, 116062. [CrossRef]
30. Sandhi, K.K.; Szpunar, J. Analysis of Corrosion of Hastelloy-N, Alloy X750, SS316 and SS304 in Molten Salt High-Temperature Environment. *Energies* **2021**, *14*, 543. [CrossRef]
31. Kofstad, P. *High Temperature Corrosion*; Elsevier Applied Science: London, UK; New York, NY, USA, 1988; ISBN 1-85166-154-9.
32. Von Bohnstein, M.; Yildiz, C.; Frigge, L.; Ströhle, J.; Epple, B. Simulation study of the formation of corrosive gases in coal combustion in an entrained flow reactor. *Energies* **2020**, *13*, 4523. [CrossRef]
33. Tesfaye, F.; Lindberg, D.; Moroz, M.; Hupa, L. Investigation of the K-Mg-Ca Sulfate System as Part of Monitoring Problematic Phase Formations in Renewable-Energy Power Plants. *Energies* **2020**, *13*, 5366. [CrossRef]
34. Sweden Linnæus University (Project Leader); Estonia Tallinn University of Technology; Greece CERTH; Italy IVALSA-Ireland Tipperary Energy Agency. The Bioenergy System Planners Handbook-BISYPLAN. Available online: http://bisyplan.bioenarea.eu/html-files-en/ (accessed on 24 July 2016).

35. Basu, P. *Combustion and Gasification in Fluidized Beds*; CRC Press: Boca Raton, FL, USA, 2006; ISBN 9780849333965.
36. Kulokas, M.; Praspaliauskas, M.; Pedišius, N. Investigation of Buckwheat Hulls as Additives in the Production of Solid Biomass Fuel from Straw. *Energies* **2021**, *14*, 265. [CrossRef]
37. Van Loo, S.; Koppejan, J. *The Handbook of Biomass Combustion and Co-Firing*; Earthscan: London, UK, 2008; ISBN 978-1-84407-249-1.
38. Keipi, T.; Tolvanen, H.; Kokko, L.; Raiko, R. The effect of torrefaction on the chlorine content and heating value of eight woody biomass samples. *Biomass Bioenergy* **2014**, *66*, 232–239. [CrossRef]
39. Reza, M.T.; Yan, W.; Uddin, M.H.; Lynam, J.G.; Hoekman, S.K.; Coronella, C.J.; Vásquez, V.R. Reaction kinetics of hydrothermal carbonization of loblolly pine. *Bioresour. Technol.* **2013**, *139*, 161–169. [CrossRef]
40. Kambo, H.S.; Dutta, A. Comparative evaluation of torrefaction and hydrothermal carbonization of lignocellulosic biomass for the production of solid biofuel. *Energy Convers. Manag.* **2015**, *105*, 746–755. [CrossRef]
41. Śliz, M.; Wilk, M. A comprehensive investigation of hydrothermal carbonization: Energy potential of hydrochar derived from Virginia mallow. *Renew. Energy* **2020**, *156*, 942–950. [CrossRef]
42. Wilk, M.; Magdziarz, A.; Kalemba-Rec, I.; Szymańska-Chargot, M. Upgrading of green waste into carbon-rich solid biofuel by hydrothermal carbonization: The effect of process parameters on hydrochar derived from acacia. *Energy* **2020**, *202*, 117717. [CrossRef]
43. Smith, A.M.; Ekpo, U.; Ross, A.B. The influence of pH on the combustion properties of bio-coal following hydrothermal treatment of swine manure. *Energies* **2020**, *13*, 331. [CrossRef]
44. Aragón-Briceño, C.I.; Grasham, O.; Ross, A.B.; Dupont, V.; Camargo-Valero, M.A. Hydrothermal carbonization of sewage digestate at wastewater treatment works: Influence of solid loading on characteristics of hydrochar, process water and plant energetics. *Renew. Energy* **2020**, *157*, 959–973. [CrossRef]
45. Aragón-Briceño, C.I.; Ross, A.B.; Camargo-Valero, M.A. Mass and energy integration study of hydrothermal carbonization with anaerobic digestion of sewage sludge. *Renew. Energy* **2021**, *167*, 473–483. [CrossRef]
46. Surup, G.R.; Leahy, J.J.; Timko, M.T.; Trubetskaya, A. Hydrothermal carbonization of olive wastes to produce renewable, binder-free pellets for use as metallurgical reducing agents. *Renew. Energy* **2020**, *155*, 347–357. [CrossRef]
47. Ameen, M.; Zamri, N.M.; May, S.T.; Azizan, M.T.; Aqsha, A.; Sabzoi, N.; Sher, F. Effect of acid catalysts on hydrothermal carbonization of Malaysian oil palm residues (leaves, fronds, and shells) for hydrochar production. *Biomass Convers. Biorefin.* **2021**. [CrossRef]
48. Atallah, E.; Zeaiter, J.; Ahmad, M.N.; Leahy, J.J.; Kwapinski, W. Hydrothermal carbonization of spent mushroom compost waste compared against torrefaction and pyrolysis. *Fuel Process. Technol.* **2021**, *216*, 106795. [CrossRef]
49. Zhang, H.; Li, J.; Yang, X.; Guo, S.; Zhan, H.; Zhang, Y.; Fang, Y. Influence of coal ash on potassium retention and ash melting characteristics during gasification of corn stalk coke. *Bioresour. Technol.* **2018**, *270*, 416–421. [CrossRef]
50. Elled, A.L.; Davidsson, K.O.; Åmand, L.E. Sewage sludge as a deposit inhibitor when co-fired with high potassium fuels. *Biomass Bioenergy* **2010**, *34*, 1546–1554. [CrossRef]
51. Aho, M.; Yrjas, P.; Taipale, R.; Hupa, M.; Silvennoinen, J. Reduction of superheater corrosion by co-firing risky biomass with sewage sludge. *Fuel* **2010**, *89*, 2376–2386. [CrossRef]
52. Kassman, H.; Pettersson, J.; Steenari, B.M.; Åmand, L.E. Two strategies to reduce gaseous KCl and chlorine in deposits during biomass combustion—Injection of ammonium sulphate and co-combustion with peat. *Fuel Process. Technol.* **2013**, *105*, 170–180. [CrossRef]
53. Niu, Y.; Tan, H.; Hui, S. Ash-related issues during biomass combustion: Alkali-induced slagging, silicate melt-induced slagging (ash fusion), agglomeration, corrosion, ash utilization, and related countermeasures. *Prog. Energy Combust. Sci.* **2016**, *52*, 1–61. [CrossRef]
54. Wang, L.; Skreiberg, Ø.; Becidan, M.; Li, H. Investigation of rye straw ash sintering characteristics and the effect of additives. *Appl. Energy* **2016**, *162*, 1195–1204. [CrossRef]
55. Antunes, R.A.; de Oliveira, M.C.L. Corrosion in biomass combustion: A materials selection analysis and its interaction with corrosion mechanisms and mitigation strategies. *Corros. Sci.* **2013**, *76*, 6–26. [CrossRef]
56. Naganuma, H.; Ikeda, N.; Ito, T.; Satake, H.; Matsuura, M.; Ueki, Y.; Yoshiie, R.; Naruse, I. Control of ash deposition in solid fuel fired boiler. *Fuel Process. Technol.* **2013**, *105*, 77–81. [CrossRef]
57. Pophali, A.; Emami, B.; Bussmann, M.; Tran, H. Studies on sootblower jet dynamics and ash deposit removal in industrial boilers. *Fuel Process. Technol.* **2013**, *105*, 69–76. [CrossRef]
58. Kruczek, H. Effect of staged combustion in PC boilers on rate of fire-side corrosion. In Proceedings of the 7th European Conference on Industrial Furnaces and Boilers, Porto, Portugal, 18–21 April 2006.
59. Hardy, T.; Kakietek, S.; Janda, T. A tool for on-line control of high-temperature corrosion hazard in steam boilers. *Arch. Combust.* **2017**, *37*, 107–118.
60. Pronobis, M.; Litka, R. Rate of corrosion of waterwalls in supercritical pulverised fuel boilers. *Chem. Process. Eng.-Inz. Chem. I Proces.* **2012**, *33*, 263–277. [CrossRef]
61. Wejkowski, R.; Kalisz, S.; Hardy, T.; Sarapata, B.; Kubiczek, H.; Janda, T. The control of high temperature corrosion phenomenon in boundary layer using different measurement methods. In *Modern Energy Technologies, Systems and Units*; Taler, J., Ed.; Publishing House of Kraków University of Technology: Krakow, Poland, 2013; pp. 289–300.

62. Modlinski, N.J. Computational modelling of a tangentially fired boiler with deposit formation phenomena. *Chem. Process. Eng.-Inz. Chem. I Proces.* **2014**, *35*, 361–368. [CrossRef]
63. Weber, R.; Mancini, M.; Schaffel-Mancini, N.; Kupka, T. On predicting the ash behaviour using Computational Fluid Dynamics. *Fuel Process. Technol.* **2013**, *105*, 113–128. [CrossRef]
64. Davis, K.A.; Linjewile, T.M.; Valentine, J.; Swensen, D.; Shino, D.; Letcavits, J.J.; Sheidler, R.; Cox, W.M.; Carr, R.N.; Harding, N.S. A multi-point corrosion monitoring system applied in a 1300 MW coal-fired boiler. *Anti-Corros. Methods Mater.* **2004**, *51*, 321–330. [CrossRef]
65. Hardy, T.; Modlinski, N. Development of Prognostic Tool for High-temperature Corrosion Risk Assessment in Pulverised Coal Boilers Based on the Waterwalls Boundary Layer Monitoring System and CFD Simulations. *Energy Procedia* **2017**, *105*, 1833–1838. [CrossRef]
66. Modlinski, N.; Hardy, T. Development of high-temperature corrosion risk monitoring system in pulverized coal boilers based on reducing conditions identification and CFD simulations. *Appl. Energy* **2017**, *204*, 1124–1137. [CrossRef]
67. Gruber, T.; Schulze, K.; Scharler, R.; Obernberger, I. Investigation of the corrosion behaviour of 13CrMo4-5 for biomass fired boilers with coupled online corrosion and deposit probe measurements. *Fuel* **2015**, *144*, 15–24. [CrossRef]
68. Zhang, Y.; Fang, Y.; Jin, B.; Zhang, Y.; Zhou, C.; Sher, F. Effect of Slot Wall Jet on Combustion Process in a 660 MW Opposed Wall Fired Pulverized Coal Boiler. *Int. J. Chem. React. Eng.* **2019**, *17*, 1–13. [CrossRef]
69. Vavilov, V.P.; Burleigh, D.D. Review of pulsed thermal NDT: Physical principles, theory and data processing. *NDT E Int.* **2015**, *73*, 28–52. [CrossRef]
70. Vakhguelt, A.; Kapayeva, S.D.; Bergander, M.J. Combination Non-Destructive Test (NDT) Method for Early Damage Detection and Condition Assessment of Boiler Tubes. *Procedia Eng.* **2017**, *188*, 125–132. [CrossRef]
71. Ho, M.; El-Borgi, S.; Patil, D.; Song, G. Inspection and monitoring systems subsea pipelines: A review paper. *Struct. Health Monit.* **2020**, *19*, 606–645. [CrossRef]
72. Shi, Y.; Zhang, C.; Li, R.; Cai, M.; Jia, G. Theory and application of magnetic flux leakage pipeline detection. *Sensors* **2015**, *15*, 31036–31055. [CrossRef]
73. Lu, X.; Liu, G.; Luan, S. The development of the boiler water wall tube inspection. In Proceedings of the 2008 3rd International Conference on Electric Utility Deregulation and Restructuring and Power Technologies, Nanjing, China, 6–9 April 2008; pp. 2415–2420. [CrossRef]
74. Hariyadi, A.; Kis Agustin, H.C.; Wijayanti, I.D. Metallography Investigation of Dry Corrosion Boiler Tube. *Appl. Mech. Mater.* **2016**, *836*, 72–77. [CrossRef]
75. Meribout, M.; Mekid, S.; Kharoua, N.; Khezzar, L. Online monitoring of structural materials integrity in process industry for I4.0: A focus on material loss through erosion and corrosion sensing. *Meas. J. Int. Meas. Confed.* **2021**, *176*, 109110. [CrossRef]
76. Vakhguelt, A.; Jo, R.S.; Bergander, M.J. Electromagnetic acoustic boiler tubes inspection with robotic device. *Vibroeng. Procedia* **2017**, *15*, 115–118. [CrossRef]
77. Bergander, M.J. EMAT thickness measurement for tubes in coal-fired boilers. *Appl. Energy* **2003**, *74*, 439–444. [CrossRef]
78. Kogia, M.; Gan, T.-H.; Balachandran, W.; Livadas, M.; Kappatos, V.; Szabo, I.; Mohimi, A.; Round, A. High Temperature Shear Horizontal Electromagnetic Acoustic Transducer for Guided Wave Inspection. *Sensors* **2016**, *16*, 582. [CrossRef]
79. Klepacki, F. Ocena stanu rur kotłowych, których grubość ścianek uległa zmniejszeniu wskutek erozji lub korozji. *Energetyka* **2005**, 886–888. Available online: https://www.pronovum.pl/static/upload/store/Ocena_stanu_rur_kotlowych_ktorych_gruboisc_scianek_ulega_zmniejszeniu_wskutek_erozji_lub_korozji.pdf (accessed on 30 September 2021).
80. Innospection SLOFEC TM—Fast Corrosion Screening Technique. Available online: http://www.ndttechnologies.com/Brochures/TankScanning/SLOFECTechnique.pdf (accessed on 1 September 2021).
81. Cheng, W. Pulsed eddy current testing of carbon steel pipes' wall-thinning through insulation and cladding. *J. Nondestruct. Eval.* **2012**, *31*, 215–224. [CrossRef]
82. Fernandez, N.M.; Charlton, P.C.; Granville, R.K. Pulsed Eddy Current CUI Simulation with Coupled Electric Circuit Analysis using Quickfield FEA. *e-J. Nondestruct. Test.* **2019**. Available online: https://www.ndt.net/article/ndtnet/papers/Pulsed-Eddy-Current-CUI-Simulation-with-Coupled-Electric-Circuit-Analysis-using-Quickfield-FEA.pdf (accessed on 30 September 2021).
83. Cheng, W.; Komura, I. Pulsed Eddy Current Testing of Carbon Steel Pipes' Wall-thinning. In Proceedings of the 9th International Conference on NDE in Relation to Structural Integrity for Nuclear and Pressurized Components, Seattle, DC, USA, 22–24 May 2012; pp. 336–342.
84. Katagiri, T.; Chen, G.; Yusa, N.; Hashizume, H. Demonstration of detection of the multiple pipe wall thinning defects using microwaves. *Meas. J. Int. Meas. Confed.* **2021**, *175*, 109074. [CrossRef]
85. Boateng, A.; Danso, K.A.; Dagadu, C.P.K. Non-Destructive Evaluation Of Corrosion On Insulated Pipe Using Double Wall Radiographic Technique. *Chem. Mater. Res.* **2013**, *3*, 73–83.
86. Boateng, A.; Danso, K.A.; Dagadu, C.P.K. Non-Destructive Evaluation Of Corrosion On Insulated Pipe Using Tangential Radiographic Technique. *Int. J. Sci. Technol. Res.* **2013**, *2*, 7–13.
87. Edalati, K.; Rastkhah, N.; Kermani, A.; Seiedi, M.; Movafeghi, A. The use of radiography for thickness measurement and corrosion monitoring in pipes. *Int. J. Press. Vessel. Pip.* **2006**, *83*, 736–741. [CrossRef]
88. MISTRAS Automated Thickness Measurement Using Line Scanning Thermography (LST)TM. Available online: http://www.mistrasgroup.com/services/company/publications/Line_Scanning_Thermography.pdf (accessed on 1 July 2018).

89. Al-mayouf, A.M.; Al-shalwi, M.N. Galvanic sensor for detecting corrosion during acid cleaning of magnetite in steam boilers. *Metals* **2021**, *11*, 343. [CrossRef]
90. Badodkar, D.N.; Singh, N.K.; Dalal, N.S.; Singh, M. EMAT Integrated with Vertical Climbing Robot for Boiler Tube Inspection. In Proceedings of the National Seminar & Exhibition on Non-Destructive Evaluation, Tiruchirappalli, India, 10–12 December 2009; pp. 32–36.
91. Vajpayee, A.; Russell, D. Automated Condition Assessment of Boiler Water Wall Tubes Using Remote Field Technology. A Revolution Over Traditional and Existing Techniques. In Proceedings of the 10th International Conference of the Slovenian Society for Non-Destructive Testing 2009, Application of Contemporary Non-Destructive Testing in Engineering, Ljubljana, Slovenia, 1–3 September 2009; pp. 523–530.
92. Caprari, G.; Breitenmoser, A.; Fischer, W.; Hürzeler, C.; Tâche, F.; Siegwart, R.; Schoeneich, P.; Rochat, F.; Mondada, F.; Moser, R. Highly compact robots for inspection of power plants. In Proceedings of the 1st International Conference on Applied Robotics for the Power Industry (CARPI), Montreal, QC, Canada, 5–7 October 2010. [CrossRef]
93. Rafajłowicz, E.; Wietrzych, J.; Rafajłowicz, W. A computer vision system for evaluation of high temperature corrosion damages in steam boilers. In *Intelligent Systems in Technical and Medical Diagnostics. Advances in Intelligent Systems and Computing*; Korbicz, J., Kowal, M., Eds.; Springer: Berlin/Heidelberg, Germany, 2014; Volume 230. [CrossRef]

Article

Application of the Mechanical and Pressure Drop Tests to Determine the Sintering Temperature of Coal and Biomass Ash

Karol Król and **Dorota Nowak-Woźny** *

Department of Energy Technologies, Turbines and Modelling of Thermal and Fluid Flow Processes, Wroclaw University of Science and Technology, 50-370 Wrocław, Poland; karol.krol@pwr.edu.pl
* Correspondence: dorota.nowak-wozny@pwr.edu.pl

Abstract: The aim of this paper is to investigate the mechanical properties of coal and biomass ash during the sintering process. For this study, bituminous coal, lignite, wheat straw, barley straw, and rye straw were selected. The proximate, ultimate, and oxide analyses were performed. The ash from these fuels was prepared in a special way that ensured the physicochemical invariability of the initial state of the mineral matter of coal and biomass. The purpose of this selection was to obtain widely available and clearly diversified materials. Based on the results of ash composition and ultimate analysis the most common ash deposition, indices were determined. Certain conflict of index indications was observed. Then, the mechanical test and pressure drop test were performed. During the mechanical test, the fracture stress as a function of sintering temperature was measured. During the pressure drop test, the pressure before and behind the sample was measured as a function of sintering temperature. Both tests showed that the characteristic changes (the occurrence of a maximum on the pressure drop curve and the inflection point at the mechanical curve) dependencies were at nearly the same temperatures. These results were compared with the initial deformation temperature (IDT) from the standard Leitz method. A linear relationship between sintering temperatures determined by the mechanical test, pressure drop test, and IDT Leitz test was obtained. The obtained results are promising in terms of the application of the mechanical methods (fracture stress test and pressure drop test) as methods of the early stage prediction of slagging/fouling risks.

Keywords: biomass ash; coal ash; sintering; mechanical test; pressure drop test; slagging; fouling

Citation: Król, K.; Nowak-Woźny, D. Application of the Mechanical and Pressure Drop Tests to Determine the Sintering Temperature of Coal and Biomass Ash. *Energies* **2021**, *14*, 1126. https://doi.org/10.3390/en14041126

Academic Editor: Robert Król

Received: 12 January 2021
Accepted: 17 February 2021
Published: 20 February 2021

Publisher's Note: MDPI stays neutral with regard to jurisdictional claims in published maps and institutional affiliations.

1. Introduction

The efficiency of the combustion system producing the energy depends on processes such as ash adhesion and ash removal. These processes lead to ash deposits creation on the heat exchanger tubes and in the gas circuit (slagging and fouling process). The formation of ash deposits on the surface of boilers and heat exchangers reduces the heat transfer rate and therefore the efficiency of the whole system. For this reason, it is important to study the slagging and fouling processes. These processes are mainly caused by reactive alkaline and alkaline earth compounds found in coal and in a relatively large amount of biomass.

The amount of alkaline elements significantly influences the mechanism of mineral matter transformation during the coal and biomass combustion process. These problems have been widely described and discussed by many authors, in particular during biomass combustion [1–4] and during coal combustion [5–8].

Slagging and fouling processes are closely associated with the ash fusion behavior—mainly with the melting of ash. The ash sintering process takes place on the microstructure level of the ash. But from a microscopic point of view, it is a very inhomogeneous material. The individual particles of the ash have different shapes, different chemical compositions, and different crystallographic structures. Then, it is very difficult to find a method that can predict the temperature at which this ash starts to be viscous and starts, therefore, to make the fouling or slagging on the surfaces of heat exchangers. The main mechanism of the ash sintering process is the melting of the surface of individual ash particles. The ash particles with melted surface coalescent as a result of minimization of the surface tension [6,9,10]. The formation of ash deposits—slagging in the radiative region and fouling in the convection region of the combustion system can cause consequences such as deterioration of heat transfer, damage of boiler surfaces, and even shutdown of the boiler [11,12]. It is interesting that the ash deposit buildup is rather related to ash physicochemical properties, while its strength development depends rather on ash sintering characteristics [12,13].

They are some widely used methods of assessing the slagging and fouling hazard. The ash fusion test (AFT) is widely used. This test determines temperatures at which change of samples' shape is observed—deformation temperature (DT), spherical/softening temperature (ST), hemispherical temperature (HT), and fluid temperature (FT), which is a part of mineral matter transformation processes [3,5,14–16]. The other widely used method is based on the statistical observations of the relationship between the amount of basic and acid oxides and the tendency to slagging and fouling [17–20]. In addition to these two methods, there are others based on the observation of the change of physical properties of the ash. To detect the shrinkage behavior of the ash sample during the sintering process, the Thermomechanical Analysis (TMA) method can be implemented. That is originally applied in metallurgy [10]. The other interesting methods are thermal conductivity analysis [21], compressing strength analysis [22–24], electrical resistivity measurements [23], thermogravimetric analysis (TGA) [18,25], thermodynamic FactSage analysis [18], optical heating stage microscopy (OSHM) [26], pressure drop test [9,10,27,28], and other techniques presented in detail [29].

These methods are constantly applied to different types of solid fuels in order to better understand the process of mineral matter transformation of solid fuels and to develop a reliable and objective method for the assessment of operational hazards.

The interesting results of the ash deposits on tube strength in the convection section of pulverized coal-fired boilers were presented [22]. Based on the tested compressive strengths of sintered ash at three temperatures (750 °C, 850 °C, and 950 °C), it was found that the compressive strength was higher for higher temperatures and lower for larger ash particles. That was explained by the Frenkel sintering mechanism and the chemical reactions of the formation of calcium aluminosilicates [22].

Our article focuses on the study of the nonstandard mechanical method based on the measurement of the fracture stress of sintered ash samples in comparison with the results of the pressure drop test, oxide indices, and the standard AFT test. The aim of our study is to test the proposed mechanical method of the relationship between the fracture stress and the temperature of sintered samples for two types of solid fuels (coal and biomass). Two types of coal (bituminous and lignite) and three types of biomass (with relatively high Na and Ca content) were used as the experimental material. Due to the fact that the content of sodium and calcium determines the early stage of the ash sintering process [23], this choice will allow us to test the proposed nonstandard mechanical method—sensitive to the initial stage of the ash sintering process.

The proposed mechanical test has a large usage potential as an objective method for assessing the slagging and fouling hazards for coal ash and biomass. Interestingly, rapid changes in the breaking stress during the ash sintering process were observed in the same temperature range as the rapid changes in the pressure drop test. That observation shows that the breaking stress is very sensitive at the beginning stage of the ash sintering process. In turn, the study of the initial stage of the ash sintering process is very important in identifying the mechanisms of transformation of the mineral substance of solid fuels during combustion under different conditions [22,30].

2. Materials and Methods

The study was carried out on three selected types of crops (rye, barley, and wheat straw) and, as a reference, on two types of coals from Polish mines (lignite and bituminous coal). For all the tested materials, the mechanical test and pressure drop test were conducted to determine their technological suitability, i.e., technical and elemental analysis. Each tested biomass and coal sample was analyzed in terms of moisture content, ash content and volatile matter, the content of solid combustible parts, the content of elements such as C, H, O, S, N, and additionally, the heating value was determined.

The technical analysis was carried out in accordance with Polish Standards for solid fuels, i.e., determination of moisture content by the gravimetric method [31], ash content by the gravimetric method [32], the heat of combustion and calculation of calorific value by methods of calorimetry [33], and volatile matter content by the gravimetric method [34]. The results refer to the analytical state, i.e., the state of fuel with moisture and ash content like the sample brought to the level of equilibrium with the surrounding atmosphere. Fixed carbon (FC) is the solid combustible flammable residue that remains in fuel after moisture, volatile matter, and ash are expelled. FC content divided by the content of volatile matter is called the fuel ratio (FR). According to their fuel ratios, coals have been classified as anthracite with FR of at least 10; semi–anthracite with FR of 6 to 10; semi–bituminous with FR of 3 to 6; bituminous with FR of 3; and as alternative fuels with FR of 1 or less. The composition of the ashes was determined by the method of a Thermo iCAP 6500 Duo ICP plasma spectrometer using ASCRM–010 as a reference substance.

The ash samples with 200 μm fraction were tested, i.e., fraction sizes with polydispersed dust consistencies that agree with generally applied principles used for hard and brown coals, according to the norm [35].

Fuels of which ash samples were prepared were dried, grounded, and fractionated to a grain size of less than 200 μm. Subsequently, all of the tested fuels were initially degassed in a vertical oven (approx. 5 h time). Degassed samples were then sintered at 500 °C to achieve complete combustion of combustible components remaining in fuels after degassing (approx. 20–30 h). The temperature of fuel sintering (500 °C) was selected in the way that the melting point of mineral matter in the ashes was avoided. Therefore, these prepared ash samples were fractionated to grains below 100 μm. In the next step, the cylindrical samples from this form of ash are prepared.

In the strength method for forming the cylindrical samples, a special matrix was used. The specific quantity of tested ash was put into the special matrix (1 g) without any addition of binders. That sample was then compressed at constant pressure (1 Mpa). Formation of the samples occurs as a result of ash compaction. Dimensions of the samples were 8–8.5 mm in diameter and 10–12 mm in height. These obtained ash samples were then isochronally sintered for four hours in different temperatures ranging from 500 °C to 1200 °C. After being isochronally sintered, each sample was then mechanically tested in the following way: each sample was put under devastating stress using UCT—5882 device according to Figure 1.

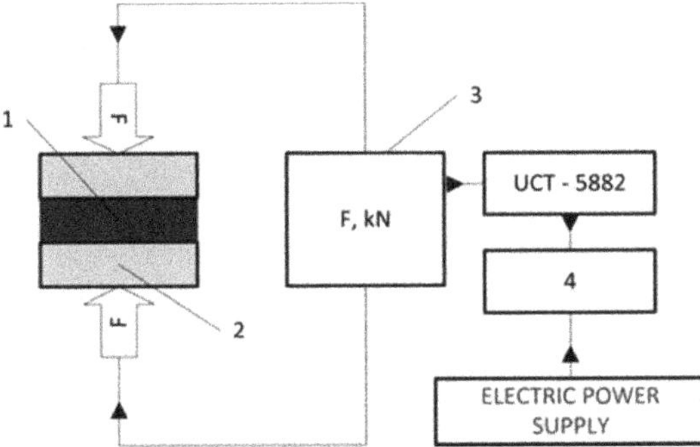

Figure 1. Schematic diagram of the experimental system for the mechanical test (1—sample, 2—hydraulic press with strain gauge, 3—compression force measuring system, 4—data processing system).

The sample preparation for the pressure test method is similar to the mechanical one. The only difference is that the cylindrical shape of the samples is not formed in the matrix but directly in the measuring tube. This tube has special steel pins (upper and lower ones) fitted loosely in the tube. The lower pin serves as a base when forming cylindrical samples followed by the pressure on the sample of the upper pin using the hydraulic press under the same pressure as in the strength method (1.0 MPa). Samples obtained in the described process have a diameter of 10 mm and an approximate height of 1.5 mm.

Determination of the fouling temperature according to the pressure method is based on measuring the gradient of the pressure of compressed air flowing through a heated ash sample. The pressure drop system (Figure 2) includes (1) measuring tube, (2) ash sample, (3) electric furnace, (4) autotransformer with a temperature controller, (5) radiator, (6) thermocouple (6), (7) electronic pressure gauge, (8) electronic temperature gauge, (9) hoses with compressed air, (10) fittings, (11) recording system, and (12) rotameter.

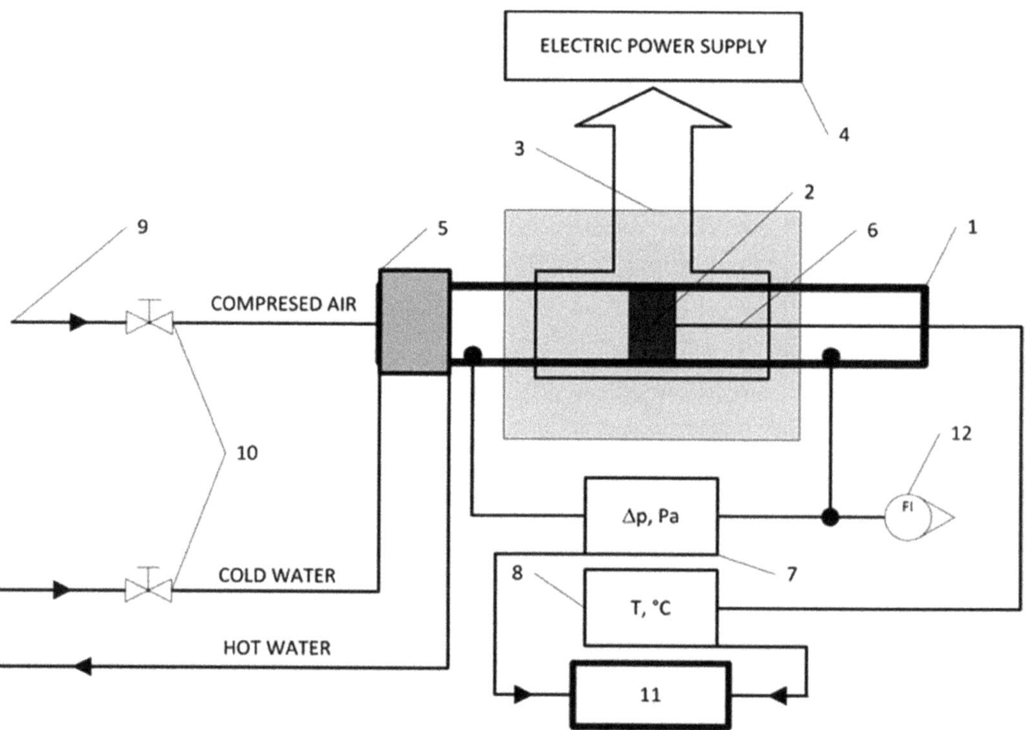

Figure 2. Schematic diagram of the experimental setup for the pressure drop test.

Each mechanical and pressure measurement was repeated on five samples made of the same ash. The results presented in Figures 3–5 are the arithmetic mean and the uncertainty is the mean standard deviation.

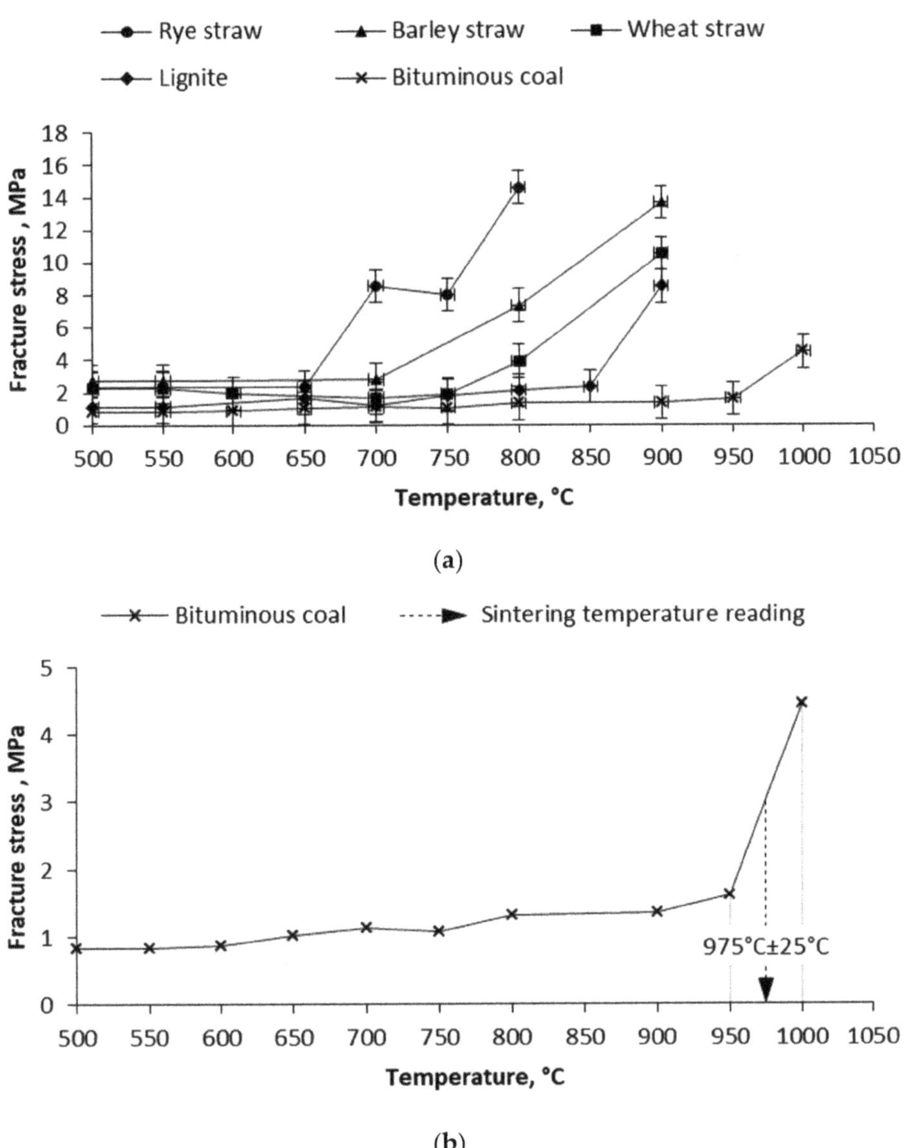

(a)

(b)

Figure 3. (**a**) Fracture stress versus annealing temperature curves for tested ashes and (**b**) method for determining the sintering temperature.

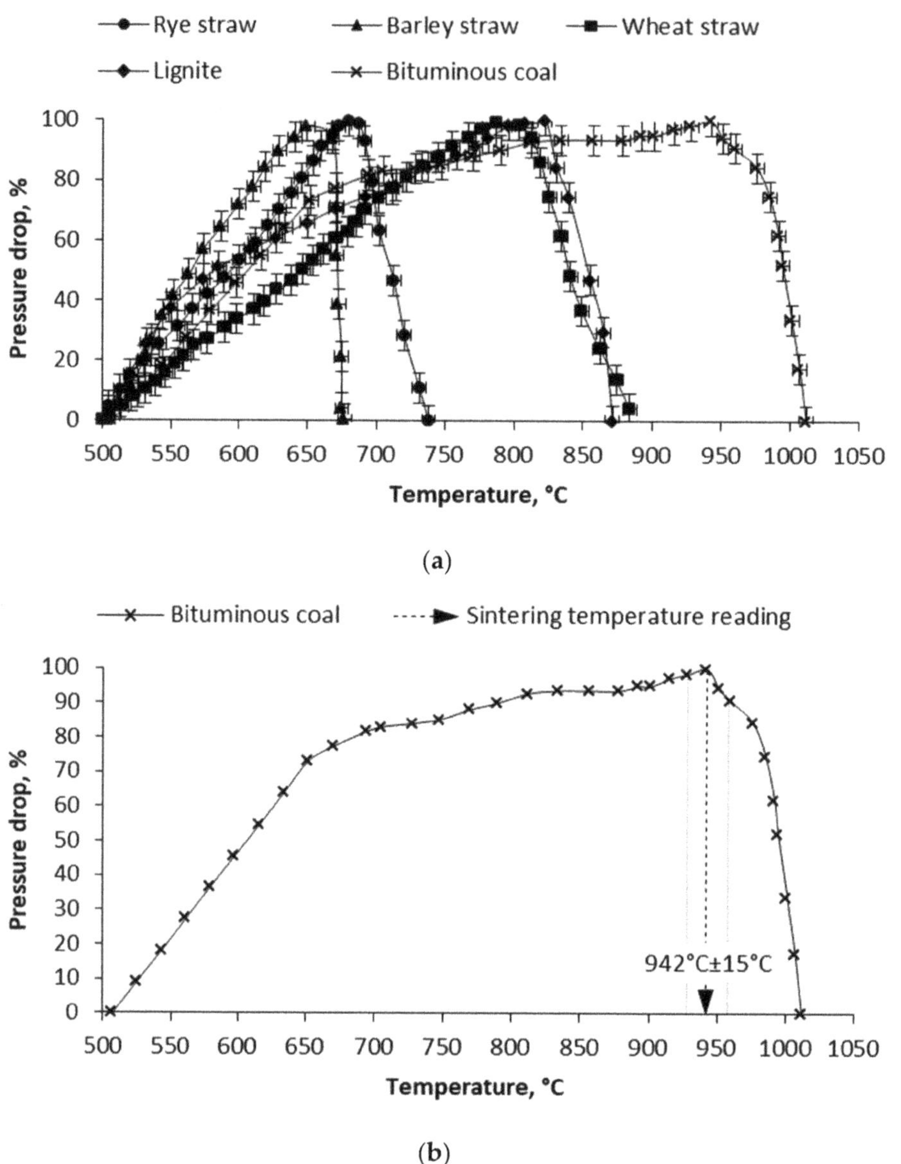

Figure 4. (**a**) Pressure drop versus annealing temperature for tested ash samples and (**b**) method for determining the sintering temperature.

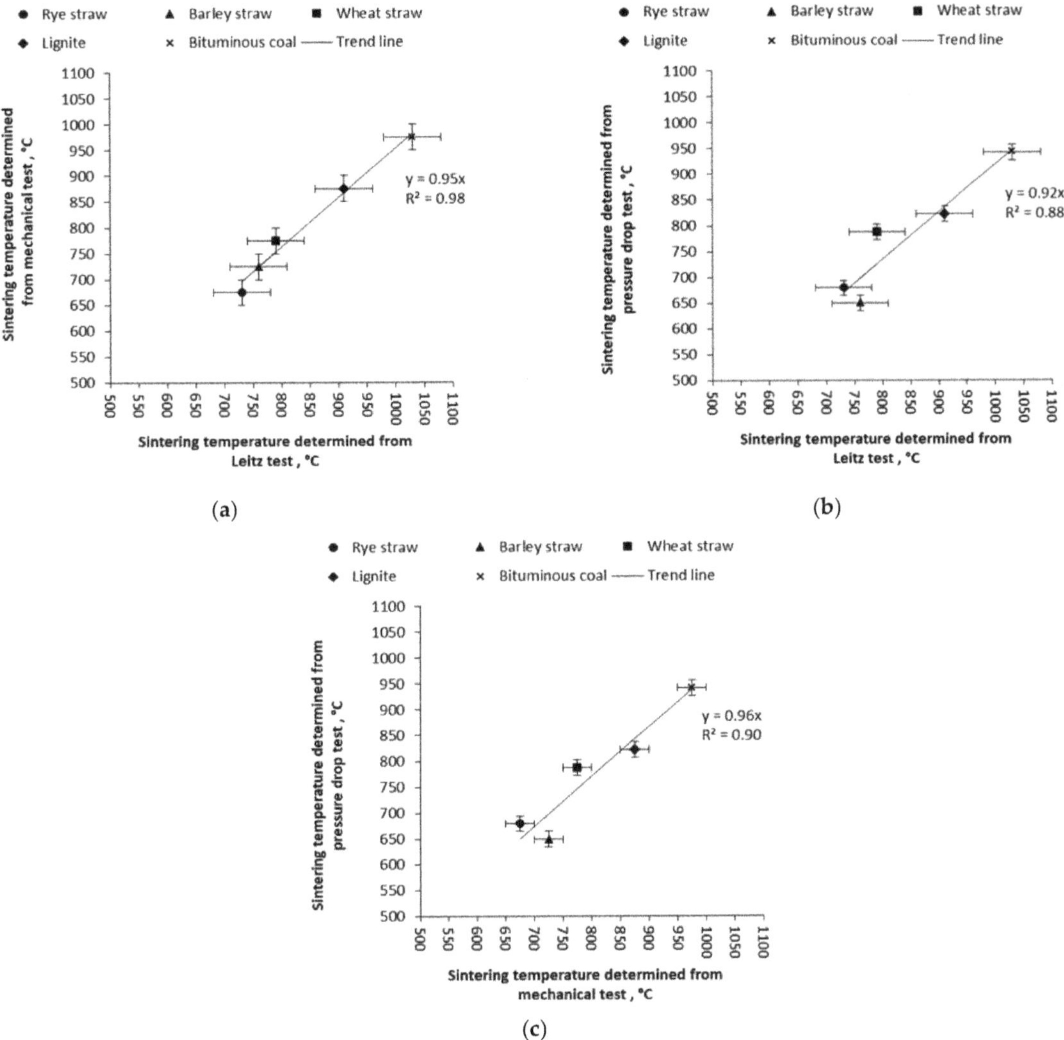

Figure 5. Relationships between sintering temperatures determined by the mechanical test/Leitz test (**a**), pressure drop test/Leitz test (**b**), and pressure drop test/mechanical test (**c**).

3. Results and Discussion

3.1. Samples Analysis

The results of proximate and elemental analysis of the selected biomass in comparison with the bituminous coal (BC) and lignite (LC) from Polish mines with data concerning the heating value (HHV) are provided in Table 1.

Table 1. Basic properties of the biomass and coal samples (on air dried basis).

Samples	Proximate Analysis					Ultimate Analysis						Heating Values
	M	A	VM	FC	FR	C	H	N	S	O	Cl	HHV
	wt.%	wt.%	wt.%	wt.%	–	wt.%	wt.%	wt.%	wt.%	wt.%	wt.%	MJ/kg
Wheat straw	5.7	3.7	69.3	21.3	0.31	45.4	5.9	0.6	0.08	38.4	0.21	17.3
Barley straw	8.5	8.2	62.7	20.6	0.33	47.8	5.7	0.4	0.09	28.9	0.45	16.1
Rye straw	6.3	3.6	68.7	21.4	0.31	46.5	6.1	0.6	0.08	36.4	0.42	16.8
Bituminous coal	2.0	17.3	26.9	53.8	2.00	65.4	3.7	1.3	1.2	8.6	0.51	27.8
Lignite coal	10.7	19.5	41.6	28.2	0.68	43.5	4.9	0.7	2.6	17.2	0.88	16.6

The results presented in Table 1 show some similarity (HHV, C, FC) of the basic physical–chemical properties of the examined biomasses and lignite samples. The heating values (HHV) and fixed carbon (FC) values are very close and are about twice lower than in the bituminous coal sample. Analytical moisture content is in the range from 2 wt.% (BC) to 10.7 wt.% (LC). The water content of biomass that comes from selected crops can vary between 5.7 wt.% and 8.5 wt.%. Increased moisture content results in an additional amount of water vapor in the exhaust. This leads to an increase in the volume of exhaust gases per unit of energy. Too high moisture content in fuel can, among other factors, impede ignition and lower the combustion temperature due to heat demand for water evaporation. Reduced combustion temperature affects the composition of combustion products and therefore also the level of hazardous emissions [4].

The tested biomass materials showed a high share of volatile matter (VM), ranging from 62.7 wt.% for barley to 69.3 wt.% for wheat biomass, in comparison with lignite (41.6 wt.%) and bituminous coal (26.9 wt.%). Due to the high content of volatiles, biomass can ignite at relatively low temperatures, and combustion proceeds quickly. Therefore, the combustion process should be adequately controlled [36].

The FR index value [37] for selected biomasses is nearly at the same level (0.31 for barley and rye biomass and 0.33 for wheat biomass) and is about twice lower than for lignite coal and about seven times lower than for bituminous coal.

The tested biomass material was characterized by low ash content—ranging from 3.6 wt.% (rye) to 8.2 wt.% (wheat) in comparison with the coal. Despite it, biomass ash is very dangerous in terms of operational hazards such as fouling and slagging because its chemical composition (high content of alkali metals like sodium, calcium, and potassium) increases exploitation hazards such as slagging, fouling, and corrosion. The ash in fuel has negative effects on the heating surfaces, increasing their erosive wear and causing the formation of slag and ash deposits [36].

The elemental carbon content in the tested biomass (45.5–47.8 wt.%) does not differ significantly from the content of this element in lignite (65.4 wt.%) and is lower than in bituminous coal (65.4 wt.%). The content of nitrogen and sulfur in the analyzed samples determines their potential to pose risks related to the emission of nitrogen oxides and sulfur dioxide. The nitrogen content determined in the studied biomass (from 0.4 wt.% for barley to 0.6 wt.% for wheat) is comparable with the value for lignite (0.7 wt.%) and about twice lower than for bituminous coal (1.3 wt.%). Sulfur content in tested biomasses (0.081–0.085 wt.%) is much lower than in bituminous coal (1.2 wt.%) and lignite (2.6 wt.%). Therefore, during combustion, the addition of selected biomass to the coal reduces nitrogen and sulfur emission.

The composition of the tested materials' ash is presented in Table 2.

Table 2. Chemical composition of biomass and coal ash.

Samples	Wheat Straw	Barley Straw	Rye Straw	Bituminous Coal	Lignite Coal
Oxides	wt.%	wt.%	wt.%	wt.%	wt.%
SiO_2	65.8	67.5	30.7	51.4	29.1
Al_2O_3	0.9	0.2	0.3	25.5	17.6
Fe_2O_3	0.5	0.2	0.3	6.6	11.3
CaO	8.0	4.9	12.8	4.6	3.0
MgO	2.0	1.5	5.6	3.3	2.9
Na_2O	0.4	0.4	0.1	1.0	4.5
K_2O	18.6	21.0	38.3	2.6	1.5
P_2O_5	2.5	2.7	6.4	0.5	0.1
TiO_2	–	–	–	1.1	2.1
SO_3	1.2	1.5	5.6	3.2	27.8
Mn_3O_4	–	–	–	0.1	0.1
BaO	–	–	–	0.2	–
SrO	–	–	–	0.1	–

The content of elemental carbon, which is important because of its influence on HHV value, and nitrogen in selected biomasses is comparable with that in lignite and smaller than in bituminous coal. The content of elemental sulfur in selected biomasses is much lower than in both coals, whereas the content of oxygen is almost twice as high in lignite and about four times higher than in bituminous coal.

It can be seen that the potassium content is significantly higher in the biomass than in both coals. The opposite is the case of iron and aluminum oxides. Additionally, selected bituminous coal has a very high content of sulfur.

Various predictive factors for ash characterization have been developed at present deposition tendencies, such as silica content (SiO_2), the chlorine content of the fuel (Cl), basic to acidic oxides (B/A), bed agglomeration index (BAI), Babcock index (Rs), fouling index (F_u), fouling factor (R_f), slag viscosity index (Sr) and initial deformation temperature (IDT). Indicators and the formulas used for the calculations are extensively described in [17,38]. Patterns used for this determination are listed below (1–6).

$$\frac{B}{A} = \frac{Fe_2O_3 + CaO + MgO + Na_2O + K_2O + P_2O_5}{SiO_2 + Al_2O_3 + TiO_2}, \tag{1}$$

$$BAI = \frac{Fe_2O_3}{Na_2O + K_2O}, \tag{2}$$

$$Rs = \left(\frac{B}{A}\right) \cdot S^d, \tag{3}$$

$$F_u = \left(\frac{B}{A}\right) \cdot (Na_2O + K_2O), \tag{4}$$

$$R_f = \left(\frac{B}{A}\right) \cdot (Na_2O + 0.659 \cdot K_2O), \tag{5}$$

$$Sr = \frac{SiO_2 \cdot 100}{Fe_2O_3 + SiO_2 + CaO + MgO}, \tag{6}$$

Applicability of these indicators is limited because they are mainly established to characterize coal rather than biomass ash properties [17,38]. Ranges of slagging and fouling indices are presented in Table 3 and marked in greyscale to highlight the difference in values to improve its visibility. The same Table 3 shows indices calculated for fuels under analysis.

Table 3. Predictive ash deposition indices with their ranges [17,38] and results of ash deposition indices of fuels under analysis.

Parameter/Symbol		Value				Fuel					Unit
		Low	Medium	High	Extremely High	Wheat Straw	Barley Straw	Rye Straw	Bituminous Coal	Lignite Coal	
Silica content in ash	SiO_2	<20	20–25	>25		65.82	67.46	30.73	51.35	29.12	wt.%
Chlorine content in fuel	Cl^d	<0.2	0.2–0.3	0.3–0.5	>0.5	0.22	0.49	0.45	0.52	0.99	wt.%
Basic to acidic compounds ratio	B/A	<0.5	0.5–1.0	1.0–1.75		0.48	0.45	2.05	0.24	0.48	–
Bed agglomeration index	BAI			<0.15		0.03	0.01	0.01	1.83	1.89	–
Babcock index	Rs	<0.6	0.6–2.0	2–2.6	>2.6	0.04	0.04	0.17	0.29	1.39	–
Fouling index	F_u	0.6–2.0		2.0–40	>40	9.15	9.75	78.63	0.86	2.86	–
Fouling factor	R_f	<0.2	0.2–0.5	0.5–1.0	>1.0	6.10	6.49	51.91	0.64	2.61	–
Slag viscosity index	Sr	>72	65–72	<65		86.19	91.04	62.20	77.96	62.86	–
Initial deformation temperature	IDT	>1100	900–1100	<900		790	760	730	1030	910	°C

^{d} in dry state.

Oxide indices are widely used to characterize ash deposition tendencies. This method was based on the statistical observation of ash behavior. It was first proposed for coal and adopted for biomass. The ash deposition coefficients for tested coal and biomass are presented in Table 3. All tested fuels have a high potential of ash deposition tendency because of the high content of silica (>25 wt.%), with the lowest for lignite (29.12 wt.%) and the highest for barley straw (67.46 wt.%). Silica plays an important role because it reacts with sodium and potassium and creates eutectics with a low melting point. The chlorine content is extremely high for both coals (0.52 wt% for bituminous and 0.99 wt% for lignite), high for rye and barley straw (0.45 wt.%, 0.49 wt.%, respectively), and medium for wheat straw (0.22 wt.%). Chlorine is responsible for potassium chloride (KCl) formation and formation of ash deposits on low-temperature heating surfaces. In contrast, the basic to acid ratio, which describes the general melting behavior, is low for all samples, except for rye straw for which this ratio is high. The B/A ratio is the lowest for bituminous coal (0.24), but the values for lignite, barley straw, and wheat straw are similar (0.45–0.48). BAI index describes the tendency to agglomerate the bed during fluidized bed combustion—for all tested biomass it is below the limit, and for both coals, it is above the limit. The Babcock index (Rs) is medium for lignite (1.39) and low for other tested fuels, with the lowest for wheat and barley straw (0.04) and the higher for rye straw (0.17) and bituminous coal (0.29). This index provides some information about K_2SO_4, $K_2Ca(SO_4)_2$, and $K_3Na(SO_4)_2$ formation that may increase the deposition tendency on high-temperature heating surfaces. It means that tested biomass and bituminous coal are characterized by a low risk of ash deposition. Fouling index (Fu) is extremely high (>40) for rye straw (78.63), high for barley straw (9.75), wheat straw (9.15), and lignite (2.86) but is in its low range for bituminous coal (0.86). Fouling factor follows these tendencies. Slag viscosity index (Sr) is in the range of low value for barley straw (91.04), wheat straw (86.19), and bituminous coal (77.96). In contrast, for rye straw (62.20) and lignite (62.86), it is in a high range. The IDT temperature is the highest for bituminous coal (1030 °C) and lowers for lignite (910 °C), wheat straw (790 °C), barley straw (760 °C), and rye straw (730 °C). The data collected in Table 3 show a certain conflict of index indications. Therefore, it is difficult to unequivocally determine the degree of hazards associated with the slagging and fouling process solely on the basis of the chemical composition of solid fuels.

3.2. Mechanical Test and Pressure Drop Test

In the mechanical method, the fracture stress for the ash samples is measured. The correlation between the change of the slop of the fracture stress versus the sintering temperature curve was observed. The results of the series of measurements are shown in Figure 3. The measuring procedure consists of the series of measurements, during which the cylindrical ash sample is isochronally sintered at selected temperatures (500–1000 °C) in an electric furnace, then dropped on the metal plate at room temperature for quenching and then compressed at room temperature up to the fracture stress.

For all tested fuels, a rapid increase in fracture stress depending on the temperature was observed (Figure 3). The value of the temperature for which the change in the slope of the fracture stress versus temperature took place was taken as the sintering temperature. This ash sintering temperature was the lowest for rye straw (675 °C), higher for barley straw (725 °C), wheat straw (775 °C), and brown coal (875 °C), and the highest for hard coal (975 °C). This tendency is in line with the IDT results of the Leitz standard method (Table 4), but the sintering temperature values are slightly lower.

Table 4. The sintering temperatures as determined by Leitz, mechanical, and pressure drop methods.

Sample	Sintering Temperature Determined According to the Method:		
	Leitz, DT	Mechanical test	Pressure drop test
Wheat straw	790 °C ± 50 °C	775 °C ± 25 °C	787 °C ± 15 °C
Barley straw	760 °C ± 50 °C	725 °C ± 25 °C	649 °C ± 15 °C
Rye straw	730 °C ± 50 °C	675 °C ± 25 °C	679 °C ± 15 °C
Bituminous coal	1030 °C ± 50 °C	975 °C ± 25 °C	942 °C ± 15 °C
Lignite coal	910 °C ± 50 °C	875 °C ± 25 °C	822 °C ± 15 °C

This procedure was repeated for each previously sintered sample. In that way, the dependency between fracture stress and the temperature was obtained. The temperature at which the evident change of the slope of that dependency can be observed was determined as the sintering temperature.

According to the pressure method, the fouling temperature determination is based on measuring the changes in the gradient of the pressure of the gas flowing by the ash sample that was previously positioned in the tube (low coefficient of expansion) with increasing temperature of the furnace.

For all tested fuels, a rapid decrease of pressure drop in relation to the temperature was observed (Figure 4). The value of the temperature at which the maximum of the relation between the pressure drop and temperature took place was taken as the sintering temperature. This ash sintering temperature was the lowest for barley straw (649 °C), higher for rye straw (679 °C), wheat straw (787 °C), and brown coal (822 °C), and the highest for hard coal (942 °C). This tendency is in accordance with the IDT results of the standard Leitz method and the results of mechanical tests (Table 4), except for the results for rye straw (higher than for barley straw).

The nature of the pressure drop dependency on the sample temperature is due to two main processes on a microscopic scale. The first process is the thermal expansion of the ash sample with increasing temperature. This leads to a better contact of the cylindrical sample with the walls of the tube and thus to an increase in pressure upstream in front of the sample—rising part of the curve in Figure 4. The second mechanism is due to the shrinkage of the ash sample as a result of chemical reactions and physical processes on a microscopic scale, such as phase change and minimization of surface tension. Melting the surface of individual ash particles causes the grains to stick to each other. The progressing sintering process causes further melting of the surface of single ash grains and the reduction of the viscosity of the already melted layer. The next step is the minimization of the surface tension of the sticked particles and forming the larger ones. In a macroscopic scale, it results in shrinkage of the ash sample.

The values of sintering temperatures determined by presented methods (mechanical test and pressure drop test) and by the standard Leitz method as reference are presented as linear relations (linear regression) in Figure 5.

The relation of type was obtained as follows:

$$T_{sintering}(mechanical) = 0.95 \cdot T_{sintering}(Leitz), \tag{7}$$

$$T_{sintering}(pressure) = 0.92 \cdot T_{sintering}(Leitz), \tag{8}$$

$$T_{sintering}(pressure) = 0.96 \cdot T_{sintering}(mechanical), \tag{9}$$

It is clear (Equations (7) and (8)) that the sintering temperatures determined by tested nonstandard methods (mechanical test and pressure drop test) are lower than those determined by the standard Leitz method. It is obvious due to the visual character of the Leitz method (the sintering temperature is subjectively determined by the observer observing the change in the shape of the sample). The sintering temperature determined by the pressure drop test is slightly lower than that determined by the mechanical method (Equation (9)). The reason for that is the character of these two methods. A pressure drop test is a test conducted in situ. In this type of test, the measurements (the pressure drop) are continuously recorded during the sample's heating, when the microstructure of the ash sample is in a non-stable state. In contrast, in the mechanical method, the tested samples are in a stable state (measurement at room temperature after some stage of the sintering process). It means that the results of the pressure drop test are more sensitive to the early stage of the sintering process.

Due to the relatively non-objective method of IDT determination using the standard Leitz method, this measurement is burdened with a fairly large error that can reach a value of up to 100 °C. The presented results of mechanical and pressure drop tests show the great potential of these two methods to assess the slagging/fouling hazards during coal and biomass combustion. These tests seem to be more adequate than the standard method because it directly shows the material shrinkage and increase of fracture stress due to changes in the microstructure of the ash mineral substance (pressure drop test) [27,28].

The observed increase of the fracture stress and decrease of the pressure drop is caused by viscous flow. However, we suggest that the viscous flow mechanism leads to the two macroscopic states of the ash. The processes that lead to these states occur simultaneously and are both controlled by viscous flow and the minimization of the surface tensile strength. One of these processes is the porosity change (decrease of closed pores and increase of open pores) of the coal ash proposed by Al-Ottom et al. [6]. The proposed mechanical test (fracture stress versus temperature) is rather slightly sensitive to other mechanisms of sintering. It is governed by the coalescent process of the individual ash particles (as a result of minimization of the surface tensile strength of the molten surface of ash grains) and the change of the microstructure of the ash sample. This microstructure change influences the increase of fracture stress.

Similar mechanical studies were carried out by Schimpke et al. [39] for lignite and hard coal ash. Slightly lower sintering temperature values obtained from the pressure drop test than from the mechanical test for the tested coal ash samples are consistent with the results presented in the literature [39]. The authors [39] noticed that the sintering initiating mass transport can take place already several Celsius degrees below the sintering temperatures obtained from the mechanical test. However, in the presented study, this effect was not observed for the tested biomass. It is still an open issue due to the very complicated microstructure of coal ash and biomass ash. From a physical point of view, ash is a very difficult material because it is composed of a lot of individual grains. Each of them is different in shape, have different crystallographic structure, different grain interfaces, and different chemical compositions. For this reason, the properties of the ash can be influenced by many different physicochemical processes. For this reason, the behavior of ash during its sintering process is the result of processes such as phase changes (e.g., melting), change of the crystallographic structure, chemical reactions, single grains sticking to each other, the coalescence of individual grains surrounded by molten layer, and the change of ash porosity. The obtained results are very promising, although they need further research.

4. Conclusions

Our paper presents the results of the pressure drop test and mechanical test of hard coal and brown coal, in addition to rye, barley, and wheat straw. The tested fuels differ in the proximate and ultimate analysis and the ash composition. Oxides indices for the tested ash samples were calculated. The conflict of these indices was found. Two nonstandard tests were performed—mechanical test (fracture stress as a function of temperature) and the pressure drop test (pressure drop as a function of temperature). It was observed that the characteristic rapid increase of fracture stress and rapid decrease of pressure drop test are at nearly the same temperature. This temperature was taken as the ash sintering temperature. This temperature was compared with the results of the Leitz IDT temperature. The linear relationship between the sintering temperature determined by mechanical test and Leitz method, by pressure drop test and Leitz method, and by pressure drop test and mechanical test were obtained. These two methods (mechanical test and pressure drop test) have a good potential for usage, especially as an ash diagnostic method because of their sensitivity in the early stage of the sintering process. However, from a physical point of view, ash is a very difficult material (a lot of individual grains, each with different shapes, different crystallographic structures, different boundaries, and different chemical compositions. For this reason, the exact identification of the processes taking place in the sintered ash requires further systematic research.

Author Contributions: Conceptualization, K.K. and D.N.-W.; methodology, K.K. and D.N.-W.; validation, K.K. and D.N.-W.; formal analysis, K.K. and D.N.-W.; investigation, K.K. and D.N.-W.; resources, K.K. and D.N.-W.; data curation, K.K. and D.N.-W.; writing—original draft preparation, K.K. and D.N.-W.; writing—review and editing, K.K. and D.N.-W.; supervision, D.N.-W. All authors have read and agreed to the published version of the manuscript.

Funding: This research received no external funding.

Institutional Review Board Statement: Not applicable.

Informed Consent Statement: Not applicable.

Data Availability Statement: Not applicable.

Conflicts of Interest: The authors declare no conflict of interest.

References

1. Tchapda, A.H.; Pisupati, S.V. A Review of Thermal Co-Conversion of Coal and Biomass/Waste. *Energies* **2014**, *7*, 1098–1148. [CrossRef]
2. Niu, Y.; Tan, H.; Hui, S. Ash-related issues during biomass combustion: Alkali-induced slagging, silicate melt-induced slagging (ash fusion), agglomeration, corrosion, ash utilization, and related countermeasures. *Prog. Energy Combust. Sci.* **2016**, *52*, 1–61. [CrossRef]
3. Vassilev, S.V.; Baxter, D.; Vassileva, C.G. An overview of the behavior of biomass during combustion: Part II. Ash fusion and ash formation mechanisms of biomass types. *Fuel* **2014**, *117*, 152–183. [CrossRef]
4. Nunes, L.; Matias, J.; Catalão, J. Biomass combustion systems: A review on the physical and chemical properties of the ashes. *Renew. Sustain. Energy Rev.* **2016**, *53*, 235–242. [CrossRef]
5. Vassilev, S.V.; Kitano, K.; Takeda, S.; Tsuru, T. Influence of mineral and chemical composition of coal ashes on their fusibility. *Fuel Process. Technol.* **1995**, *45*, 27–51. [CrossRef]
6. Al-Otoom, A.Y.; Elliott, L.K.; Wall, T.F.; Moghtaderi, B. Measurement of the Sintering Kinetics of Coal Ash. *Energy Fuels* **2000**, *14*, 994–1001. [CrossRef]
7. Wall, T.; Creelman, R.; Gupta, R.; Gupta, S.; Coin, C.; Lowe, A. Coal ash fusion temperatures—New characterization techniques, and implications for slagging and fouling. *Prog. Energy Combust. Sci.* **1998**, *24*, 345–353. [CrossRef]
8. Anthony, E.J.; Jia, L. Agglomeration and strength development of deposits in CFBC boilers firing high-sulfur fuels. *Fuel* **2000**, *79*, 1933–1942. [CrossRef]
9. Jing, N.; Wang, Q.; Luo, Z.; Cen, K. Effect of different reaction atmospheres on the sintering temperature of Jincheng coal ash under pressurized conditions. *Fuel* **2011**, *90*, 2645–2651. [CrossRef]
10. Shi, W.; Bai, J.; Kong, L.; Li, H.; Bai, Z.; Vassilev, S.V.; Li, W. An overview of the coal ash transition process from solid to slag. *Fuel* **2021**, *287*, 119537. [CrossRef]
11. Bryers, R.W. Fireside slagging, fouling and high-temperature corrosion of heat-transfer surface due to impurities in stem rising fuels. *Prog. Energy Combust. Sci.* **1996**, *22*, 29–120. [CrossRef]
12. Raask, E. *Mineral Impurities in Coal Combustion: Behavior, Problems and Remedial Measures*; Hemisphere Publishing Corporation: Washington, DC, USA, 1985; Chapter 10; pp. 137–160.
13. Luan, C.; You, C.; Zhang, D. Composition and sintering characteristics of ashes from co–firing of coal and biomass in a la-boratory-scale drop tube furnace. *Energy* **2014**, *69*, 562–570. [CrossRef]
14. Polish Standard PN-ISO 540:2001 Solid Fuels—Determination of Ash Fusibility at High Temperature Using the Pipe Method. Available online: https://sklep.pkn.pl/pn-iso-540-2001p.html (accessed on 19 February 2021).

15. Du, S.; Yang, H.; Qian, K.; Wang, X.; Chen, H. Fusion and transformation properties of the inorganic components in biomass ash. *Fuel* **2014**, *117*, 1281–1287. [CrossRef]
16. Li, Q.; Zhang, Y.; Meng, A.; Li, L.; Li, G. Study on ash fusion temperature using original and simulated biomass ashes. *Fuel Process. Technol.* **2013**, *107*, 107–112. [CrossRef]
17. Jagodzińska, K.; Gądek, W.; Pronobis, M.; Kalisz, S. Investigation of ash deposition in PF boiler during combustion of torrefied biomass. In Proceedings of the IOP Conference Series: Earth and Environmental Science, Krakow, Poland, 14–17 November 2017; Volume 214, p. 012080.
18. Magdziarz, A.; Wilk, M.; Gajek, M.; Nowak-Woźny, D.; Kopia, A.; Kalemba-Rec, I.; Koziński, J. Properties of ash genarated during sewage sludge combustion: A multifaceded analysis. *Energy* **2016**, *113*, 85–94. [CrossRef]
19. Pronobis, M. Evaluation of the influence of biomass co-combustion on boiler furnace slagging by means of fusibility correlations. *Biomass. Bioenergy* **2005**, *28*, 375–383. [CrossRef]
20. Pronobis, M. The influence of biomass co-combustion on boiler fouling and efficiency. *Fuel* **2006**, *85*, 474–480. [CrossRef]
21. Al-Otoom, A.Y.; Bryant, G.W.; Elliott, L.K.; Skrifvars, B.J.; Hupa, M.; Wall, T.F. Experimental Options for Determining the Temperature for the Onset of Sintering of Coal Ash. *Energy Fuels* **2000**, *14*, 227–233. [CrossRef]
22. Jung, B.; Schobert, H.H. Viscous sintering of coal ashes. 1. Relationships of sinter point and sinter strength to particle size and composition. *Energy Fuels* **1991**, *5*, 555–561. [CrossRef]
23. Jung, B.; Schobert, H.H. Viscous sintering of coal ashes. 2. Sintering behavior at short residence times in a drop tube furnace. *Energy Fuels* **1992**, *6*, 59–68. [CrossRef]
24. Moroń, W.; Rybak, W. Strength method for determining the ash sintering temperature. In Proceedings of the Scientific and Technical Conference: Energy 2002, Wroclaw, Poland, 6–8 November 2002; pp. 503–508.
25. Febrero, L.; Granada, E.; Perez, C.; Patiño, D.; Arce, E.; Arce-Fariña, M.E. Characterisation and comparison of biomass ashes with different thermal histories using TG-DSC. *J. Therm. Anal. Calorim.* **2014**, *118*, 669–680. [CrossRef]
26. Qin, Y.-H.; Feng, M.-M.; Zhao, Z.-B.; Vassilev, S.V.; Feng, J.; Vassileva, C.G.; Li, W.-Y. Effect of biomass ash addition on coal ash fusion process under CO_2 atmosphere. *Fuel* **2018**, *231*, 417–426. [CrossRef]
27. Płaza, P.; Moroń, W.; Król, K.; Ferens, W. Sintering of ashes from coal and biofuels. *Syst. J. Transdiscipl. Syst. Sci.* **2006**, *11*, 525–532.
28. Płaza, P.; Hercog, J.; Hrycaj, G.; Król, K.; Rybak, W. Predicting Ash Deposit Formation During Co-Firig of Coal with Biomass, Success and Visions for Bioenergy. In *Thermal Processing of Biomass for Bioenergy, Biofuels and Bioproducts*; Bridgwater, A.V., Ed.; European Commission: Salzburg, Austria, 2007; p. 10.
29. Vassilev, S.V.; Baxter, D.; Vassileva, C.G. An overview of the behavior of biomass during combustion: Part I. Phase-mineral transformations of organic and inorganic matter. *Fuel* **2013**, *112*, 393–449. [CrossRef]
30. Wu, J.; Yu, D.; Liu, F.; Yu, X.; Zeng, X.; Han, J.; Yu, G.; Xu, M. Impact of Oxy-fuel Combustion on Ash Properties and Sintering Strength Development. *Energy Fuels* **2018**, *32*, 4315–4322. [CrossRef]
31. ISO. *Polish Standard PN-EN ISO 18134-1:2015-11 Solid Biofuels—Determination of Moisture Content—Oven Dry Method—Part 1: Total Moisture—Reference Method (ISO 18134-1:2015)*; ISO: London, UK, 2015.
32. ISO. *Polish Standard PN-EN ISO 18122:2016-01 Solid Biofuels—Determination of Ash Content (ISO 18122:2015)*; ISO: London, UK, 2015.
33. ISO. *Polish Standard PN-ISO 1928:2020-05 Solid Mineral Fuels—Determination of Gross Calorific Value by the Bomb Calorimetric Method and Calculation of Net Calorific Value*; ISO: London, UK, 2020.
34. ISO. *Polish Standard PN-EN ISO 18123:2016-01 Solid Biofuels—Determination of the Content of Volatile Matter (ISO 18123:2015)*; ISO: London, UK, 2015.
35. ISO. *Polish Standard PN-G-04502:2014-11 Hard and Brown Coals—Sampling and Preparation of Samples for the Laboratory Tests—Primary Methods*; ISO: London, UK, 2014.
36. Abuelnuor, A.A.A.; Wahid, M.A.; Hosseini, S.E.; Saat, A.; Saqr, K.M.; Sait, H.H.; Osman, M. Characteristics of biomass in flameless combustion: A review. *Renew. Sustain. Energy Rev.* **2014**, *33*, 363–370. [CrossRef]
37. Król, K.; Iskra, K.; Ferens, W.; Miodoński, J.M. Testing properties of sewage sludge for energy use. *Environ. Prot. Eng.* **2019**, *45*, 61–73. [CrossRef]
38. Garcia-Maraver, A.; Mata-Sanchez, J.; Carpio, M.; Perez-Jimenez, J.A. Critical review of predictive coefficients for biomass ash deposition tendency. *J. Energy Inst.* **2017**, *90*, 214–228. [CrossRef]
39. Schimpke, R.; Klinger, M.; Krzack, S.; Meyer, B. Determination of the initial ash sintering temperature by cold compression strength tests with regard to mineral transitions. *Fuel* **2017**, *194*, 157–165. [CrossRef]

Article

Gasification of Coal by CO_2: The Impact of the Heat Transfer Limitation on the Progress, Reaction Rate and Kinetics of the Process

Krzysztof M. Czajka

Department of Energy Conversion Engineering, Wroclaw University of Science and Technology, 50-370 Wroclaw, Poland; krzysztof.czajka@pwr.edu.pl; Tel.: +48-71-320-23-24

Abstract: This paper presents the impact of thermal lag on the progress of different coal types' gasification by CO_2. The analysis was performed using thermogravimetry and numerical modeling. Experiments were carried out at a heating rate of 1–50 Kmin^{-1} and a temperature ranging from 383 to 1173 K. The developed numerical model enabled the determination of a true sample temperature considering the gasification process to consist of two single-step consecutive reactions. Analysis revealed that the average thermal lag in CO_2 is about 11% greater than that in N_2, which is related to the properties of CO_2 itself and the occurrence of the char–CO_2 reaction. The onset temperature of the reverse Boudouard reaction depends on the type of fuel; however, no simple relationship with the coal rank was found. Thermal lag has an impact on the kinetic parameter $A_{\alpha 0.5}$ describing devolatilization, up to 19.8%, while in the case of the char–CO_2 reaction, this influence is expected to be even greater. The performed analysis proved that disregarding thermal lag may significantly hinder the interpretation of the analyzed processes; thus, TG experiments should be carried out with a low heating rate, or at the post-processing stage, a thermal lag model needs to be employed.

Keywords: thermal lag; fossil fuels; pyrolysis; TG; thermal analysis

Citation: Czajka, K.M. Gasification of Coal by CO_2: The Impact of the Heat Transfer Limitation on the Progress, Reaction Rate and Kinetics of the Process. *Energies* **2021**, *14*, 5569. https://doi.org/10.3390/en14175569

Academic Editor: Mejdi Jeguirim

Received: 9 August 2021
Accepted: 1 September 2021
Published: 6 September 2021

Publisher's Note: MDPI stays neutral with regard to jurisdictional claims in published maps and institutional affiliations.

1. Introduction

Nowadays, two thirds of coal is consumed by the electricity sector [1]. Although the consumption of coal for energy production in 2020 decreased by approximately 8% compared to 2019, coal-fired power plants are still the dominant source of energy production, yielding ca. 36% of the global electrical power [1]. Bearing in mind that the use of fossil fuels raises numerous concerns related to the environmental cost [2], international organizations and governments of different countries are undertaking numerous initiatives such as the "Green Deal", "Clean Power Plan" or "Carbon Neutrality" that are supposed to reduce the use of coal for energy purposes. Nevertheless, the world's forecasted growing electricity demand is going to require increases in all generation sectors, including renewables, coal and nuclear; thus, it may be expected that coal will continue to play a significant role in power generation [3].

One of the ways to meet the growing electricity needs while maintaining concern for the natural environment is the concept of the development of clean coal technology (CCT). CCT covers a wide range of topics, from coal mining, through its use for heat and power production, up to the utilization of coal-based wastes, and comprises a variety of pollutant abatement techniques. The most recent challenges in CCT are related to the need to tackle the rising CO_2 emissions, extending CCT to include carbon capture and sequestration techniques (also called carbon capture and storage, both abbreviated as CCS) or carbon capture, utilization and storage methods (CCUS) [4,5].

Among the many clean coal technologies integrated with carbon capture, utilization and storage, the concept of gasification of fossil fuels in reactors in which CO_2 is the gasifying agent deserves special attention. This technology enables the highly efficient production of electricity from coal fuels while reducing the negative impact on the environment. It is part of the circular carbon economy concept (CCE), thanks to which the carbon cycle could be effectively closed and the rising CO_2 could be reused within different processes [6].

Unlike steam and air gasification, which are quite established and well documented, most CO_2 gasification studies are limited to the lab scale and are quite far from industrial implementation. According to the author's knowledge, no commercial CO_2 gasifier is in operation at present. Patent research from the past 17 years indicated that until 2021, less than 190 patents concerning CO_2 gasifiers were recognized worldwide [7]. As stated in the comprehensive review papers of [7–9], nowadays, the main challenge to be overcome is to scale up results obtained at laboratory reactors to the industrial scale.

Analysis of the gasification process in CO_2 is usually carried out using conventional methods such as conduction and convection heating. Laboratory-scale experimental apparatuses such as thermogravimetric analyzers (TG) usually employ electrical resistances to heat up and precisely control the parameters of operation. The use of TG is common, although it has its limitations [10] and has been criticized for its inability to obtain high temperatures and high heating rates [11]. Nevertheless, as indicated in [12], kinetic parameters determined employing TG can be successfully used to predict fuel behavior on an industrial scale.

To successfully apply kinetic parameters to describe the course of the process in industrial conditions, laboratory measurements should be performed in the absence of mass transfer limitations [13]. According to the recommendations of the International Thermal Analysis and Calorimetry Society (ICTAC) Kinetic Committee, it is suggested to analyze a thin layer of the sample, use low heating rates or limit the temperature of the process [14]. Nevertheless, good practice during the experimental stage may decrease the mass transfer limitations but not eliminate them. It is impossible to avoid inaccuracies arising from, for example, the occurrence of endothermic/exothermic reactions or the finite heat transfer rate between the gaseous environment and the analyzed sample.

As indicated in [15–17], to increase the accuracy of determination of the true sample temperature (T_s), it is worth using a model that considers the nature of the heat transfer inside the TG reactor. The numerical analysis presented in [15], supported by experimental results, showed that in the case of the pyrolysis process in nitrogen, in the heating rate range from 1 to 80 $Kmin^{-1}$, the discrepancy between the T_s and the temperature measured in the close surroundings of the sample (T_c), known as thermal lag, might be from 4.8 to 56.2 K. In the work of [18], it was shown that for the heating rate of 40 $Kmin^{-1}$, the discrepancies reach approximately 35 K, while in [17], it was indicated that for the heating rate of 10 $Kmin^{-1}$, the maximum thermal lag can be from 51 to over 200 K. Unfortunately, in the case of the CO_2 gasification process, despite a detailed literature review, it was not possible to identify studies devoted to the determination of the impact of thermal lag on the accuracy of the TG measurement.

The shown inaccuracies in temperature determination affect the reliability of the measured reactivity of the analyzed samples and their kinetic parameters, which may lead to a misinterpretation of the nature of the investigated process. According to the information presented in [18], the discrepancies in the kinetic parameters determined before and after considering the limited heat transfer may be from 10% to 20%, and according to [15], they can even reach 40–50%. Please note that the above results are shunted to pyrolysis in the nitrogen atmosphere. In the case of the carbon dioxide atmosphere, the author is not aware of any works on this issue, but it seems that due to the different properties of CO_2 compared to N_2, including its higher density and heat capacity [19], the real differences may be even greater.

Therefore, this study aimed to analyze and assess to what extent the heat limitation affects the measurement of the coal gasification process with CO_2 in TG conditions. For this study, a thermal lag model which considers both the nature of the gasifying agent and the physicochemical properties of the analyzed sample was developed. Experimental and model studies were aimed at explaining to what extent the measured temperature of the sample differs from the true temperature and what the consequences of these differences are in the context of analyzing the reaction rate, the amount of converted fuel and the kinetic parameters describing the gasification process. Particular attention was paid to the reaction of the remaining char with CO_2, the so-called reverse Boudouard or Boudouard–Bell [20] reaction, which will play a key role in terms of the need to scale laboratory results to an industrial scale if the gasification process is planned to be carried out at temperatures above 1273 K [21].

2. Experimental Setup and Tests

2.1. Facility and Instrumentation

Thermogravimetric analysis was carried out with a SETARAM Setsys Evolution TGA. This device can continuously measure weight loss as a function of time and temperature. It consists of a microbalance connected to a tray, on which lay two crucibles. The investigated sample filled one of the crucibles, while the reference crucible remained empty and was used to determine the temperature drift of the device. Both crucibles were located inside a cylindrical furnace with a programmable control temperature. The device was equipped with three thermocouples (as shown in Figure 1), of which one was located below the sample crucible and provided the crucible temperature (T_c), the second one measured the temperature of the reference crucible and the third measured the temperature of the surrounding gaseous atmosphere.

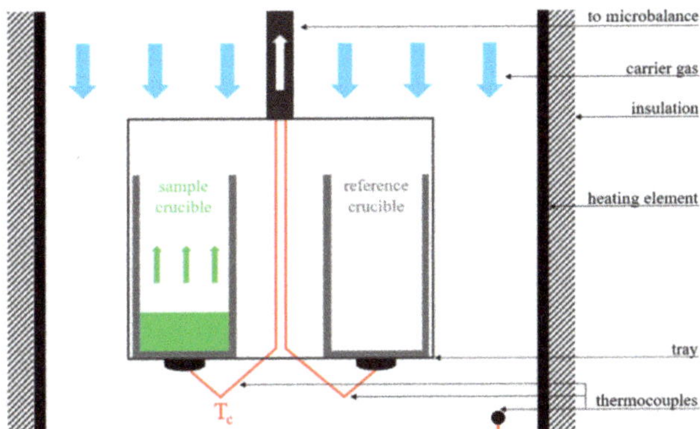

Figure 1. Scheme of the TG device.

The mass of the sample was limited to ca. 10^{-6} kg, in order to mitigate the impact of self-heating/self-cooling [14]. At the first stage of the experiments, the sample was dried at 383 K until its weight became constant. Afterwards, to simulate the gasification process, samples were heated at a rate of 1, 5, 15, 30 and 50 Kmin^{-1} (to simplify the notation further, these rates are referred to as HR1, HR5, etc.), with a flow of 32 mL min^{-1}, in an atmosphere containing 100% CO_2, at a temperature ranging from 383 to 1173 K. The reference measurements were conducted according to a similar heating program, except that the atmosphere contained 100% N_2.

2.2. Sample Properties

The samples analyzed in the experiments were as follows:

-   Lignites, LB supplied by Bełchatów Coal Mine (Poland) and LT supplied by Turów Coal Mine (Poland);
-   Hard coals, HJ supplied by Janina Coal Mine located in Libiąż (Poland), HS supplied by Sobieski Coal Mine located in Jaworzno (Poland) and HZ supplied by Ziemowit Coal Mine located in Lędziny (Poland);

- Anthracite, AI mined in Ibbenbüren Coal Mine (Germany).

The samples were air dried, pulverized to a size below 10 µm and homogenized. The high fineness of the coal resulted from the requirement to ensure a kinetic gasification regime and was directly related to the need to obtain the appropriate Biot and external pyrolysis numbers (for more details, please see 3.2). The necessity to grind the fuel to a similarly fine fraction is confirmed, among others, in [22]. The remaining determinations were conducted under the relevant standards, i.e., moisture content was determined according to [23], volatile matter according to [24] and ash according to [25]. The ultimate analysis was conducted through LECO TruSpec Micro CHNS following [26,27], while the higher heating value was analyzed with IKA WERKE C2000, according to [28]. The obtained results are summarized in Table 1.

Table 1. Proximate and ultimate analysis of samples.

Smp.	M	V	FC	A	FR	C	H	N	S	O *	HHV
		wt% ar			-			wt% ar			MJ·kg^{-1}
LB	4.4	44.4	35.0	16.2	0.79	55.2	4.5	0.7	1.8	17.2	18.9
LT	2.0	46.7	33.8	17.5	0.72	59.0	4.8	0.5	1.3	14.9	21.7
HJ	2.1	32.7	55.6	9.6	1.70	75.7	4.3	1.2	1.2	5.9	24.7
HS	3.7	33.0	52.9	10.4	1.60	76.0	4.1	1.3	1.6	2.9	25.8
HZ	2.4	34.9	55.0	7.7	1.58	77.1	4.6	1.2	1.1	5.9	27.8
AI	1.7	4.9	90.3	3.1	18.4	89.0	3.5	1.2	0.5	1.0	33.2

Where: M—moisture, V—volatiles, FC—fixed carbon (=100%—M/V/A), A—ash, FR—fuel ratio (=FC/V), C—carbon, H—hydrogen, N—nitrogen, S—sulfur, O—oxygen, ar—on an as-received basis, *—calculated by difference.

3. Modeling Study

3.1. Thermal Lag Model

This model is a development of works devoted to the pyrolysis of solid fuels in nitrogen [15–17] and allows determining the thermal lag accompanying the gasification reaction. Although the gasification process is very complex, and its mechanism is sometimes described using up to several hundred chemical reactions [29–31], experimental studies carried out with the use of TG (described in detail in Section 4.1) have shown that, in some simplifications, the process can be viewed as two consecutive stages. In the first stage, the volatile matter evolves, while in the second stage, the remaining carbonized residue reacts with CO_2 to produce CO, namely:

$$\text{coal} \xrightarrow[\text{(devolatilization)}]{k_a} \text{char + volatiles 1} \xrightarrow[\substack{\text{(reverse}\\ \text{Boudouard reaction)}}]{k_b} \text{volatiles 2 + ash} \tag{1}$$

The rate of the process can be written as

$$\frac{d\alpha}{dt} = x_{vol}\frac{d\alpha_a}{dt} + x_{ch}\frac{d\alpha_b}{dt} \tag{2}$$

where x_{vol} and x_{ch} satisfy the relationship

$$x_{vol} + x_{ch} + x_{ash} = 1 \tag{3}$$

and the rate of devolatilization and the carbon residue reaction with CO_2 can be reflected by first-order reactions:

$$\frac{d\alpha_a}{dt} = (1 - \alpha_a)k_{0,a}e^{\frac{-E_a}{RT_m}} \tag{4}$$

$$\frac{d\alpha_b}{dt} = (\alpha_a - \alpha_b)k_{0,b}e^{\frac{-E_b}{RT_m}} \tag{5}$$

The energy balance equation was derived taking into consideration the conditions of the heat transfer inside the TG furnace, in which the sample was spread thinly and uniformly in the crucible. The stream of the carrier gas flowed from the top of the crucible to the bottom, sweeping away evolved gases. The heat flux absorbed by the sample (raw fuel), char, ash and the crucible was assumed to be equal to the difference between the heat flux supplied to the system, the heat flux accompanying the devolatilization reaction and the heat flux accompanying the reaction of the carbon residue with CO_2:

$$[(1 - x_{ash})(1 - x_{ch})(1 - \alpha)m_0 C_{Ps} + x_{ch}m_0 C_{Pch} + x_{ash}m_0 C_{Pash} + m_c C_{Pc}]\frac{dT_m}{dt}$$
$$= hA_c(T_r - T_m) - m_0\Delta H_{dev}x_{vol}\frac{d\alpha_a}{dt} - m_0\Delta H_{gas}x_{ch}\frac{d\alpha_b}{dt} \tag{6}$$

The change in the sample temperature during the gasification process was obtained taking into account Equations (4)–(6) and is described as

$$\frac{dT_m}{dt} = \frac{hA_c(T_r - T_m) - m_0\Delta H_{dev}x_{vol}(1 - \alpha_a)k_{0,a}e^{\frac{-E_a}{RT_m}} - m_0\Delta H_{gas}x_{ch}(\alpha_a - \alpha_b)k_{0,b}e^{\frac{-E_b}{RT_m}}}{(1 - x_{ash})(1 - x_{ch})(1 - \alpha)m_0 C_{Ps} + x_{ch}m_0 C_{Pch} + x_{ash}m_0 C_{Pash} + m_c C_{Pc}} \tag{7}$$

After considering the following boundary conditions, Equation (7) was computed numerically using Mathcad to determine the sample temperature:

$$T_m = T_0 \text{ and } \alpha = 0 \text{ at } t = 0 \tag{8}$$

The overall heat transfer coefficient h, included in Equation (7), describes the amount of heat transferred to the sample and is affected by, e.g., the sample temperature (T_m) and temperature-dependent parameters of the gas, such as λ_g, v_g or C_{Pg}. To determine its value, the energy balance equation for the reference crucible was proposed:

$$\frac{dT_m}{dt} = \frac{A_c h(T_r - T_m)}{m_c C_{Pc}} \tag{9}$$

where the surface area A_c was assumed to be time-invariant.

Equation (9) was solved numerically with the Mathcad software, taking into account the same boundary conditions as previously used (Equation (8)).

The heat supplied to the sample was assumed to be transferred by convection and radiation:

$$h = h_c + h_r \tag{10}$$

Equation (11) presents the formula applied to determine the heat transferred by convection (h_c), which was derived for the flow around the cylinder with the ratio of the diameter to height of 1:2 [32]:

$$h_c = \frac{Nu\,\lambda_g}{d_c} = \frac{\lambda_g\left(1.2564 + 0.6592Re^{2/5}Pr^{1/3}\right)}{d_c} \tag{11}$$

The heat transferred by radiation (h_r) was calculated according to

$$h_r = \varepsilon\sigma(T_r + T_m)\left(T_r^2 + T_m^2\right) \tag{12}$$

The values of the specific heat of the sample, char, ash and crucible, used in Equations (6)–(9), were dependent on the temperature, according to formulas [33–36]:

$$C_{Pc} = 1.30684 \cdot 10^{-6} T_m^3 - 0.00363 T_m^2 + 3.52606 T_m + 20.29987 \tag{13}$$

$$C_{Pch} = -218 + 3.807 T_m - 0.001758 T_m^2 \tag{14}$$

$$C_{Pash} = 912 + 2.235 T_m - 0.001464 T_m^2 \tag{15}$$

$$C_{Ps} = x_{ch} C_{Pch} + (x_{vol} - 0.1)(728 + 3.391 T_m) + 0.1(2273 + 2.554 T_m) \tag{16}$$

The properties of CO_2 constituting the furnace atmosphere, such as the thermal conductivity (λ_g), the density (ρ_g), the specific heat (C_{Pg}) or the kinematic viscosity (ν_g), required to calculate the number of Nusselt (Nu), Prandtl (Pr) or Reynolds (Re), were calculated based on the mini-REFPROP [37] and CoolProp [38] databases. The remaining parameters used for the model are summarized in Table 2.

Table 2. Parameters applied for the thermal lag model.

l_c, m	d_c, m	d_f, m	m_0, kg	m_c, kg	μ	σ, Wm^{-2}K^{-4}	ΔH_{dev}, Jkg^{-1}	ΔH_{gas}, Jkg^{-1}	h [d], Wm^{-2}K^{-1}
10^{-2}	$5 \cdot 10^{-3}$	$5 \cdot 10^{-2}$	10^{-6}	$1.8 \cdot 10^{-4}$	0.15 [a]	$5.67 \cdot 10^{-8}$	$400 \cdot 10^{3}$ [b]	$604 \cdot 10^{4}$ [c]	17.2–79.0

Where: [a]—[39], [b]—the heat of reaction is not the same as the heat measured directly by DSC, but it is reduced by the heat of the char formation following [40], [c]—the average heat of Boudouard reaction in a temperature range of 298–1173 K [21], [d]—computed from Equation (9).

3.2. Kinetic Regime

To enable a direct scale-up of kinetic parameters determined at the laboratory to the industrial scale, the measurements should be carried out in conditions where the rate of reaction is controlled by chemical reactions, in the so-called *regime I*. To meet these requirements, the Biot (Bi) and external pyrolysis (Py') numbers need to be analyzed.

The Bi provides the ratio between heat convection at the surface of the layer and heat conduction within the layer. For a thin layer filling a cylindrical crucible, it takes the form [41]

$$Bi = \frac{h d_c l_b}{2 \lambda_s (2 d_c + l_b)} \tag{17}$$

where λ_s (Wm^{-1}K^{-1}) is the thermal conductivity of the sample, assumed to be equal to [42]

$$\lambda_s = 1.071 - 0.00884 T_m + 3.2939 \cdot 10^{-5} T_m^2 - 5.8083 \cdot 10^{-8} T_m^3 + 4.7875 \cdot 10^{-11} T_m^4 - 1.4143 \cdot 10^{-14} T_m^5 \tag{18}$$

The Py' is the ratio between heat convection and the reaction rate of a particle, and for a layer, it is given by [41]

$$Py' = \frac{2 h_c (2 d_c + l_b)}{\lambda_s \rho_s C_{Ps} d_c l_b} \tag{19}$$

A thermally thin regime is ensured when the Bi number is much smaller than one. In such conditions, the heat conduction within the layer is a lot faster than the heat convection away from its surface; thus, the sample temperature can be assumed to be essentially uniform. Low Py' numbers correspond to control by external heat transfer, whilst a number larger than unity represents conditions where heat convection to the particle is a lot faster than the chemical reactions taking place. To ensure that the process is controlled purely by kinetics, both requirements need to be met: $Bi \ll 1$ and $Py' \gg 1$ [43]. Kinetic parameters determined in such conditions are often called intrinsic or true, in order to distinguish them from apparent kinetics determined under other regimes.

3.3. Reaction Kinetics

The kinetic analysis was carried out in two variants, model-based and experimental-based. The model-based variant assumed that the true temperature is equal to the temperature calculated using the model ($T_s = T_m$), while the experimental variant assumed that the true temperature follows the temperature measured under the crucible ($T_s = T_c$). Kinetic results obtained for both variants were compared to illustrate to what extent the measurement uncertainty related to the determination of the true temperature affects the kinetic parameters of gasification.

As stated in Section 3.1, the gasification process was divided into two stages, i.e., devolatilization, and the char–CO_2 reaction. Each of these steps can be reflected by a first-order reaction according to Equations (4) and (5). To determine the kinetic parameters describing each of the stages, such as the activation energy ($E_{a/b}$) and the pre-exponential factor ($k_{0,a/b}$), a model fitting approach based on a single non-isothermal TG measurement was used. The arguments for the choice of the model fitting approach were, inter alia, the shape of the curves suggesting two single-step consecutive reactions, the non-variability of the kinetic parameters depending on the range of analyzed temperatures [44] and the experiences of other authors in applying this type of approach for analyzing thermal processes [45–47].

For the model fitting purpose, a single first-order reaction model (SFOR) [48] using the simplifying temperature integral function with the Doyle integral approximate [49] was employed.

$$ln[-ln(1 - \alpha)] = ln\left(\frac{k_{0,a/b}E_{a/b}}{\beta R}\right) - 1.0518\frac{E_{a/b}}{RT_s} - 5.33 \tag{20}$$

Plotting $ln[-ln(1 - \alpha)]$ vs. $1/T_s$ returns the activation energy and the pre-exponential factor from the slope and the intercept, respectively. The results derived at low temperatures concern devolatilization and were denoted with index a, while the high-temperature kinetics of the char–CO_2 reaction were denoted with index b.

4. Results and Discussion

4.1. Progress and Reaction Rate

As mentioned in the introduction, measurement with TG is inextricably subject to measurement uncertainty related to the heat transfer limitation, which results from the inability to directly measure the temperature of the sample inside the TG reactor using a thermocouple. Installing a thermocouple in direct contact with the sample would result in errors due to, among others, conduction of heat along the thermocouple wire, convection in the boundary layer around the thermocouple and radiation between the wires and the external medium [50,51]. Hence, commercially available TG analyzers typically measure the temperature in the vicinity of the fuel-containing crucible (T_c), and this temperature is generally taken as the true sample temperature. Figure 2 shows a comparison of the measured temperature T_c with the calculated actual temperature T_m. The difference between both temperatures, marked as a gray area, is the so-called thermal lag, denoted as $\Delta T = T_c - T_m$.

As it can be seen, the thermal lag phenomenon accompanies the gasification process practically in the entire range (ΔT is equal to zero only for the 0 s of the experiment). The analysis carried out for LT at HR50 indicates that in the considered measurement range, the average thermal lag was 64.0 K, while the maximum thermal lag even reached 91.4 K (309 s of the process). The nature of the changes in ΔT was complex, which is well illustrated by the continuous violet line (thermal lag in CO_2) shown in Figure 3.

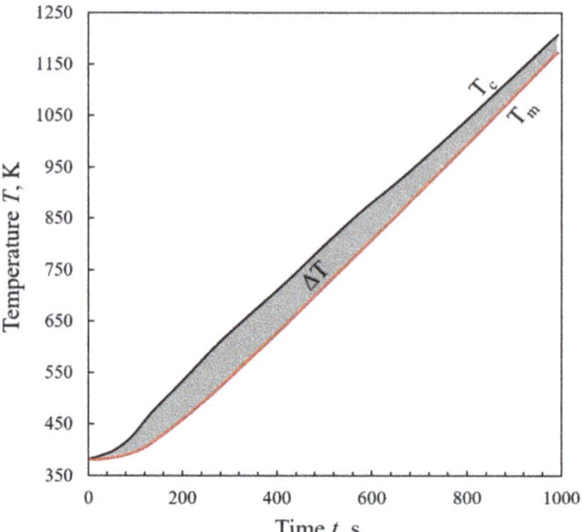

Figure 2. Thermal lag determined for LT at HR50.

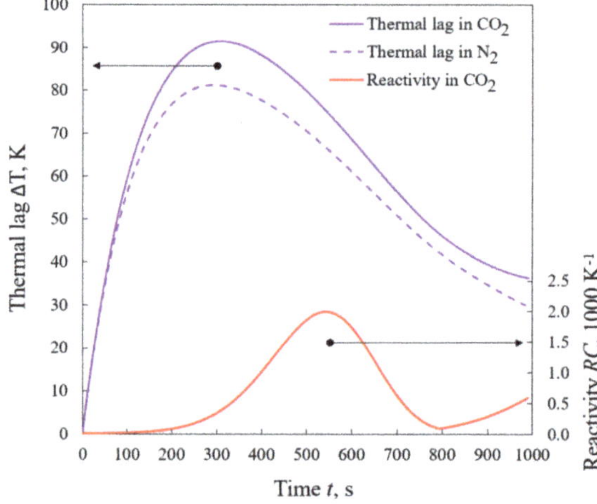

Figure 3. Comparison of thermal lag determined in CO_2 and N2 (LT, HR50).

In the first stage (up to approximately 300 s), the thermal lag determined during the experiments in CO_2 increased with time. After reaching the maximum, the ΔT values decreased until reaching 800 s of the experiment. Interestingly, the influence of the endothermic devolatilization on the increase in ΔT was almost invisible. This observation differs from the observations for the pyrolysis of cellulose in nitrogen presented in [15]. In the author's opinion, the revealed differences are the result of both a lower proportion of volatile parts in lignite than in cellulose (46.7% vs. ca. 87–89% [15]) and a different heat capacity of the surrounding gas atmosphere, raw fuel, char and ash. After reaching 800 s, a third stage should be distinguished in which ΔT decreased slower than previously. This phenomenon was directly related to the beginning of the char–CO_2 reaction. This observation is confirmed by the correlation between the thermal lag curve in CO_2 and the fuel reactivity curve in CO_2 (red line). As it can be seen from the figure, the reactivity of the fuel increases during the 800 s as a result of the reverse Boudouard reaction, in which an increase directly corresponds to the increase in the thermal lag of the temperature measurement. Moreover, the influence of this reaction on an increase in thermal lag is indirectly proved by the shape of the thermal lag curve determined in the nitrogen atmosphere (dashed violet line). In this atmosphere, the reverse Boudouard reaction does not take place, and in the range of times and temperatures in which it would occur, a systematic decrease in the thermal lag value (unlike for the CO_2 atmosphere) is observed. Such a significant, compared to devolatilization, influence of the char–CO_2 reaction on thermal lag is a direct consequence of the highly endothermic nature of this process, which absorbs about 15 times more heat than devolatilization ($\Delta H_{dev} = 400 \times 10^3$ J·kg^{-1} vs. $\Delta H_{gas} = 604 \times 10^4$ J·kg^{-1}).

Interesting information is also provided by the comparison of the thermal lag curve in N_2 and CO_2. In addition to the already mentioned differences related to the char–CO_2 reaction, the thermal lag values determined in both atmospheres for similar times also differ. As mentioned earlier, in the case of CO_2, the average thermal lag was 64.0 K, and the maximum thermal lag was even 91.4 K, while for the N_2 atmosphere, these values were, respectively, 57.6 K and 81.2 K. These differences are directly related to the occurrence of the char–CO_2 reaction and the properties of both gases, such as the thermal conductivity (λ_g), the density (ρ_g), the specific heat (C_{Pg}) or the kinematic viscosity (ν_g). This observation explains why, in some papers [52,53], certain differences in the TG profiles recorded during heating of the same fuels in atmospheres of N_2 and CO_2 were recognized. Consideration of the differences in thermal lag indicates that at low temperatures, CO_2 is an inert gas and does not significantly affect the mechanism of coal pyrolysis, which confirms the reports presented in [54,55]. However, it should be noted that the results obtained do not exclude that for some materials characterized by properties other than coals [56,57], CO_2 may influence the reaction progress even at the early stages of devolatilization. In the light of the obtained data, it seems valuable to take into account the influence of thermal lag on the precision of the obtained results when assessing the influence of CO_2 on the progress of the gasification reaction.

Figure 4 shows the influence of the heating rate on thermal lag. To better visualize the behavior of a set of curves, on the x-axis, a logarithmic scale is proposed (which explains the differences in shape between the HR50 curves presented in Figures 3 and 4).

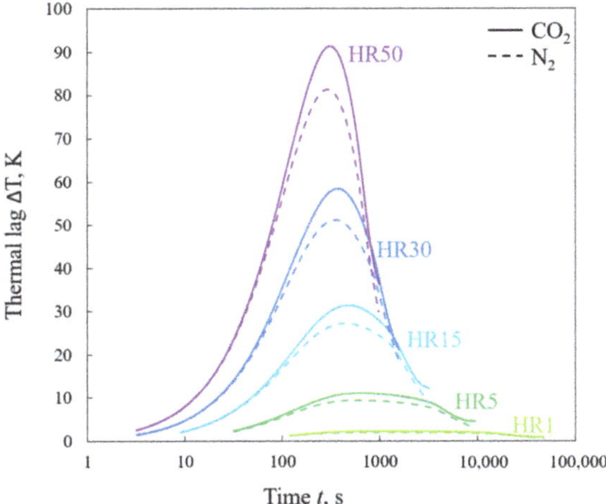

Figure 4. Thermal lag determined for LT at different heating rates.

The conducted analyses indicate that the uncertainty of the temperature measurement in TG increased with the heating rate. In the case of gasification of LT with HR1, the average value of ΔT was 1.46 K; for HR5, it was 7.17 K; for HR15, it was 20.8 K; for HR40, it was 44.0 K; and for HR50, it was 64.0 K. Regardless of the heating rate, ΔT determined in the nitrogen atmosphere was always lower by approximately 11–13 percentage points than that determined for the CO_2 atmosphere.

The results presented thus far were obtained for one type of fuel, i.e., LT. Figure 5 shows the influence of the fuel type on the thermal lag. The results are presented for the selected heating rate, namely, HR5.

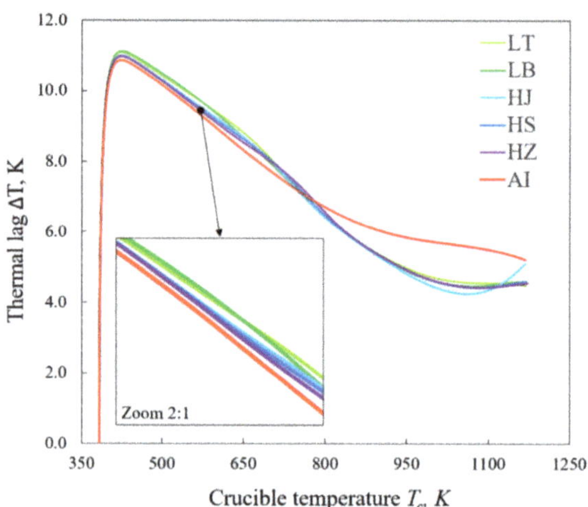

Figure 5. Thermal lag determined for different fuels at HR5.

The value of the thermal lag in the temperature range up to ca. 800 K is quite similar for all analyzed fuels and ranges, on average, from 8.81 to 9.30 K. As expected, a slightly higher value of ΔT was obtained in this range for fuels with a high content of volatiles (LB: 9.36 K, and LT: 9.07 K) than for those with a low content (AI: 8.81 K), which is a consequence of the endothermic nature of the devolatilization. The influence of the fuel type on the value of the thermal lag becomes more visible at high temperatures (above 800 K), i.e., in the range where the char–CO_2 reaction is observed. In this respect, fuels with a higher content of fixed carbon were characterized by higher thermal lag values, e.g., for the AI sample, containing 90.3% fixed carbon, the thermal lag was 5.72 K, i.e., 14–20 percentage points more than the thermal lag determined for the other fuels (containing from 33.8% to 55.6% of fixed carbon).

According to Equation (7), thermal lag influences the shape of the conversion curve. Figure 6 compares the conversion curves obtained for the LT sample, gasified with HR1–50. Following the adopted nomenclature, in this and the following figures, the curves determined with the assumption that the temperature T_c corresponds to the true fuel temperature are marked with a solid line, and the curves established for the assumption that T_m corresponds to the true test temperature are marked with a dashed line.

As it can be seen, for HR1, i.e., in conditions where thermal lag played an almost insignificant role (average ΔT equal to 1.46 K), both fuel conversion curves (light green lines) have an almost identical shape. Nevertheless, as the heating rate increased, the differences in the shape of the conversion curves increased. In the case of HR5, the curve based on the assumption that the actual test temperature was equal to T_c indicated that, e.g., 30% of the fuel conversion occurred at 680.0 K, and 60% of the fuel conversion occurred at 1149.1 K. Assuming that the true sample temperature was equal to T_m, the same conversion rates were achieved for the temperatures of 688.3 K and 1153.7 K, respectively. For HR15, 30% of fuel conversion was recorded at 717.0 K and 740.8 K, and for HR30, this was at 745.7 K and 782.3 K, while for HR50, this was at 792.5 K and 855.0 K.

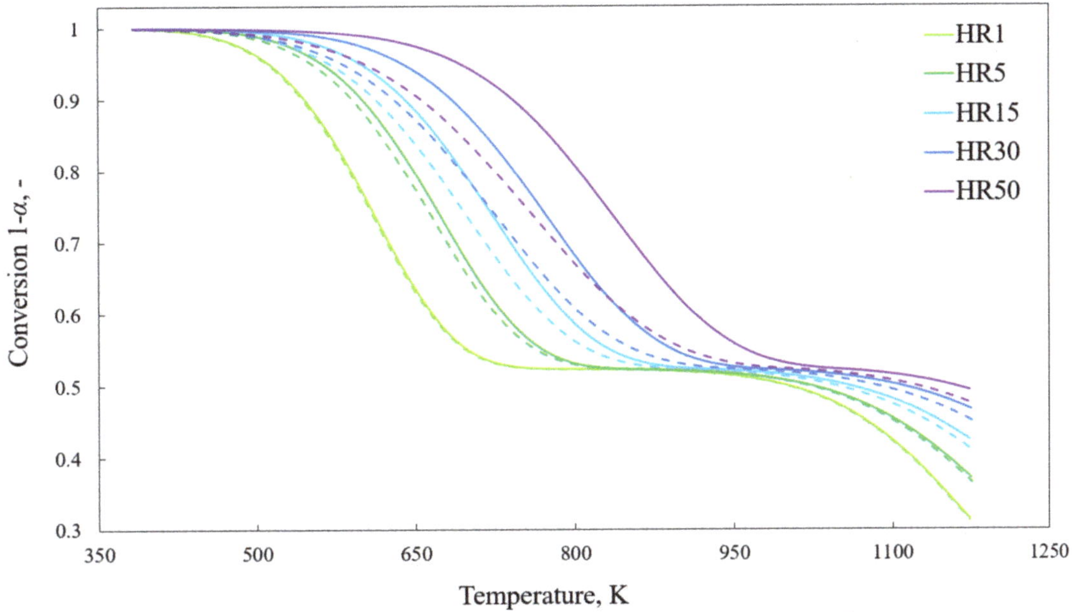

Figure 6. The impact of thermal lag on the conversion curves, determined for LT at different heating rates.

As a consequence of the different shapes of the curves, there were also differences in the total amount of converted fuel. The experimental results indicate that with the increase in the heating rate, the amount of reacted fuel was 68.5%, 62.7%, 57.6%, 53.2% and 50.6%, respectively. The model results obtained for the assumption of measurements carried out in the conditions of no heat transfer limitations indicate that the amount of converted fuel should be 68.7%, 63.3%, 58.8%, 54.8% and 52.5%. As it can be seen, the discrepancies increase with the increase in the heating rate, and for HR50, they reach up to 1.8%.

The obtained results indicate that when analyzing the LT gasification in CO_2, the use of heating rates higher than about a few degrees per minute results in significant discrepancies in the shape of the fuel conversion curves. Hence, it is recommended to use the lowest possible heating rates or to recalculate the results obtained at a high HR with a model that allows the determination of the true sample temperature.

To illustrate how the fuel properties affect the shape of the conversion curve, the results obtained for LT were compared with those obtained for other fuels. Figure 7a–e show the profiles obtained during the research with HR30. The profiles are presented on separate graphs to increase the readability of the presented data. The profile for LT is not shown as it was previously presented in Figure 6.

As evidenced by the obtained results, for each of the analyzed fuels, a significant effect of thermal lag on the shape of the conversion curve was found. Comparing the discrepancies between the temperature characterizing the 30% degree of fuel conversion determined experimentally (solid line) and with the model (dashed line), it should be noted that they ranged from 30.5 (AI) to 45.8 K (LT). The discrepancies in the total amount of fuel reacted significantly differed, depending on the type of fuel analyzed, and ranged from ca. 1% to even 5.3% (in the case of AI).

The results obtained for LB, HJ, HS, HZ and AI confirm the observation based on LT that carrying out the gasification process with an HR equal to several degrees per minute results in a thermal lag of several dozen degrees, which may significantly hinder the interpretation of the analyzed phenomena. As in the case of LT, running the process with a heating rate equal to a few degrees per minute guaranteed low thermal lag values, equal to 5.39–8.37 K (at the 30% degree of fuel conversion).

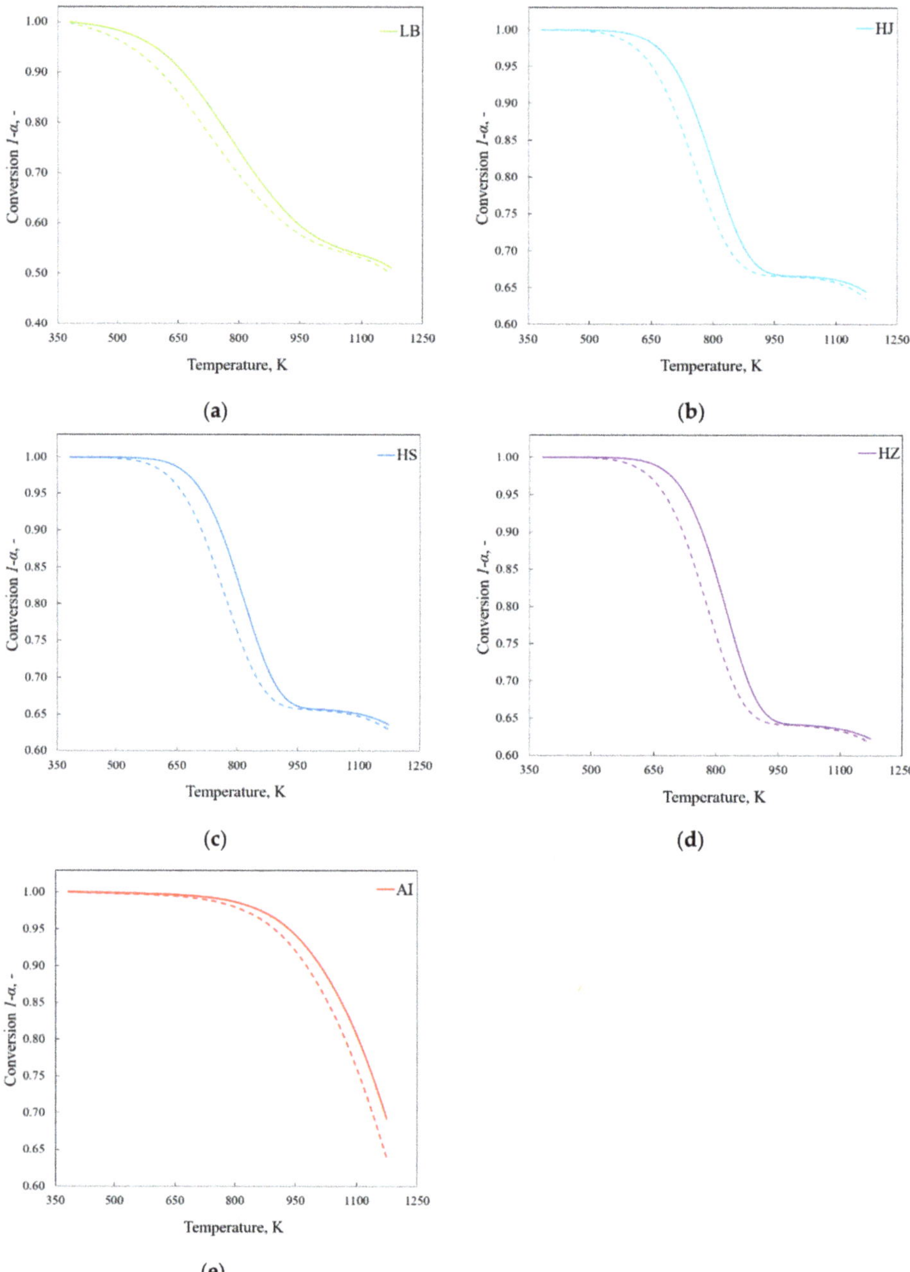

Figure 7. The impact of thermal lag on the conversion curve, determined for different fuels at HR30. (**a**) LB; (**b**) HJ; (**c**) HS; (**d**) HZ; (**e**) AI.

The conversion curves determined employing TG (presented in Figures 6 and 7) are commonly used to determine the reactivity of the fuel. According to the definition, reactivity is considered as the rate of mass loss over time or temperature [58–60]. Figure 8 shows the reactivity curves obtained for LT at HR1–50. As in the case of the previous figures, the experimental results are marked with a solid line, and the model results are marked with a dashed line.

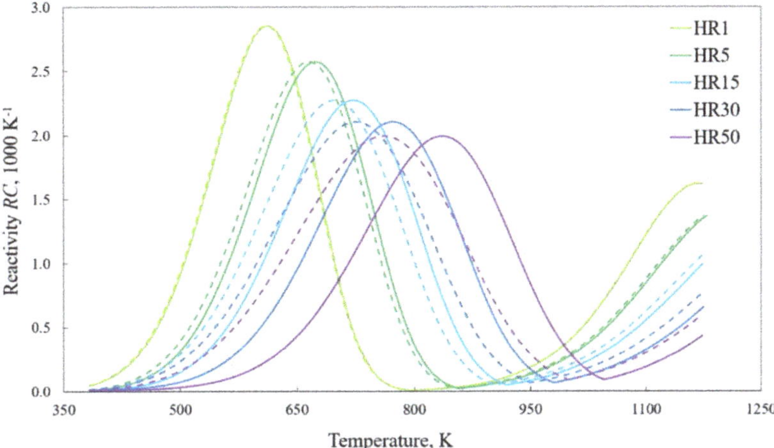

Figure 8. The impact of thermal lag on the reactivity curves, determined for LT at different heating rates.

The obtained curves show two characteristic peaks: the first occurring at lower temperatures, illustrating the susceptibility of a sample to undergo devolatilization, and the second occurring at higher temperatures, showing the susceptibility of the remaining char to react with CO_2. As it can be seen, the second peak is not shown in full because the measurement was stopped as the temperature increased over 1173 K and the process approached the transitional regime, in which the rate of reaction was affected by diffusion phenomena. Work in this area was undesirable from the point of view of the possibility of further extrapolation of the obtained results (more information is presented in Section 3.2).

It should be noted that the discrepancy between the experimental and model reactivity curves can be significant, especially for high heating rates. The comparison of the temperature at which the maximum reactivity was recorded during devolatilization shows that, for example, for HR50, the discrepancies may be as high as 75.2 K (T_{Rmax} *(model)* = 761.1 K vs. T_{Rmax} *(experiment)* = 836.3 K). The difference between the T_{Rmax} parameters determined experimentally for HR1 and HR50, denoted as ΔT_{Rmax}, was 227.3 K, and the same value determined by the model was 147.1 K. On this basis, it should be concluded that the discrepancies between T_{Rmax} determined at different heating rates are, to some extent, affected by the finite heat transfer rate. However, even after taking into account thermal lag, the T_{Rmax} parameter recorded for different heating rates differs significantly, meaning it cannot be treated as universal, and when using it, it should be specified under what measurement conditions it was determined.

The analysis of the peak responsible for the fuel susceptibility to the reaction of char with CO_2 leads to an interesting conclusion. Depending on the heating rate, this reaction starts at the following temperatures: 800.5 K (HR1), 859.4 K (HR5), 910.7 K (HR15), 949.8 K (HR30) and 997.0 K (HR50). It is generally assumed in the literature that the temperature range of the reverse Boudouard reaction starts at 953 [61]–973 K [62]. In the case of HR1–15, the obtained values are lower than those allowed by the literature sources [61,62], which does not mean that they are incorrect. The possibility of the reaction between char and CO_2 at lower temperatures was confirmed by [63], in which the presented equilibrium calculations indicate that, theoretically, the Boudouard reaction can start at temperatures as high as approximately 673 K. Moreover, it was experimentally proved in [64] that under isothermal conditions, the Boudouard reaction can be observed at temperatures as high as 773 K.

Figure 9 shows the experimental and model curves of reactivity of the different fuels, determined at HR30.

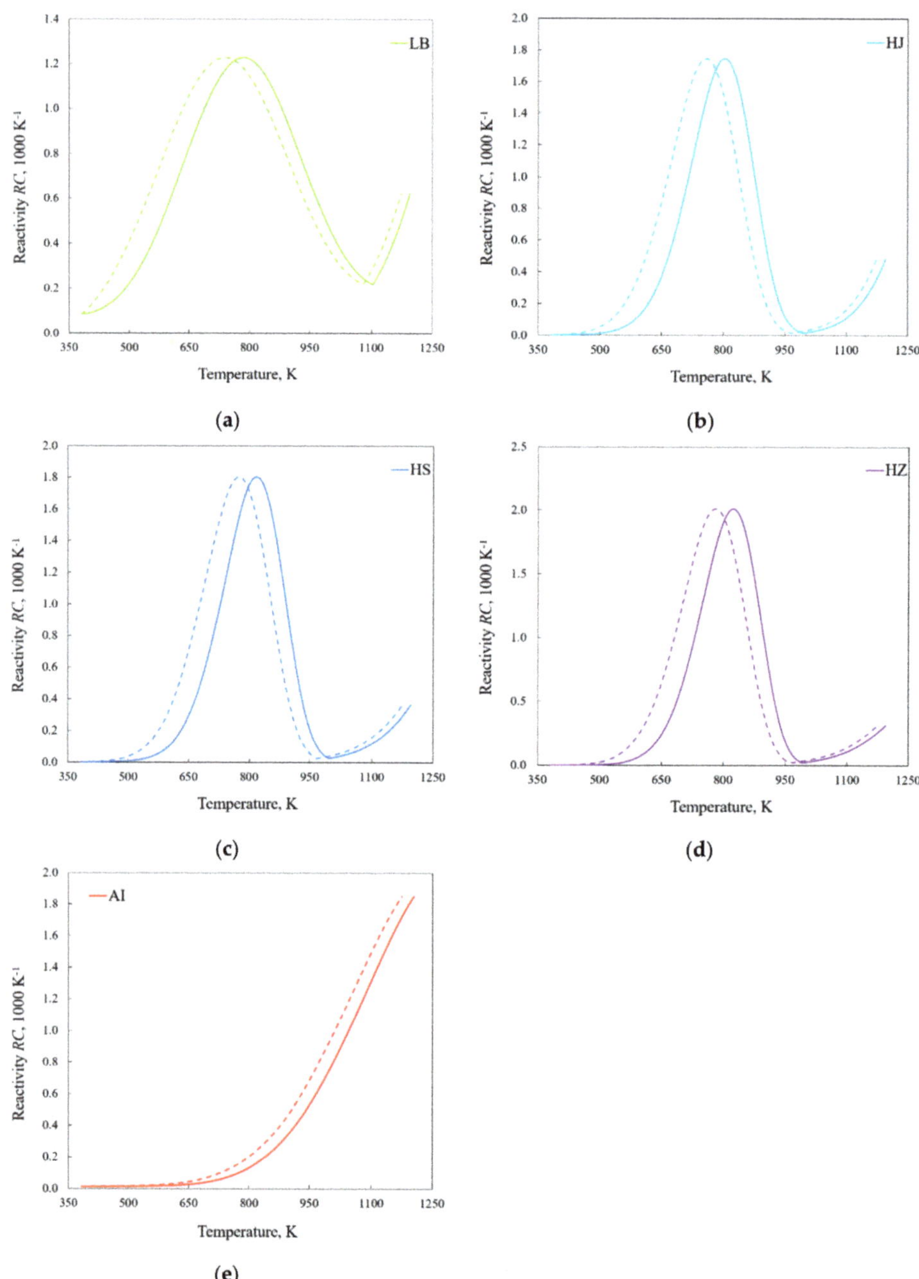

Figure 9. The impact of thermal lag on the reactivity curve, determined for different fuels at HR30. (**a**) LB; (**b**) HJ; (**c**) HS; (**d**) HZ; (**e**) AI.

The obtained results indicate that the initial temperature of the char–CO_2 reaction for LB was 1080.0 K; for HJ, it was 960.9 K; for HS, it was 964.4 K; for HZ, it was 966.2 K; and for AI, it was below 923 K. The discrepancies in the results obtained for the different fuels indicate that the properties and origins of the fuels influence the progress of the reverse Boudouard reaction. However, no simple relationship between the basic parameters characterizing the fuels, such as the fixed carbon content or elemental carbon content, with the parameters characterizing the reverse Boudouard reaction has been demonstrated. In the author's opinion, the progress of this reaction may be influenced by other physicochemical parameters of the fuel, such as the distribution and shape of pores, the presence of mineral matter or the presence of surface functional groups [65].

4.2. Kinetics

To further investigate the influence of thermal lag on the coal gasification process, a kinetic analysis was performed. As mentioned in Section 3.2, to scale up the kinetic results to industrial conditions, laboratory experiments should be carried out under the conditions of the kinetic regime. The Bi number and the Py' number, determined according to Equations (17) and (19), ranged from $3.19 \cdot 10^{-4}$ to $1.02 \cdot 10^{-3}$ and from 1.23 to 9.19. As the Bi number was much lower than unity, and the Py' number was higher than 1, it was found that the selected measurement parameters ensured that the gasification process was controlled by chemical reactions [22].

A kinetic study was carried out following the procedure presented in Section 3.3, separately analyzing the processes of devolatilization and the reaction between char and CO_2. The obtained results are presented in Tables 3 and 4. The results called "experimental" were calculated assuming that the true sample temperature was equal to T_c, and the so-called "model" results were calculated assuming that it was equal to T_m.

Table 3. Results of a kinetic study of LT.

Sample	HR	Approach	Devolatilization			Char–CO_2 Reaction		
			E_a kJ mol^{-1}	$k_{o,a}$ s^{-1}	$A_{\alpha0.5}$ s^{-1}	E_b kJ mol^{-1}	$k_{o,b}$ s^{-1}	$A_{\alpha0.1}$ s^{-1}
LT	1	Mod.	41.42	6.23×10^{-1}	1.57×10^{-4}	126.4	73.6	2.56×10^{-5}
		Exp.	41.86	6.67×10^{-1}	1.57×10^{-4}	126.9	76.5	2.56×10^{-5}
	5	Mod.	43.43	2.03	7.02×10^{-4}	117.4	85.9	1.18×10^{-4}
		Exp.	45.53	2.72	7.17×10^{-4}	119.2	99.8	1.19×10^{-4}
	15	Mod.	39.94	1.90	1.76×10^{-3}	112.2	92.6	3.23×10^{-4}
		Exp.	45.66	4.16	1.87×10^{-3}	116.9	1.35×10^2	3.25×10^{-4}
	30	Mod.	36.88	1.50	2.98×10^{-3}	112.3	1.12×10^2	6.00×10^{-4}
		Exp.	48.03	6.74	3.41×10^{-3}	122.4	2.35×10^2	6.09×10^{-4}
	50	Mod.	36.46	1.61	4.51×10^{-3}	117.3	2.17×10^2	9.77×10^{-4}
		Exp.	53.30	13.1	5.41×10^{-3}	130.7	5.91×10^2	1.00×10^{-3}

Table 4. Results of a kinetic study for HR5.

Sample	HR	Approach	Devolatilization			Char–CO_2 Reaction		
			E_a kJ mol^{-1}	$k_{o,a}$ s^{-1}	$A_{\alpha0.5}$ s^{-1}	E_b kJ mol^{-1}	$k_{o,b}$ s^{-1}	$A_{\alpha0.1}$ s^{-1}
LB		Mod.	26.18	5.93×10^{-2}	4.37×10^{-4}	135.6	5.32×10^2	1.23×10^{-4}
		Exp.	27.58	7.35×10^{-2}	4.47×10^{-4}	137.7	6.24×10^2	1.24×10^{-4}
HJ		Mod.	53.08	12.4	8.36×10^{-4}	184.2	5.81×10^4	1.59×10^{-4}
	HR5	Exp.	55.49	17.3	8.52×10^{-4}	186.7	6.88×10^4	1.61×10^{-4}
HS		Mod.	58.59	23.6	8.57×10^{-4}	124.9	96.8	1.13×10^{-4}
		Exp.	61.10	32.8	8.73×10^{-4}	126.7	1.11×10^2	1.14×10^{-4}
HZ		Mod.	66.23	57.5	8.90×10^{-4}	117.8	42.9	1.08×10^{-4}
		Exp.	68.84	80.1	9.06×10^{-4}	119.5	48.9	1.08×10^{-4}
AI		Mod.	nd	nd	nd	61.58	1.26×10^{-1}	6.95×10^{-5}
		Exp.	nd	nd	nd	62.95	1.44×10^{-1}	7.04×10^{-5}

The performed kinetic analysis shows that the activation energy determined based on the experimental curves for the devolatilization of LT ranged from 41.86 to 53.30 kJ mol^{-1}. The obtained values are similar to those presented in the literature for other lignites, e.g., for the Menzen sample (33.9–40.4 kJ mol^{-1} [66]). Moreover, it was noted that the activation energy value increased with the heating rate. There is no unanimity in the literature on the influence of the heating rate on the value of the activation energy. A similar tendency to that described above was noted, among others, in [67] and [68] (for coal Afsin Elbistan). The reverse observation was shown in [66,69,70], and an observation indicating a lack of a clear correlation between the activation energy and the heating rate was shown, e.g., in [71–73]. The author of this paper does not attempt to explain the essence of the relationship between the activation energy and the heating rate but would like to emphasize that after taking into account the thermal lag, the values of E for different heating rates came closer to each other and amounted to 36.46–41.42 kJ mol^{-1}. Moreover, contrary to the trend observed for the experimental curves, the activation energies calculated based on the model curves decreased with the heating rate. The fairly slight discrepancies between the activation energy obtained for the different heating rates may indicate, unlike in the case of the analysis of experimental data, that after considering thermal lag, the mechanism of the observed reaction seems not to change over the studied range of heating rates. Moreover, the results prove that taking into account the limited heat transfer may be of key importance for the interpretation and understanding of the mechanism of the reactions under study.

The activation energy determined for the remaining fuels (at HR5) is similar to the data presented in the literature. In the case of lignite LB, this value is similar to, e.g., the Soma sample (36.4 kJ mol^{-1}) [68], while E for hard coals was comparable to that determined for the bituminous coals of Azdavay and Karadon (65.0 and 74.2 kJ mol^{-1} [74]). As anthracite contained only 4.9% of volatile matter and its evolution process was difficult to identify unequivocally, it was not decided to determine the kinetic parameters characterizing it. A comparison of the model and experimental results for various pyrolyzed fuels with HR5 shows that at such a low heating rate, considering the measurement uncertainty related to the determination of the actual temperature of the sample results in a reduction in the activation energy by 1.40–2.62 kJ mol^{-1}, i.e., 4–5 percentage points relative to the experimental value.

As indicated in [15], the comparison of only one kinetic parameter may lead to erroneous conclusions due to the existence of the so-called compensation effect [48], i.e., the interrelation between E and k_0. To avoid misunderstandings, and to unequivocally define the impact of thermal lag on kinetic parameters, the A_α parameter was introduced. The parameter relates E and k_0 according to the relation

$$A_\alpha = k_{0,a/b} e^{\left(\frac{-E_{a/b}}{RT_\alpha}\right)} \tag{21}$$

This parameter was determined for the temperature corresponding to the stage at which conversion of an exact amount of fuel took place; thus, for example, the parameter determined at the conversion of 50% was denoted as $A_{\alpha 0.5}$. A comparison of the A_α values obtained for the devolatilization of LT with the different heating rates shows (Table 3) that not considering thermal lag when studying the kinetics of HR1 leads to an error of 0.4%. In the case of HR5, the error is 2.1%, and for HR15, it is 6.3%, while for HR30 and HR50, it is 14.4% and 19.8%, respectively. The comparison of A_α for various fuels, determined for HR5, shows that it is, to some extent, dependent on the type of fuel; however, the obtained discrepancies between A_α remain quite similar, i.e., from 1.7% for HZ to 2.4% for LB. In light of the above, it seems reasonable to recommend, when analyzing the coal devolatilization process, limiting the heating rate to a few degrees per minute.

The analysis of the kinetic data obtained for the LT char–CO_2 reaction (at different HRs) indicates that the results obtained based on the experimental data ranged from 116.9 to 130.7 kJ·mol^{-1}. The inclusion of the thermal lag model narrowed the scope of the obtained results to 112.2–126.4 kJ·mol^{-1}, while no correlation between E and the heating rate was recognized. The differences between the A_α parameter determined for the char–CO_2 reaction, based on the model and experimental curves, were smaller than in the case of the devolatilization process and amounted to a maximum of 2.7%. However, it should be remembered that in the case of devolatilization, the A_α parameter was determined for the conversion degree of 50%, and in the case of the char–CO_2 reaction, it was determined for the conversion degree of 10%. If the parameter $A_{\alpha 0.5}$ was used in the case of the reverse Boudouard reaction, the obtained results indicate that the uncertainty in determining the kinetic parameters would be greater than in the case of devolatilization. In the case of the reaction between char and CO_2, the differences in activation energies for the different fuels turned out to be significant. In the case of HR5, the activation energy of the reverse Boudouard reaction determined based on the model was about 61.58 kJ·mol^{-1} for AI, 117.4–117.8 kJ·mol^{-1} for HZ and LT, 124.9 kJ·mol^{-1} for HS, 135.6 kJ·mol^{-1} for LB and 184.2 kJ·mol^{-1} for HJ. The obtained values are similar to those presented in the literature, that is, 109.5 [75]–220 kJ·mol^{-1} [76]. They indicate that the influence of the properties of the coal structure on the progress of the reverse Boudouard reaction is significant; however, they do not confirm the statement contained in [77] that the activation energy for the char–CO_2 reaction generally decreases proportionally to the coal rank. The attempt to correlate the basic physicochemical parameters of the fuel with an activation energy of the reverse Boudouard reaction did not bring satisfactory results, which indicates that the progress of this reaction will depend on other parameters, e.g., the pore distribution, the content of mineral matter or the presence of individual functional groups on the surface.

The comparison of the obtained kinetic parameters with the content of selected elements in the mineral matter leads to an interesting conclusion. The composition of the mineral matter of the investigated materials is cited from [78] and presented in Table 5.

Table 5. The content of selected elements in the mineral matter [78].

Sample	K_2O	CaO	Al_2O_3 wt%	SiO_2	P_2O_5
LB	0.11	23.0	17.2	29.1	0.13
LT	0.48	19.1	22.0	32.4	0.17
HJ	1.92	4.81	26.9	40.6	0.71
HS	1.47	5.31	23.0	44.6	0.93
HZ	0.88	6.97	27.2	36.3	0.55

There is a certain unanimity in the literature [79,80] that potassium (K) is an element having a positive effect on the gasification rate. The results obtained for lignites show that in the case of LT, containing about four times more K_2O than LB, a higher rate of the char–CO_2 reaction was observed (i.e., lower kinetic parameters). No such dependence could be observed in the case of hard coals. Nevertheless, concerning the low degree of fuel conversion, it can be assumed that potassium was not concentrated enough to enhance the catalytic effect [80,81].

Noteworthy is the effect of calcium (Ca) on the rate of the char–CO_2 reaction. The share of CaO in the investigated hard coals was equal to 6.97% (HZ), 5.31% (HS) and 4.81% (HJ). The CaO content seems to fit the trend in the activation energy required to initiate the reaction between surfaces of chars derived from hard coals and CO_2 (HJ > HS > HZ). The observation confirms the remark made in [77] that the catalytic effect of CaO in low-rank coal causes an increase in the number of active sites and hence the reactivity. Nevertheless, it should be noted that the influence of CaO on the gasification process is debatable, and, for example, in the works on gasification of biomass, some authors [81–83] observed an inverse, i.e., inhibiting, effect of CaO on the gasification rate.

Other elements commonly considered to have a significant impact on the rate of gasification are, among others, aluminum (Al), silicon (Si) and phosphorus (P). It is believed that by inhibition of the catalysis from other elements, they decrease the rate of the process. The obtained results do not show any visible correlation between their content and the rate of the char–CO_2 reaction. The author would like to emphasize that, despite the above-mentioned relationship between the content of ash components and the reactivity of the investigated samples, the impact of the mineral matter needs to be deeply analyzed and further clarified.

5. Conclusions

The impact of the heat transfer limitation on the gasification progress of different types of coals was investigated using both experimental and modeling approaches. The major findings and conclusions are as follows:

- The nature of the changes in thermal lag was complex. Unlike in the case of a nitrogen atmosphere, the endothermic devolatilization process does not significantly affect it, while the impact of the char–CO_2 reaction was visible.
- As it was expected, the average thermal lag in carbon dioxide was greater than that in nitrogen (64.0 K vs. 57.6 K), which is related to both the properties of CO_2 itself and the occurrence of the reverse Boudouard reaction.
- At low temperatures, CO_2 is an inert gas and does not significantly affect the mechanism of coal pyrolysis.
- The slightly higher reactivity of the fuel during devolatilization in CO_2 than in N_2 may be attributed to the properties of gases but not to the change in the process mechanism.
- Fuels having a higher volatiles content were characterized by a slightly higher thermal lag at the stage of devolatilization, while fuels rich in fixed carbon were characterized by this at the stage of the char–CO_2 reaction.
- Thermal lag influences the shape of the conversion curve, e.g., curves determined using the model show higher reactivity than those obtained experimentally.
- Discrepancies between T_{Rmax} determined at different heating rates are, to some extent, affected by the finite heat transfer rate; however, the T_{Rmax} parameter cannot be treated as independent of the heating rate.
- The onset temperature of the reverse Boudouard reaction depends on the type of fuel; however, no simple relationship with the basic physicochemical parameters characterizing fuels was found.
- Disregarding thermal lag may significantly hinder the interpretation of the analyzed phenomena, e.g., after considering the limitations of heat transfer, the values of the activation energy describing devolatilization became more similar for different heating rates, which suggests that the mechanism of the observed reaction seems not to change as much as previously supposed.
- Thermal lag has an impact on the kinetic parameter $A_{\alpha0.5}$ describing devolatilization, up to 19.8%. In the case of the char–CO_2 reaction, this influence is expected to be even greater.
- There is an influence of the properties of the coal structure on the progress of the reverse Boudouard reaction; however, it does not decrease proportionally to the coal rank.
- A relationship between the content of CaO and reactivity [77] was observed for hard coals but not for lignites.

Summarizing, the performed analysis indicates that to obtain trustworthy results, it is desirable to carry out TG experiments with a low heating rate or to recalculate results obtained at high heating rates with a model that allows the determination of the true sample temperature.

Funding: The thermal lag model development was co-funded by the National Centre for Research and Development of Poland, National Research Foundation of South Africa and Eskom Specialisation Centre under joint research and development project *Environomical impact of enhanced coal power plant operating flexibility (EPPOF)*, grant no. *PL-RPA2/09/EPPOF/2019*, and by the Wrocław Academic Hub, grant no. *BWU-16/2019/M8*.

Institutional Review Board Statement: Not applicable.

Informed Consent Statement: Not applicable.

Data Availability Statement: Not applicable.

Conflicts of Interest: The author declares no conflict of interest.

Nomenclature

$A_{\alpha 0.1}$	Reaction rate coefficient at the conversion of 10%, s^{-1}
$A_{\alpha 0.5}$	Reaction rate coefficient at the conversion of 50%, s^{-1}
A_c	The outer surface area of the crucible, m^2
Bi	Biot number
Cp_{ash}	Specific heat of ash, $J \cdot kg^{-1} \cdot K^{-1}$
Cp_c	Specific heat of crucible material, $J \cdot kg^{-1} \cdot K^{-1}$
Cp_{ch}	Specific heat of char, $J \cdot kg^{-1} \cdot K^{-1}$
Cp_g	Specific heat of gas atmosphere surrounding the crucible, $J \cdot kg^{-1} \cdot K^{-1}$
Cp_s	Specific heat of specimen, $J \cdot kg^{-1} \cdot K^{-1}$
d_c	Crucible diameter, m
d_f	Furnace diameter, m
d_s	Sample grain diameter, m
E_a	The activation energy of devolatilization, $J \cdot mol^{-1}$
E_b	The activation energy of char–CO_2 reaction, $J \cdot mol^{-1}$
h	Overall heat transfer coefficient, $W \cdot m^{-2} \cdot K^{-1}$
h_c	Convection heat transfer coefficient, $W \cdot m^{-2} \cdot K^{-1}$
h_r	Radiation heat transfer coefficient, $W \cdot m^{-2} \cdot K^{-1}$
$k_{0,a}$	The pre-exponential factor for devolatilization, s^{-1}
$k_{0,b}$	The pre-exponential factor for char–CO_2 reaction, s^{-1}
l_b	Height of a layer of sample in a crucible, m
l_c	Crucible height, m
m_0	The initial mass of the specimen, kg
m_c	Mass of the crucible, kg
Nu	Nusselt number
Pr	Prandtl number
Py'	External pyrolysis number
R	Universal gas constant, $J \cdot K^{-1} \cdot mol^{-1}$
RC	Sample reactivity, K^{-1}
Re	Reynolds number
t	Time, s
T_α	The temperature corresponding to the conversion of an exact amount of fuel, K
T_c	Crucible temperature measured by the temperature sensor, K
T_m	Sample temperature computed using model, K
T_r	The reference temperature, K
T_{RCmax}	Maximum reactivity temperature, K
T_s	True sample temperature, K
x_{ash}	Share of ash
x_{ch}	Share of char
x_{vol}	Share of volatiles
Greek symbols	
α	Degree of sample conversion
α_a	Degree of sample conversion during devolatilization
α_b	Degree of sample conversion during the char–CO_2 reaction
β	Heating rate, $K \cdot s^{-1}$
ΔH_{dev}	The heat of pyrolysis, $J \cdot kg^{-1}$ (assumes positive values when the reaction is endothermic)
ΔH_{gas}	The heat of the char–CO_2 reaction, $J \cdot kg^{-1}$ (assumes positive values when the reaction is endothermic)
ΔT	Thermal lag, K
ΔT_{Rmax}	The difference between T_{Rmax} determined at HR1 and HR50
μ	Emissivity
λ_g	Thermal conductivity of gas atmosphere surrounding the crucible, $W \cdot m^{-1} \cdot K^{-1}$

λ_s	Thermal conductivity of the sample
ν_g	Kinematic viscosity of gas atmosphere surrounding the crucible, $m^2 \cdot s^{-1}$
ρ_g	The density of gas atmosphere surrounding the crucible, $kg \cdot m^{-3}$
ρ_s	Density of sample
σ	Stefan–Boltzmann constant, equal to $5.670367 \cdot 10^{-8}\ W \cdot m^{-2} \cdot K^{-4}$

References

1. International Energy Agency. Global Energy Review 2020: The Impacts of the Covid-19 Crisis on Global Energy Demand and CO_2 Emissions. 2020. Available online: https://iea.blob.core.windows.net/assets/7e802f6a-0b30-4714-abb1-46f21a7a9530/Global_Energy_Review_2020.pdf (accessed on 8 August 2021).
2. Jakob, M.; Steckel, J.C.; Jotzo, F.; Sovacool, B.K.; Cornelsen, L.; Chandra, R.; Edenhofer, O.; Holden, C.; Löschel, A.; Nace, T.; et al. The future of coal in a carbon-constrained climate. *Nat. Clim. Chang.* **2020**, *10*, 704–707. [CrossRef]
3. Massachusetts Institute of Technology. The Future of Coal: Options in a Carbon-Constrained World. 2007. Available online: https://energy.mit.edu/wp-content/uploads/2007/03/MITEI-The-Future-of-Coal.pdf (accessed on 8 August 2021).
4. Greig, C.; Uden, S. The value of CCUS in transitions to net-zero emissions. *Electr. J.* **2021**, *34*, 107004. [CrossRef]
5. Miller, B.G. *Clean Coal Engineering Technology*; Elsevier: Amsterdam, The Netherlands, 2011.
6. Mauerhofer, A.; Fuchs, J.; Müller, S.; Benedikt, F.; Schmid, J.; Hofbauer, H. CO_2 gasification in a dual fluidized bed reactor system: Impact on the product gas composition. *Fuel* **2019**, *253*, 1605–1616. [CrossRef]
7. Chan, Y.H.; Rahman, S.N.F.S.A.; Lahuri, H.M.; Khalid, A. Recent progress on CO-rich syngas production via CO_2 gasification of various wastes: A critical review on efficiency, challenges and outlook. *Environ. Pollut.* **2021**, *278*, 116843. [CrossRef] [PubMed]
8. Roncancio, R.; Gore, J.P. CO_2 char gasification: A systematic review from 2014 to 2020. *Energy Convers. Manag. X* **2021**, *10*, 100060. [CrossRef]
9. Irfan, M.F.; Usman, M.R.; Kusakabe, K. Coal gasification in CO_2 atmosphere and its kinetics since 1948: A brief review. *Energy* **2011**, *36*, 12–40. [CrossRef]
10. Czajka, K.M. Proximate analysis of coal by micro-TG method. *J. Anal. Appl. Pyrolysis* **2018**, *133*, 82–90. [CrossRef]
11. Gonzalo-Tirado, C.; Jiménez, S.; Ballester, J. Kinetics of CO_2 gasification for coals of different ranks under oxy-combustion conditions. *Combust. Flame* **2013**, *160*, 411–416. [CrossRef]
12. Czajka, K.; Modliński, N.; Kisiela-Czajka, A.M.; Naidoo, R.; Peta, S.; Nyangwa, B. Volatile matter release from coal at different heating rates –experimental study and kinetic modelling. *J. Anal. Appl. Pyrolysis* **2019**, *139*, 282–290. [CrossRef]
13. Illanes, A. Immobilized Biocatalysts. *Compr. Biotech.* **2011**, *2*, 25–39.
14. Vyazovkin, S.; Chrissafis, K.; Di Lorenzo, M.L.; Koga, N.; Pijolat, M.; Roduit, B.; Sbirrazzuoli, N.; Suñol, J.J. ICTAC Kinetics Committee recommendations for collecting experimental thermal analysis data for kinetic computations. *Thermochim. Acta* **2014**, *590*, 1–23. [CrossRef]
15. Czajka, K.M. The impact of the thermal lag on the interpretation of cellulose pyrolysis. *Energy* **2021**, *236*, 121497. [CrossRef]
16. Stenseng, M.; Jensen, A.; Dam-Johansen, K. Investigation of biomass pyrolysis by thermogravimetric analysis and differential scanning calorimetry. *J. Anal. Appl. Pyrolysis* **2001**, *58–59*, 765–780. [CrossRef]
17. Narayan, N.; Antal, M.J. Thermal Lag, Fusion, and the Compensation Effect during Biomass Pyrolysis. *Ind. Eng. Chem. Res.* **1996**, *35*, 1711–1721. [CrossRef]
18. Burnham, A.; Braun, R.L.; Gregg, H.R.; Samoun, A.M. Comparison of methods for measuring kerogen pyrolysis rates and fitting kinetic parameters. *Energy Fuels* **1987**, *1*, 452–458. [CrossRef]
19. Wall, T.; Liu, Y.; Spero, C.; Elliott, L.; Khare, S.; Rathnam, R.; Zeenathal, F.; Moghtaderi, B.; Buhre, B.; Sheng, C.; et al. An overview on oxyfuel coal combustion—State of the art research and technology development. *Chem. Eng. Res. Des.* **2009**, *87*, 1003–1016. [CrossRef]
20. Mianowski, A.; Robak, Z.; Tomaszewicz, M.; Stelmach, S. The Boudouard–Bell reaction analysis under high pressure conditions. *J. Therm. Anal. Calorim.* **2012**, *110*, 93–102. [CrossRef]
21. Gomez Quevedo, R.A. *Fundamental Kinetic Studies of CO_2 and Steam Gasification (Doctoral Dissertation)*; University of Calgary: Calgary, AB, Canada, 2015.
22. Dupont, C.; Boissonnet, G.; Seiler, J.-M.; Gauthier-Maradei, P.; Schweich, D. Study about the kinetic processes of biomass steam gasification. *Fuel* **2007**, *86*, 32–40. [CrossRef]
23. International Organisation for Standardisation. *Solid Mineral Fuels–Hard Coal: Determination of Moisture in the General Analysis Test Sample by Drying in Nitrogen*; ISO 11722:2013; ISO: Geneva, Switzerland, 2013.
24. International Organisation for Standardisation. *Hard Coal and Coke: Determination of Volatile Matter*; ISO 562:2010; ISO: Geneva, Switzerland, 2010.
25. International Organisation for Standardisation. *Solid Mineral Fuels: Determination of Ash*; ISO 171:2010; ISO: Geneva, Switzerland, 2010.
26. Polish Committee for Standardization. *Paliwa Stałe: Oznaczanie Zawartości Siarki Całkowitej i Popiołowej Automatycznymi Analizatorami*; PN-G-04571:1998; Polish Committee for Standardization: Warsaw, Poland, 1998. (In Polish)
27. Polish Committee for Standardization. *Paliwa Stałe: Oznaczanie Zawartości Siarki Popiołowej Metodą Spalania w Wysokiej Temperaturze*; PN-G-04581:1999/Az1:2002; Polish Committee for Standardization: Warsaw, Poland, 1999. (In Polish)

28. International Organisation for Standardisation. *Solid Biofuels: Determination of Calorific Value*; ISO 18125:2017; ISO: Geneva, Switzerland, 2017.

29. Mularski, J.; Pawlak-Kruczek, H.; Modlinski, N. A review of recent studies of the CFD modelling of coal gasification in entrained flow gasifiers, covering devolatilization, gas-phase reactions, surface reactions, models and kinetics. *Fuel* **2020**, *271*, 117620. [CrossRef]

30. Mularski, J.; Modliński, N. Entrained-Flow Coal Gasification Process Simulation with the Emphasis on Empirical Char Conversion Models Optimization Procedure. *Energies* **2021**, *14*, 1729. [CrossRef]

31. Mularski, J.; Modliński, N. Impact of Chemistry–Turbulence Interaction Modeling Approach on the CFD Simulations of Entrained Flow Coal Gasification. *Energies* **2020**, *13*, 6467. [CrossRef]

32. Hadad, Y.; Jafarpur, K. Laminar forced convection heat transfer from isothermal cylinders with active ends and different aspect ratios in axial air flows. *Heat Mass Transf.* **2011**, *47*, 59–68. [CrossRef]

33. Ginnings, D.C.; Corruccini, R.J. Enthalpy, Specific Heat, and Entropy of Aluminum Oxide from 0° to 900° C. *J. Res. Nat. Bur. Stand.* **1947**, *38*, 593–600. [CrossRef]

34. Leśniak, B.; Słupik, Ł.; Jakubina, G. The determination of the specific heat capacity of coal based on literature data. *Chemik* **2013**, *67*, 560–571.

35. Eisermann, W.; Johnson, P.; Conger, W. Estimating thermodynamic properties of coal, char, tar and ash. *Fuel Process. Technol.* **1980**, *3*, 39–53. [CrossRef]

36. Richardson, M. The specific heats of coals, cokes and their ashes. *Fuel* **1993**, *72*, 1047–1053. [CrossRef]

37. Lemmon, E.W.; Huber, M.L.; McLinden, M.O. *NIST Standard Reference Database 23: Reference Fluid Thermodynamic and Transport Properties-REFPROP*; National Institute of Standards and Technology: Gaithersburg, MD, USA, 2013.

38. Bell, I.H.; Wronski, J.; Quoilin, S.; Lemort, V. Pure and Pseudo-pure Fluid Thermophysical Property Evaluation and the Open-Source Thermophysical Property Library CoolProp. *Ind. Eng. Chem. Res.* **2014**, *53*, 2498–2508. [CrossRef] [PubMed]

39. Farag, I.H.; Allam, T. Carbon dioxide standard emissivity by mixed gray-gases model. *Chem. Eng. Commun.* **1982**, *14*, 123–131. [CrossRef]

40. Milosavljevic, I.; Oja, A.V.; Suuberg, E.M. Thermal Effects in Cellulose Pyrolysis: Relationship to Char Formation Processes. *Ind. Eng. Chem. Res.* **1996**, *35*, 653–662. [CrossRef]

41. Uhrig, A. Determination of Pyrolysis Reaction Kinetics of Raw and Torrefied Biomass. Bachelor's Thesis, University of Twente, Enschede, The Netherlands, 2014.

42. Patisson, F.; LeBas, E.; Hanrot, F.; Ablitzer, D.; Houzelot, J.-L. Coal pyrolysis in a rotary kiln: Part I. Model of the pyrolysis of a single grain. *Met. Mater. Trans. A* **2000**, *31*, 381–390. [CrossRef]

43. Prins, M.J.; Ptasinski, K.J.; Janssen, F.J. Torrefaction of wood: Part 1. Weight loss kinetics. *J. Anal. Appl. Pyrolysis* **2006**, *77*, 28–34. [CrossRef]

44. Vyazovkin, S.; Burnham, A.K.; Favergeon, L.; Koga, N.; Moukhina, E.; Pérez-Maqueda, L.A.; Sbirrazzuoli, N. ICTAC Kinetics Committee recommendations for analysis of multi-step kinetics. *Thermochim. Acta* **2020**, *689*, 178597. [CrossRef]

45. Weerachanchai, P.; Tangsathitkulchai, C.; Tangsathitkulchai, M. Comparison of pyrolysis kinetic models for thermogravimetric analysis of biomass. *J. Sci. Technol.* **2010**, *17*, 387–400.

46. Chen, H.; Liu, N.; Fan, W. Two-step Consecutive Reaction Model of Biomass Thermal Decomposition by DSC. *Acta Physico-Chim. Sin.* **2006**, *22*, 786–790. [CrossRef]

47. Guo, J.; Lua, A. Kinetic study on pyrolytic process of oil-palm solid waste using two-step consecutive reaction model. *Biomass Bioenergy* **2001**, *20*, 223–233. [CrossRef]

48. Czajka, K.; Kisiela, A.; Moroń, W.; Ferens, W.; Rybak, W. Pyrolysis of solid fuels: Thermochemical behaviour, kinetics and compensation effect. *Fuel Process. Technol.* **2016**, *142*, 42–53. [CrossRef]

49. Doyle, C.D. Estimating isothermal life from thermogravimetric data. *J. Appl. Polym. Sci.* **1962**, *6*, 639–642. [CrossRef]

50. Garnier, B.; Lanzetta, F.; Lemasson, P.H.; Virgone, J. Lecture 5A: Measurements with contact in heat. Metti 5 Spring School, Roscoff, 13–18 June 2011. Available online: https://www.sft.asso.fr/Local/sft/dir/user-3775/documents/actes/Metti5_School/Lectures&Tutorials-Texts/Text-L5A-Garnier-Lanzetta.pdf (accessed on 8 August 2021).

51. Falsetti, C.; Kapulla, R.; Paranjape, S.; Paladino, D. Thermal radiation, its effect on thermocouple measurements in the PANDA facility and how to compensate it. *Nucl. Eng. Des.* **2021**, *375*, 111077. [CrossRef]

52. Jalalabadi, T.; Moghtaderi, B.; Allen, J. Thermochemical Conversion of Biomass in the Presence of Molten Alkali-Metal Carbonates under Reducing Environments of N_2 and CO_2. *Energies* **2020**, *13*, 5395. [CrossRef]

53. Yang, M.; Luo, B.; Shao, J.; Zeng, K.; Zhang, X.; Yang, H.; Chen, H. The influence of CO_2 on biomass fast pyrolysis at medium temperatures. *J. Renew. Sustain. Energy* **2018**, *10*, 013108. [CrossRef]

54. Jamil, K.; Hayashi, J.-I.; Li, C.-Z. Pyrolysis of a Victorian brown coal and gasification of nascent char in CO_2 atmosphere in a wire-mesh reactor. *Fuel* **2004**, *83*, 833–843. [CrossRef]

55. Duan, L.; Zhao, C.; Zhou, W.; Qu, C.; Chen, X. Investigation on Coal Pyrolysis in CO_2 Atmosphere. *Energy Fuels* **2009**, *23*, 3826–3830. [CrossRef]

56. Jindarom, C.; Meeyoo, V.; Rirksomboon, T.; Rangsunvigit, P. Thermochemical decomposition of sewage sludge in CO_2 and N_2 atmosphere. *Chemosphere* **2007**, *67*, 1477–1484. [CrossRef]

57. Kim, Y.J.; Kim, M.I.; Yun, C.H.; Chang, J.Y.; Park, C.R.; Inagaki, M. Comparative study of carbon dioxide and nitrogen atmospheric effects on the chemical structure changes during pyrolysis of phenol–formaldehyde spheres. *J. Colloid Interface Sci.* **2004**, *274*, 555–562. [CrossRef] [PubMed]

58. Bilkic, B.; Haykiri-Acma, H.; Yaman, S. Combustion reactivity estimation parameters of biomass compared with lignite based on thermogravimetric analysis. *Energy Sources Part A* **2020**, 1–14. [CrossRef]

59. Pilatau, A.; Czajka, K.; Filho, G.P.; Medeiros, H.S.; Kisiela, A.M. Evaluation Criteria for the Assessment of the Influence of Additives (AlCl3 and ZnCl2) on Pyrolysis of Sunflower Oil Cake. *Waste Biomass-Valoriz.* **2017**, *8*, 2595–2607. [CrossRef]

60. Rybak, W.; Moroń, W.; Czajka, K.M.; Kisiela, A.M.; Ferens, W.; Jodkowski, W.; Andryjowicz, C. Co-combustion of unburned carbon separated from lignite fly ash. *Energy Procedia* **2017**, *120*, 197–205. [CrossRef]

61. Speight, J.G. *Heavy Oil Recovery and Upgrading*; Gulf Professional Publishing: Laramie, WY, USA, 2019; p. 576.

62. Wang, T.; Stiegel, G. *Integrated Gasification Combined Cycle (IGCC) Technologies*; Woodhead Publishing: Amsterdam, The Netherlands, 2017; p. 10.

63. Tangstad, M.; Baukes, J.P.; Steenkamp, J.D.; Ringdalen, E. Coal-based reducing agents in ferroalloys and silicon production. In *New Trends in Coal Conversion*; Suarez-Ruiz, I., Rubiera, F., Diez, M.A., Eds.; Woodhead Publishing: Amsterdam, The Netherlands, 2019; pp. 405–438.

64. Riley, J.; Siriwardane, R.; Tian, H.; Benincosa, W.; Poston, J. Experimental and kinetic analysis for particle scale modeling of a CuO-Fe2O3-Al2O3 oxygen carrier during reduction with H_2 in chemical looping combustion applications. *Appl. Energy* **2018**, *228*, 1515–1530. [CrossRef]

65. Kisiela, A.M.; Czajka, K.M.; Moron, W.; Rybak, W.; Andryjowicz, C. Unburned carbon from lignite fly ash as an adsorbent for SO_2 removal. *Energy* **2016**, *116*, 1454–1463. [CrossRef]

66. Guldogan, Y.; Evren, V.; Durusoy, T.; Bozdemir, T. Effects of Heating Rate and Particle Size on Pyrolysis Kinetics of Mengen Lignite. *Energy Sources* **2001**, *23*, 337–344.

67. Guo, Z.; Zhang, L.; Wang, P.; Liu, H.; Jia, J.; Fu, X.; Li, S.; Wang, X.; Li, Z.; Shu, X. Study on kinetics of coal pyrolysis at different heating rates to produce hydrogen. *Fuel Process. Technol.* **2013**, *107*, 23–26. [CrossRef]

68. Ozbas, K.E.; Kok, M.V. Effect of Heating Rate on Thermal Properties and Kinetics of Raw and Cleaned Coal Samples. *Energy Sources* **2003**, *25*, 33–42. [CrossRef]

69. Urych, B. Determination of Kinetic Parameters of Coal Pyrolysis to Simulate the Process of Underground Coal Gasification (UCG). *J. Sustain. Min.* **2014**, *13*, 3–9. [CrossRef]

70. Dwivedi, K.K.; Chatterjee, P.K.; Karmakar, M.K.; Pramanick, A.K. Pyrolysis characteristics and kinetics of Indian low rank coal using thermogravimetric analysis. *Int. J. Coal Sci. Technol.* **2019**, *6*, 102–112. [CrossRef]

71. Dingcheng, L.; Qiang, X.; Guangsheng, L.; Junya, C.; Jun, Z. Influence of heating rate on reactivity and surface chemistry of chars derived from pyrolysis of two Chinese low rank coals. *Int. J. Min. Sci. Technol.* **2018**, *28*, 613–619. [CrossRef]

72. Song, Y.H.; She, J.M.; Lan, X.Z.; Zhou, J. Pyrolysis Characteristics and Kinetics of Low Rank Coal. *Mater. Sci. Forum* **2011**, *695*, 493–496. [CrossRef]

73. Xu, Y.; Zhang, Y.; Wang, Y.; Zhang, G.; Chen, L. Thermogravimetric study of the kinetics and characteristics of the pyrolysis of lignite. *React. Kinet. Mech. Catal.* **2013**, *110*, 225–235. [CrossRef]

74. Yakar Elbeyli, I.; Piskin, S.; Sutcu, H. Pyrolysis Kinetics of Turkish Bituminous Coals by Thermal Analysis. *Turk. J. Eng. Environ. Sci.* **2004**, *28*, 233–239.

75. Barrio, M.; Hustad, J.E. CO2 Gasification of Birch Char and the Effect of CO Inhibition on the Calculation of Chemical Kinetics. In *Progress in Thermochemical Biomass Conversion*; Wiley: Hoboken, NJ, USA, 2008; pp. 47–60.

76. Risnes, H.; Srensen, L.H.; Hustad, J.E.; Sørensen, L.H. CO2 Reactivity of Chars from Wheat, Spruce and Coal. In *Progress in Thermochemical Biomass Conversion*; Wiley: Hoboken, NJ, USA, 2008; pp. 61–72.

77. Liu, G.-S.; Tate, A.; Bryant, G.; Wall, T. Mathematical modeling of coal char reactivity with CO_2 at high pressures and temperatures. *Fuel* **2000**, *79*, 1145–1154. [CrossRef]

78. Rybak, W.; Nowak-Woźny, D.; Moroń, W.; Urbanek, B.; Hrycaj, G. Transformacja substancji mineralnej w warunkach spalania tlenowego. In *Spalanie Tlenowe Dla Kotłów Pyłowych i Fluidalnych Zintegrowanych z Wychwytem CO2: Kinetyka i Mechanizm Spalania Tlenowego Oraz Wychwytu CO2*; Nowak, W., Rybak, W., Czakiert, T., Eds.; Wydawnictwo Politechniki Częstochowskiej: Częstochowa, Poland, 2013; pp. 176–194. (In Polish)

79. Dahou, T.; Defoort, F.; Khiari, B.; Labaki, M.; Dupont, C.; Jeguirim, M. Role of inorganics on the biomass char gasification reactivity: A review involving reaction mechanisms and kinetics models. *Renew. Sustain. Energy Rev.* **2021**, *135*, 110136. [CrossRef]

80. Bennici, S.; Jeguirim, M.; Limousy, L.; Haddad, K.; Vaulot, C.; Michelin, L.; Josien, L.; Zorpas, A.A. Influence of CO_2 Concentration and Inorganic Species on the Gasification of Lignocellulosic Biomass Derived Chars. *Waste Biomass-Valoriz.* **2019**, *10*, 3745–3752. [CrossRef]

81. Vázquez, M.D.P.G.; García, R.; Gil, M.; Pevida, C.; Rubiera, F. Unconventional biomass fuels for steam gasification: Kinetic analysis and effect of ash composition on reactivity. *Energy* **2018**, *155*, 426–437. [CrossRef]

82. Delannay, F.; Tysoe, W.; Heinemann, H.; Somorjai, G. The role of KOH in the steam gasification of graphite: Identification of the reaction steps. *Carbon* **1984**, *22*, 401–407. [CrossRef]

83. Bouraoui, Z.; Dupont, C.; Jeguirim, M.; Limousy, L.; Gadiou, R. CO2 gasification of woody biomass chars: The influence of K and Si on char reactivity. *Comptes Rendus Chim.* **2016**, *19*, 457–465. [CrossRef]

MDPI
St. Alban-Anlage 66
4052 Basel
Switzerland
Tel. +41 61 683 77 34
Fax +41 61 302 89 18
www.mdpi.com

Energies Editorial Office
E-mail: energies@mdpi.com
www.mdpi.com/journal/energies